Molecular and Integrative Toxicology

Molecular and Integrative Toxicology presents state-of-the-art toxicology in a useful context. Volumes emphasize the presentation of cellular and molecular information aimed toward the protection of human or animal health or the sustainability of environmental systems.

More information about this series at http://www.springer.com/series/8792

James C. Bonner • Jared M. Brown
Editors

Interaction of Nanomaterials with the Immune System

Springer

Editors
James C. Bonner
Department of Biological Sciences
North Carolina State University
Raleigh, NC, USA

Jared M. Brown
Department of Pharmaceutical Sciences
Skaggs School of Pharmacy
University of Colorado
Anschutz Medical Campus
Aurora, CO, USA

ISSN 2168-4219 ISSN 2168-4235 (electronic)
Molecular and Integrative Toxicology
ISBN 978-3-030-33961-6 ISBN 978-3-030-33962-3 (eBook)
https://doi.org/10.1007/978-3-030-33962-3

This Springer imprint is published by the registered company Springer Nature Switzerland AG
The registered company address is: Gewerbestrasse 11, 6330 Cham, Switzerland

Preface

Nanotechnology is manipulation of matter at the atomic scale for a plethora of applications, some of which hold solutions for our most pressing challenges, such as energy and medicine.

Despite the enormous potential benefits, there is also the potential danger that the advancement of the nanotechnology industry will bring with it adverse human health effects. One recognized effect is the impact of engineered nanomaterials (ENMs) on the immune system as these materials are foreign to the human body. The focus of this book is to present an overview of the principles and basic mechanisms of immunotoxicity caused by ENMs. Human exposure to ENMs occurs occupationally at workplaces, as a result of specific biomedical or consumer applications, or after environmental contamination resulting from nanomaterials released into the air, water, and soil.

The impact of ENMs on the human immune system has yet to be determined. This is due to the relatively recent emergence of the nanotechnology industry over the past few decades. However, the evidence from global studies using rodents or cultured human cells, some of which is presented in this book, predicts that ENMs will cause some degree of immune-related diseases in humans, including but perhaps not limited to allergies, asthma, hypersensitivity reactions, autoimmune disease, fibrosis, and cancer. For this reason, we feel that this book is timely and deals with key issues for understanding ENM interaction with the immune system that will help us proactively prevent future immune-related diseases.

ENMs, like other specific types of chemicals, influence the immune system upon inhalation, ingestion, injection, and dermal exposure. However, unlike many chemicals, ENMs deserve some special attention due to their unique interactions with biological systems. For example, the term *nano-bio interface* was coined to encompass the interaction of ENMs with biomolecules, cell membranes, or intracellular components (e.g., actin, DNA). These interactions at the subcellular scale make ENMs unique, for better or worse, and emphasize the concept that size does indeed matter. It is not our intent to present information on nanomedicine applications, although some overlap with this topic is inevitable due to immunotoxic side effects of some nanotherapeutics.

Some valuable documents worth mentioning are already available on immunotoxicity caused by chemical exposure and on methods for assessing immunotoxicity. For example, the topic of immunotoxicity and chronic disease caused by chemical exposure is a topic that has been addressed in a previous volume of *Molecular and Integrative Toxicology* (Dietert and Luebke 2012). Additionally, previous Environmental Health Criteria (EHC) documents published by the World Health Organization (WHO) have addressed chemical exposure and immunotoxicity. EHC monograph 212 of the International Programme on Chemical Safety (IPCS) focused on mechanisms, clinical aspects, epidemiology, hazard identification, and risk assessment of allergy and hypersensitivity following exposure to certain chemicals (IPCS 1999), while EHC monograph 236 focused on the induction of autoimmunity associated with chemical exposure (IPCS 2006). Finally, a forthcoming EHC monograph in 2020 entitled "Principles and Methods to Assess the Risk of Immunotoxicity Associated with Exposure to Nanomaterials" will present detailed information on testing methods (ICPS in press). Therefore, it is not our intent herein to provide a duplicative effort on immunotoxicity principles and testing methods but instead to illustrate mechanistic concepts of nanoimmunotoxicology from a diverse group of experts.

Finally, we are grateful to our scientific colleagues and friends who contributed to this book. The project was inspired by a shared interest and enthusiasm with our contributing colleagues that stemmed from formal scientific sessions, as well as informal conversations, at conferences in Europe and the USA. The chapters are authored by experts in the field of nanotechnology, toxicology, and immunology from six countries (Finland, the Netherlands, Sweden, Switzerland, the UK, and the USA). It is our hope that this book will provide some thought and guidance to the next generation of immunotoxicologists who will continue to address important issues related to nanotoxicology, human health, and the environment.

Raleigh, NC, USA James C. Bonner
Aurora, CO, USA Jared M. Brown

References

Dietert RR, Luebke RW, editors. Immunotoxicity, immune dysfunction, and chronic disease. Molecular and integrative toxicology series. Totowa: Humana Press. 2012; 440 pp.

International Programme on Chemical Safety. Principles and methods for assessing allergic hypersensitization associated with exposure to chemicals. Environmental Health Criteria monograph 212. Geneva: World Health Organization; 1999.

International Programme on Chemical Safety. Principles and methods for assessing autoimmunity associated with exposure to chemicals. Environmental Health Criteria monograph 236. Geneva: World Health Organization; 2006.

IPCS (in press). Principles and methods to assess the risk of immunotoxicity associated with exposure to nanomaterials. Geneva: World Health Organization, International Programme on Chemical Safety (Environmental Health Criteria 244). Licence: CC BY-NC-SA 3.0 IGO.

Contents

Contributors

Harri Alenius Institute of Environmental Medicine, Karolinska Institute, Stockholm, Sweden

Human Microbiome Research (HUMI), Medical Faculty, University of Helsinki, Helsinki, Finland

James C. Bonner North Carolina State University, Raleigh, NC, USA

Jared M. Brown University of Colorado, Anschutz Medical Campus, Aurora, CO, USA

Melisa Bunderson-Schelvan University of Montana, Missoula, MT, USA

Hao Chen University of Florida, Gainesville, FL, USA

Barbara Drasler Adolphe Merkle Institute, University of Fribourg, Fribourg, Switzerland

Bengt Fadeel Karolinska Institutet, Stockholm, Sweden

Matthew J. Gaffrey Pacific Northwest National Laboratory, Richland, WA, USA

Raymond F. Hamilton University of Montana, Missoula, MT, USA

Andrij Holian University of Montana, Missoula, MT, USA

Sara T. Humes University of Florida, Gainesville, FL, USA

Sandeep Keshavan Karolinska Institutet, Stockholm, Sweden

John A. Lednicky University of Florida, Gainesville, FL, USA

Ho Young Lee North Carolina State University, Raleigh, NC, USA

Ryan P. Mendoza University of Colorado, Aurora, CO, USA

Seyed M. Moghimi Colorado Center for Nanomedicine and Nanosafety, University of Colorado Anschutz Medical Campus, Aurora, CO, USA

School of Pharmacy, King George VI Building, Newcastle University, Newcastle upon Tyne, UK

Division of Stratified Medicine, Biomarkers & Therapeutics, Institute of Cellular Medicine, Faculty of Medical Sciences, Newcastle University, Newcastle upon Tyne, UK

Alke Petri-Fink Adolphe Merkle Institute, University of Fribourg, Fribourg, Switzerland

Wei-Jun Qian Pacific Northwest National Laboratory, Richland, WA, USA

Barbara Rothen-Rutishauser Adolphe Merkle Institute, University of Fribourg, Fribourg, Switzerland

Tara Sabo-Attwood University of Florida, Gainesville, FL, USA

Navid B. Saleh University of Texas, Austin, TX, USA

Kai Savolainen Finnish Institute of Occupational Health, Helsinki, Finland

Dmitri Simberg Translational Bio-Nanosciences Laboratory, Department of Pharmaceutical Sciences, Skaggs School of Pharmacy and Pharmaceutical Sciences, University of Colorado, Aurora, CO, USA

Colorado Center for Nanomedicine and Nanosafety, University of Colorado Anschutz Medical Campus, Aurora, CO, USA

Brian D. Thrall Pacific Northwest National Laboratory, Richland, WA, USA

Kevin L. Trout University of Montana, Missoula, MT, USA

Henk van Loveren Maastricht University Medical Centre, Maastricht, The Netherlands

Rob J. Vandebriel Centre for Health Protection, National Institute for Public Health and the Environment, Bilthoven, The Netherlands

Dorothy J. You North Carolina State University, Raleigh, NC, USA

Tong Zhang Pacific Northwest National Laboratory, Richland, WA, USA

Chapter 1
Introduction

James C. Bonner and Jared M. Brown

Abstract The rise of nanotechnology, a new industrial revolution, is generating a wealth of novel advanced materials that are dramatically changing the fields of electronics, engineering, and medicine. It is anticipated that these changes will solve important problems in renewable energy, more efficient communication and transportation systems, bioremediation of environmental pollution, and treatment of debilitating diseases. However, the impact of nanomaterials on the immune system is a concern, since manipulation of matter with a size range on par with subcellular structures has the potential to activate or suppress cells of the innate or adaptive immune system. This chapter overviews the topics covered in this book and thereby sets the stage for understanding the complexity of immune responses to a diversity of emerging engineered nanomaterials.

Keywords Nanotechnology · Immune system · Immunotoxicity

1.1 The Rapidly Expanding World of Engineered Nanomaterials (ENMs)

Nanotechnology, by a colloquial definition, is a new field of science focused on precision engineering of objects smaller than can be seen with the human eye. In more technical terms, nanotechnology may be defined as the design and manipulation of matter at the atomic scale to develop novel advanced materials. The potential benefits of nanotechnology are numerous, and include superior advances in electronics,

J. C. Bonner (✉)
North Carolina State University, Raleigh, NC, USA
e-mail: jcbonner@ncsu.edu

J. M. Brown
Department of Pharmaceutical Sciences, Skaggs School of Pharmacy,
University of Colorado, Anschutz Medical Campus, Aurora, CO, USA

J. C. Bonner, J. M. Brown (eds.), *Interaction of Nanomaterials with the Immune System*, Molecular and Integrative Toxicology,
https://doi.org/10.1007/978-3-030-33962-3_1

highly efficient renewable energy cells that will replace and reduce conventional fossil fuel energy, nanoparticle-mediated remediation of environmental pollution, and new nano-enabled therapeutic approaches to treat deadly diseases such as cancer (Cattaneo et al. 2010). In order to maximize the success and benefits of nanotechnology, the design and synthesis of advanced nanomaterials should be considered with awareness of potential adverse effects on human health and the environment (Xia et al. 2009).

Like any new emerging technology, benefits are accompanied by risk. Some materials that are relatively innocuous at the macro- or larger scale might behave differently at the nanoscale. An example is titanium dioxide (TiO_2) that is relatively inert as a micron-sized particle, but as a nanomaterial can cause significant oxidative stress, inflammation, and tumor formation when instilled into the lungs of rodents (Oberdorster 1996). Also, some idea of relative toxicity or mode of action might be expected based on elemental composition alone, but some effects might be entirely unanticipated. For example, in elemental terms single-walled carbon nanotubes (SWCNTs) are essentially the same as graphite, a relatively nontoxic substance, but the dimensions of the nanotube structure (~1 to 4 nm width with a length >1 micrometer) are similar in scale to subcellular structures such as the cytoskeletal protein actin or a DNA molecule (Pampaloni and Florin 2008; Sargent et al. 2009). As such, there is the potential for nanotubes to interact with biomolecules with similar dimensions; e.g., wrapping or intertwining. This type of hybrid ENM–biomolecule interaction might explain some observed phenomena in immune cells; e.g., carbon bridge-like structures that form between alveolar macrophages in rats exposed to SWCNTs that do not occur upon exposure to spherical carbon nanoparticles (Mangum et al. 2006). ENMs also interact with biomolecules in the extracellular microenvironment to form a "*biocorona*" that influences the immune response (Neagu et al. 2017; Chen and Riviere 2017). In general, understanding the biophysicochemical interactions at the "*nano-bio interface*" is critical toward elucidating the impact of ENMs on cells of the immune system and whether these interactions could have biocompatible or bio-adverse outcomes on host immunity (Nel et al. 2009).

A major challenge for toxicologists and regulatory agencies in assessing risks associated with nanotechnology is the ever-increasing number and variety of ENMs. An important distinction to make at the forefront of this book is the difference between ENMs and anthropogenic nanoparticles (e.g., ambient ultrafine particles generated unintentionally as a by-product of man's activity). However, it has been acknowledged that much of what we know about ambient ultrafine particles can be applied to ENMs (Stone et al. 2017). The term *engineered nanomaterial* (ENM) may be more applicable than *engineered nanoparticle* (ENP) which refers to all dimensions and shapes <100 nm (e.g., spherical polyhedron and cube). Therefore, the abbreviation ENM is used throughout this book and includes ENPs as well as other structures with at least one dimension <100 nm (e.g., tubes and sheets).

1.2 Historical Perspective of Nanoparticle-Induced Immunotoxicity

It is not surprising that ENMs would present a challenge to our immune system. The mammalian immune system evolved to cope with foreign particles, many of them possessing nanoscale dimensions (between 1 and 100 nm), which presented a detrimental impact on the cellular and physiological well-being of the host. Foreign material entering the body, mainly through inhalation or ingestion, is identified as "not self" and dealt with accordingly. Some obvious invading "not-self" entities that accompanied the evolution of man are pathogenic microbes (e.g., viruses and bacteria). While most of these microbes are larger than 100 nm, their nano-sized components may be shed or released to trigger an immune response of the host. The same applies to nano-sized components shed from even larger organisms, such as invertebrates like house dust mites and cockroaches, which shed chitin nanoparticles from their defoliating exoskeletons. In addition to microbial and zoological sources, naturally occurring nanoparticles include organic and inorganic "dusts" from wildfires, dust storms, and volcanic activity.

The technological ability to manipulate matter at the nanoscale is relatively new. However, physicist Richard Feynman introduced the idea of manipulating individual atoms and molecules in 1959. A decade later, the term "nanotechnology" was coined by Professor Norio Taniguchi. Scientific advances such as the introduction of the scanning tunneling microscope in the early 1980s, capable of visualizing individual atoms, enabled the emergence of modern nanotechnology. The Nobel Prize was awarded to Richard Smalley in 1985 for discovery of the Buckminsterfullerene or "Buckeyball," a C60 carbon nanoparticle. This landmark event was followed by the discovery of the carbon nanotube by Sumio Iijima in 1991. These discoveries set the stage for nanotechnology companies to emerge in the 1990s. Nanotechnology gained momentum by the turn of the twenty-first century when production methods became more controllable and ENMs with adjustable properties were generated, characterized, and imaged at atomic resolution. The National Nanotechology Initiative (NNI) was launched in 2000 to coordinate federal research and development (R&D) efforts and promote US competitiveness in nanotechnology. Similar efforts were launched in the European Union (EU), China, and Japan.

While natural environmental nanoparticles continue to present a challenge to the human immune system and promote immune-mediated diseases, exposure to nanoparticles released as a by-product of anthropogenic activity has been in progress since the Industrial Revolution. Combustion associated with the burning of fossil fuels for energy and transportation contributed to an increase in air pollution particulate matter (PM); a complex mixture of agents (e.g., polycyclic aromatic hydrocarbons and transition metals) surrounding a carbon core. The emergence of new occupational diseases with an immunological basis accompanied the rise of industry. Metallurgy and welding generated metal and metal oxide particles and fumes, while mining released mineral particles like coal dust or asbestos fibers, resulting in occupational respiratory diseases such as asthma, hypersensitivity pneumonitis, pulmonary fibrosis, and lung cancer (De Vuyst and Camus 2000). Strong epidemiologic evidence linked exposure with disease outcome. Unfortunately, in many cases irreversible disease (e.g., cancer and fibrosis) had already been diagnosed in the population before such associations were confirmed. A noteworthy example is the epidemic of asbestos-related lung diseases: pulmonary fibrosis and mesothelioma. Therefore, disease prevention associated with exposure to products of the emerging nanotechnology industry seems like a logical "lesson learned" and should compel society to avoid repeating mistakes of the past with regard to occupational or consumer exposure and human health. However, advances in nanotechnology are moving forward faster than toxicology research aimed at identifying the relative risk of emerging ENMs. Moreover, the number of products on the market incorporating ENMs will inevitably continue to grow as the number of new applications increases.

1.3 Innate and Adaptive Immune Responses to ENMs

Immunotoxicity can be defined as any adverse effect on the immune system following toxicant exposure. Further, immunotoxic effects can be subdivided into five adverse event categories: (1) immunosuppression, (2) immunogenicity, (3) hypersensitivity, (4) autoimmunity, and (5) adverse immunostimulation (Hussain et al. 2012). *Immunosuppression* refers to impairment of immune cell components leading to decreased immune function, resulting in increased susceptibility to infectious diseases or tumor growth. The other four adverse event categories involve *immunostimulation*. Specifically, *immunogenicity* results in an allergic response following multiple exposures to a foreign agent. Disease outcomes such as asthma and contact dermatitis fall into this category. *Hypersensitivity* involves immune sensitization to a foreign agent that results in a severe adverse response. *Autoimmunity* refers to the immune system adversely reacting to "self" antigens, resulting in tissue destruction by host cells. *Adverse immunostimulation* refers to any antigen-nonspecific activation of the immune system. As illustrated in Fig. 1.1, ENMs may have either immunosuppressive

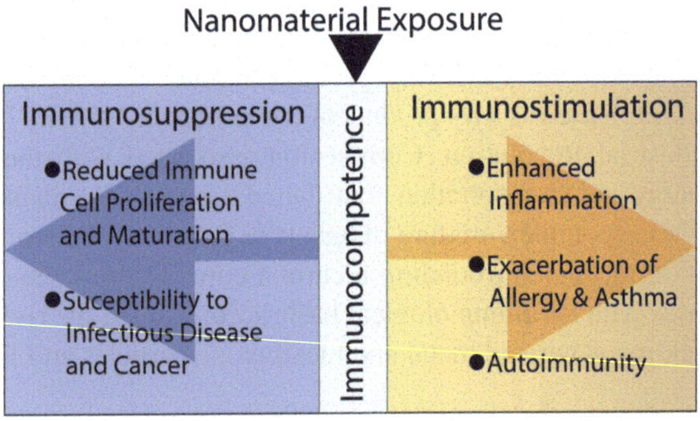

Fig. 1.1 Illustration of the basic principles of immunosuppression and immunostimulation induced by engineered nanomaterials (ENMs)

or immunostimulatory effects. Whether ENMs exert immunosuppressive or immunostimulatory effects likely depends on the timing and context of exposure, as well as susceptibility factors such as age, sex, genetics, co-exposure with other immunogenic agents, or preexisting disease (Scoville et al. 2017; Frank et al. 2017; Duke et al. 2018; Ihrie et al. 2019).

The immune system consists of a network of specific cell types and secreted biomolecules that work in a coordinated fashion to recognize "non-self" materials, communicate the invading threat, and execute defensive mechanisms. Typically, the immune system is divided into the *innate immune system* that rapidly responds to foreign invasion and the delayed but highly specific *adaptive immune system*. The adaptive immune system is endowed with immunological memory following initial response to a specific antigen, leading to an enhanced response to subsequent exposures to the same antigen. This type of immune sensitization results in the polarization and expansion of specific T lymphocyte populations that recognize and react against a specific antigen and is the underlying basis for vaccination used in the prevention of infectious diseases in humans.

Many of the immunotoxic effects of ENMs are mediated via direct interaction with the innate immune system (Farrera and Fadeel 2015). The innate immune system consists of cells in the mononuclear phagocytic system that includes tissue macrophages, peripheral blood monocytes, and granulocytes such as neutrophils. Macrophages, discussed in Chap. 2, represent the earliest response to ENM exposure and function to remove foreign material as well as serve other important functions in host defense (Hussell and Bell 2014). Neutrophils also play important roles in the innate immune response to microbes as well as ENMs and are recruited from the circulation to injured tissues by chemokines released by resident cells, including macrophages (Borregaard 2010). The diverse functions of neutrophils in the immune response to ENMs are discussed in Chap. 3.

Mast cells also play a prominent role in innate immune responses. These cells are abundant in tissues exposed to the external environment (e.g., lung, gastrointestinal tract, and skin) where they are a first responder to insult. Upon activation, they immediately release preformed mediators via degranulation such as histamine, serotonin, proteases, and lipid mediators that promote recruitment of immune cells to the surrounding tissue as well as many physiological effects such as increased vascular permeability and cross talk with the nervous system. Following degranulation, mast cells further produce a myriad of cytokines in a late phase response that further define the local immune response by recruitment of eosinophils, neutrophils, and macrophages, as well as promoting the polarization of T cells. Evidence from cell culture experiments in vitro and rodent studies in vivo indicates that ENMs can trigger mast cell activation and degranulation. For example, mast cells orchestrate adverse pulmonary and cardiovascular responses to carbon nanotubes through the interleukin 33/suppression of tumorigenicity 2 receptor (IL-33/ST2) axis (Katwa et al. 2012). The role of mast cells in mediating immunotoxicity to ENMs is discussed in Chap. 4.

While the innate immune system is the immediate defense against foreign agents, the adaptive immune response takes days or weeks to develop. However, once an adaptive immune response has been established, the immune system rapidly responds to subsequent exposure to the same foreign agent. Cells of the adaptive immune system include dendritic cells, T cells, and B cells. In general, T cells are involved in cell-mediated responses, while B cells mediate humoral immunity (antibody production). T cells can be further polarized to different T helper (Th) CD4+ lymphocyte subtypes (Th1, Th2, Th17) after antigen presentation by dendritic cells in combination with the appropriate cytokine microenvironment, giving rise to either Th1 "non-allergic" inflammation or Th2/Th17 "allergic" inflammation (Lambrecht and Hammad 2015). Regulatory T cells are required for maintenance of immune tolerance, essential for the discrimination between "self" and "non-self." Dendritic cells play a central role in the immune response by coordinating innate and adaptive immunity and serving as inducers of the adaptive immune response through antigen presentation to T lymphocytes. These cells are effective phagocytes at primary target organs exposed to ENMs (lung and intestine) and further facilitate processing and presentation of foreign antigens to T cells (Banchereau and Steinman 1998). After phagocytosis of the foreign agent, immature dendritic cells migrate from peripheral primary target tissues to lymph nodes, where they undergo maturation and stimulate T cells and B cells.

In general, exposure to foreign agents could lead to immunosuppression, wherein the immune system fails to respond in expanding specific T cell populations, or immunostimulation that potentially leads to autoimmune or allergic disease (Luster 2014; Boraschi et al. 2017). While, fewer studies have been conducted on the effects of ENMs on the adaptive immune response, a few studies indicate that ENMs may disrupt dendritic cell function, which may in turn affect T and B cells. For example, carbon nanotubes have a direct effect on dendritic cells that diminishes their ability to stimulate T cell proliferation in the spleen (Tkach et al. 2011). The role of dendritic cells in ENM-induced immunotoxicity or as a therapeutic target for immune regulation by nanoparticles is discussed in Chap. 5.

1.4 Physicochemical Properties of ENMs that Determine Immunotoxicity

Size is a key physicochemical property that determines the unique behavior of ENMs in biological systems, allowing for translocation across tissue and cellular barriers, as well as interactions with biomolecules and intracellular organelles. *Shape* is also an important feature of ENMs that could determine immunogenicity. ENMs come in a variety of shapes, including spheres, hedrons, fibers, tubes, and sheets. Some examples are shown in Fig. 1.2. For fiber or tube-shaped ENMs, the rigidity of the structure also plays a role in innate immune cell recognition. In addition to size and shape, a variety of other *physicochemical properties* determine the immunotoxic activity of ENMs. For example, some ENMs undergo partial or complete *dissolution* in aqueous fluids and release ions, while other ENMs are relatively stable in solution or suspension. ENMs can also form agglomerates that may modify

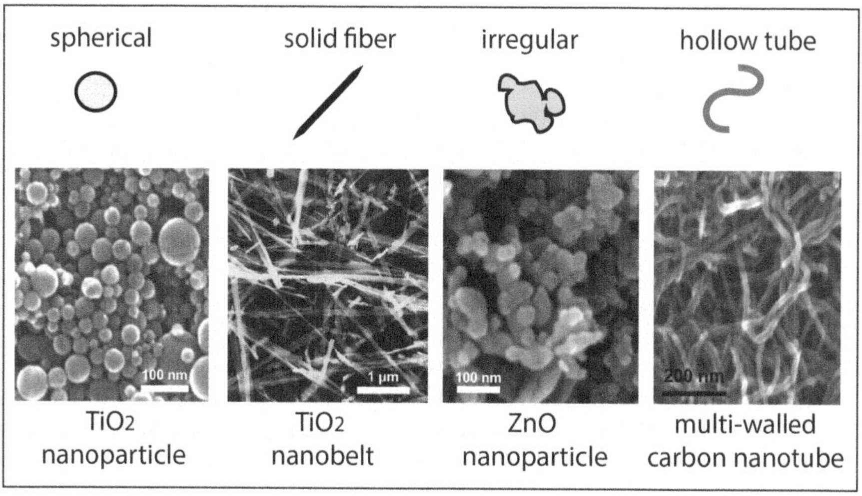

Fig. 1.2 Examples of engineered nanomaterials (Reproduced with permission from Ihrie and Bonner, 2018)

immune cell recognition and uptake or translocation across biological barriers. *Agglomeration* of ENMs, due to noncovalent interaction through van der Waals forces or electrostatic attraction, may result in the formation of micron-sized particles. However, unlike larger particles, ENM agglomerates with the same overall dimensions retain extremely high surface area and reactivity due to the loose attraction between the individual components. The *surface charge* on ENMs, referred to as the *zeta potential*, determines agglomeration as well as other interactions with biomolecules, cell membranes, and translocation across mucosal barriers such as the lining of the lungs or gastrointestinal (GI) tract. Upon interaction with biological systems, some physicochemical characteristics, such as surface charge, change as the ENMs accumulate a biocorona and move across biological barriers (discussed below).

1.5 Biocorona Formation and Recognition of ENMs by the Immune System

In biological systems, proteins, lipids, and carbohydrates adsorb to the surfaces of ENMs to form a *biocorona.* The biocorona coating on the ENM is often what the immune cell first recognizes and therefore is important in mediating these interactions. For example, the composition of the biocorona can determine whether an ENM is recognized by immune cells as biocompatible or immunotoxic. While physicochemical properties determine the synthetic identity of ENMs, the biocorona determines *biological identity*; that is to say, how immune cells recognize

ENMs (Fadeel et al. 2013). Biocorona composition depends on the extracellular microenvironment surrounding the ENM. For example, inhaled ENMs encounter a different makeup of biomolecules in the lungs (e.g., surfactant proteins and lipids) as opposed to ENMs ingested into the GI tract. The formation of the primary biocorona in the respiratory or GI tracts could modify the penetration and passage of nanoparticles across biological barriers from the lungs or intestines into the circulation, although it has been shown that an albumin biocorona around gold nanoparticles does not impede their translocation across the air–blood barrier in the lungs in rats (Konduru et al. 2017). Once in the circulation, a *secondary biocorona* forms through interaction with serum proteins such as albumin, immunoglobulins, and complement proteins. Natural antibodies may recognize epitopes of "self" proteins in the biocorona as "non-self" since proteins are known to undergo denaturation and unfold when bound to nanoparticles (Deng et al. 2011). The effect of ENMs, specifically nanoparticles, on the complement cascade and the implications for immunotoxicity versus nanomedicine applications are discussed in Chap. 6.

1.6 Translocation of ENMs across Biological Barriers

Upon exposure, ENMs directly contact the skin or may encounter the epithelial lining of the respiratory or intestinal tract after inhalation or ingestion, respectively. ENMs have been reported to translocate to different tissues and organs after exposure, whether the exposure route is inhalation, ingestion, or dermal. The keratinized epidermis of the skin generally acts as an effective barrier against ENM penetration, unless there is abrasion or damage resulting from sunburn. In contrast to skin, the epithelial lining of the respiratory and GI tract are more susceptible to injury by ENM exposure and translocation of ENMs can occur much more readily. The ability of ENMs to cross the air–blood barrier in the lung or the gut–blood barrier in the GI tract depends on the size and shape of the nanoparticle, but can also be influenced by surface charge, agglomeration status, and biocorona surrounding the ENM. For example, after inhalation, noncationic nanoparticles with a diameter of less than 34 nm that do not bind serum proteins reach lymph nodes within about 30 min (Choi et al. 2010). Translocation of ENMs across biological barriers to the organs of the immune system (lymph nodes, thymus, spleen) provides direct access to immune cells (Geiser and Kreyling 2010). It is also possible for ENMs with appropriate physicochemical characteristics to accumulate in other organ systems (liver, kidneys, heart), or cross the blood–brain barrier or the placental barrier (Kreyling et al. 2010). It is noteworthy that ENMs do not necessarily have to translocate to influence immune responses in tissues distant from the primary target organ. For example, inhaled carbon nanotubes cause release of soluble signals from the lungs (e.g., TGF-β1) that causes immunosuppression in the spleen (Mitchell et al. 2009). The translocation, bioaccumulation, and fate of ENMs in the body are discussed in Chap. 7.

1.7 Oxidative Stress in the Immune Response to ENMs

Many of the toxicological effects of inhaled particles are due to oxidative stress. For example, the oxidative stress potential of ambient air pollution particles, especially the ultrafine fraction (i.e., nanoparticles), has been attributed at least in part to transition metals (e.g., zinc, copper, and iron) that generate reactive oxygen species (ROS) either via Fenton-like reactions or by stimulating an oxidative burst when engulfed by phagocytes. For example, for nanoparticles that directly generate ROS through surface chemistry, decreasing particle diameter generally correlates with increasing toxicity. This is because an equivalent mass of smaller particles has a greater surface area per unit mass compared to larger particles. The generation of ROS in response to air pollution nanoparticles has been implicated in the pathogenesis of immune-related lung diseases in humans such as asthma and pulmonary fibrosis (Li et al. 2008). Moreover, ROS have been implicated in mediating the toxicity of ENMs in cells and experimental animals (Nel et al. 2009).

Two major mechanisms of ROS generation have been proposed. First, molecular oxygen interacts with the ENM surface, in the absence of cells, to generate free radicals such as superoxide ion and hydroxyl radical. Like ambient air pollution particles, the oxidative potential of metal oxide nanoparticles is due in part to generation of ROS via Fenton-like reactions. Second, ENMs activate inflammatory cells (e.g., macrophages and neutrophils) to stimulate the release of ROS through activation of the nicotinamide adenine dinucleotide phosphate (NADPH) oxidase system on cell membranes. For example, NADPH oxidase-derived ROS have been shown to play a central role in pulmonary fibrogenesis induced by exposure to carbon nanotubes (Shvedova et al. 2008). Intracellular ROS can also be generated from the mitochondria via disruption of the electron transport chain and ENMs can also disrupt mitochondrial homeostasis. For example, silver nanoparticles have been reported to reduce adenosine triphosphate (ATP) content in cultured glioblastoma cells and fibroblasts, causing damage to mitochondria and increasing production of ROS (Asharani et al. 2009). To counterbalance ROS generation, cells possess antioxidant systems, including *superoxide dismutase* (SOD), *catalase*, and *glutathione redox systems,* which degrade specific oxidants in specific ways. However, ENMs have the potential to inhibit these antioxidant systems. Finally, ROS can also serve as signaling intermediates to activate intracellular signaling molecules such as receptor tyrosine kinases, mitogen-activated protein (MAP) kinases, and transcription factors, leading to the expression of genes involved in immune responses (Thompson et al. 2014). The role of ROS in ENM-induced immunotoxicity is discussed in Chap. 8.

1.8 ENMs and Immune-Mediated Diseases

Immune-mediated diseases linked to toxicant exposure in the human population include asthma, hypersensitivity pneumonitis, pulmonary fibrosis, allergic contact dermatitis, and autoimmune diseases. It is also increasingly recognized that cancer has a strong immunological basis and cells of both the innate and adaptive immune system are found in the tumor microenvironment (Gajewski et al. 2013). As discussed earlier in the Sect. 1.2, there is a wealth of epidemiological evidence that links occupational and environmental exposure to toxicants (e.g., ambient particles and metals) (Cooper et al. 2002). However, human disease associated with exposure to ENMs has not been documented. Logically, one could make the assumption that this is either because ENMs will not cause human disease or it is simply too early, given the recent advent of nanotechnology. The latter is likely more plausible than the former, based on adverse effects of metals and the fact that many ENMs are composed of metals or metal oxides. However, epidemiological studies confirm that ENM exposure has occurred in workers (Basinas et al. 2018). Moreover, there is some evidence of adverse inflammatory or immune responses in workers exposed to ENMs. For example, biomarkers of pulmonary fibrosis are increased in workers exposed to multiwalled carbon nanotubes (Fatkhutdinova et al. 2016).

Much of what we know about the potential for ENMs to cause immune-related diseases is inferred from studies with rodents. For example, there is a wealth of information on ENMs in causing pulmonary fibrosis and allergic lung disease in rodents that has been recently reviewed (Duke and Bonner 2018; Ihrie and Bonner 2018). ENMs are also causes of contact dermatitis in experimental animals. The evidence for ENMs in causing or exacerbating airway inflammation and asthma or allergic dermatitis is discussed in Chap. 9.

1.9 ENM Interaction with the Microbial World and Implications for Immunity

In addition to the effects of ENMs on host immune cells, it is also possible that ENMs impact host susceptibility to microbes; bacteria and viruses that constantly pose a challenge to our immune system. It is established that exposure to ambient ultrafine particles by inhalation can enhance host susceptibility to viral and bacterial infections. Some types of ENMs also increased susceptibility to viral infection. For example, single-walled carbon nanotubes (SWCNT) increase infectivity of human epithelial cells exposed to pandemic influenza A and also increase viral titers in mice (Sanpui et al. 2014; Chen et al. 2017). Chapter 10 focuses on the impact of ENMs on host susceptibility to viral infections. While ENMs may increase host susceptibility to microbes, certain types of ENMs could also be exploited as delivery platforms to target viruses or bacteria for eradication. On the other hand, ENMs in the gut could wreak havoc on the host microbiome, which is essential for maintenance of host immunity. For example, dietary silver nanoparticles have been shown to disrupt the gut microbiome in mice (van den Brule et al. 2016).

1.10 Methods for Assessing the Immunotoxicity of ENMs

A tiered approach is commonly used for immunotoxicity testing of chemicals and for the most part, the same principles can theoretically be applied to ENMs. In the first tier, standard toxicity tests (e.g., 90-day repeated dose toxicity study) are performed in rodents to assess systemic effects of ENMs on organs of the immune system after inhalation or oral exposure. In the second tier, more specific immune function assays are utilized; for example, antibody formation or resistance against infectious agents. However, these tiered studies performed in rodents are increasingly impractical given the increasing variety of different ENMs, which is further complicated by postsynthesis surface functionalization to enhance functional properties of the materials. As a result, new and improved in vitro tests and in silico approaches are being developed that address high-throughput immunotoxicity testing, which will reduce reliance on animal testing (Nel et al. 2013). Specifically, state-of-the-art in vitro cell culture systems coupled with relevant dose–response exposure systems are increasingly being improved and refined to assess ENM immunotoxicity at the cellular level (Alépée et al. 2014; Fytianos et al. 2016; Septiadi et al. 2018). These in vitro systems for ENM immunotoxicity testing are discussed in Chap. 11. In addition, novel "organ-on-a-chip" technologies could provide more advanced in vitro assessments of immunotoxicity (Polini et al. 2019). Collectively, these alternative approaches will allow for the comparative analysis of large numbers of ENMs simultaneously and will also be important for hazard assessment at various stages of product development as well as throughout the life cycle of ENMs (Thomas et al. 2009). Finally, it is increasingly apparent that an integrated approach toward assessing the issue of nanoimmunotoxicology will be accomplished by the coordinated efforts of toxicologists, immunologists, engineers, exposure scientists, and risk assessors to foster sustainable development of nanotechnology.

References

Alépée N, Bahinski A, Daneshian M, De Wever B, Fritsche E, Goldberg A, Hansmann J, Hartung T, Haycock J, Hogberg H, Hoelting L, Kelm JM, Kadereit S, McVey E, Landsiedel R, Leist M, Lübberstedt M, Noor F, Pellevoisin C, Petersohn D, Pfannenbecker U, Reisinger K, Ramirez T, Rothen-Rutishauser B, Schäfer-Korting M, Zeilinger K, Zurich MG. State-of-the-art of 3D cultures (organs-on-a-chip) in safety testing and pathophysiology. ALTEX. 2014;31(4):441–77.

Asharani PV, Low Kah Mun G, Hande MP, Valiyaveettil S. Cytotoxicity and genotoxicity of silver nanoparticles in human cells. ACS Nano. 2009;3(2):279–90.

Banchereau J, Steinman RM. Dendritic cells and the control of immunity. Nature. 1998;392:245–52.

Basinas I, Jiménez AS, Galea KS, Tongeren MV, Hurley F. A systematic review of the routes and forms of exposure to engineered nanomaterials. Ann Work Expo Health. 2018;62(6):639–62.

Boraschi D, Fadeel B, Duschl A. Immune system. In: Fadeel B, Pietroiusti A, Shvedova A, editors. Adverse effects of engineered nanomaterials: exposure, toxicology, and impact on human health. 2nd ed: Elsevier/Academic Press (London); 2017. p. 313–37.

Borregaard N. Neutrophils, from marrow to microbes. Immunity. 2010;33(5):657–70.

Cattaneo AG, Gornati R, Sabbioni E, Chiriva-Internati M, Cobos E, Jenkins MR, Bernardini G. Nanotechnology and human health: risks and benefits. J Appl Toxicol. 2010;30(8):730–44.

Chen R, Riviere JE. Biological surface adsorption index of nanomaterials: modelling surface interactions of nanomaterials with biomolecules. Adv Exp Med Biol. 2017;947:207–53.

Chen H, Zheng X, Nicholas J, Humes ST, Loeb JC, Robinson SE, Bisesi JH Jr, Das D, Saleh NB, Castleman WL, Lednicky JA, Sabo-Attwood T. Single-walled carbon nanotubes modulate pulmonary immune responses and increase pandemic influenza a virus titers in mice. Virol J. 2017;14(1):242.

Choi HS, Ashitate Y, Lee JH, Kim SH, Matsui A, Insin N, Bawendi MG, Semmler-Behnke M, Frangioni JV, Tsuda A. Rapid translocation of nanoparticles from the lung airspaces to the body. Nat Biotechnol. 2010;28(12):1300–3.

Cooper GS, Miller FW, Germolec DR. Occupational exposures and autoimmune diseases. Int Immunopharmacol. 2002;2(2–3):303–13.

De Vuyst P, Camus P. The past and present of pneumoconioses. Curr Opin Pulm Med. 2000;6(2):151–6.

Deng ZJ, Liang M, Monteiro M, Toth I, Minchin RF. Nanoparticle-induced unfolding of fibrinogen promotes Mac-1 receptor activation and inflammation. Nat Nanotechnol. 2011;6:39–44.

Duke KS, Bonner JC. Mechanisms of carbon nanotube-induced pulmonary fibrosis: a physicochemical characteristic perspective. Wiley Interdiscip Rev Nanomed Nanobiotechnol. 2018;10(3):e1498.

Duke KS, Thompson EA, Ihrie MD, Taylor-Just AJ, Ash EA, Shipkowski KA, Hall JR, Tokarz DA, Cesta MF, Hubbs AF, Porter DW, Sargent LM, Bonner JC. Role of p53 in the chronic pulmonary immune response to tangled or rod-like multi-walled carbon nanotubes. Nanotoxicology. 2018;12(9):975–91.

Fatkhutdinova LM, Khaliullin TO, Vasil'yeva OL, Zalyalov RR, Mustafin IG, Kisin ER, Birch ME, Yanamala N, Shvedova AA. Fibrosis biomarkers in workers exposed to MWCNTs. Toxicol Appl Pharmacol. 2016;299:125–31.

Frank EA, Carreira VS, Shanmukhappa K, Medvedovic M, Prows DR, Yadav JS. Genetic susceptibility to toxicologic lung responses among inbred mouse strains following exposure to carbon nanotubes and profiling of underlying gene networks. Toxicol Appl Pharmacol. 2017;327:59–70.

Farrera C, Fadeel B. It takes two to tango: understanding the interactions between engineered nanomaterials and the immune system. Eur J Pharm Biopharm. 2015;95(Pt A):3–12.

Fadeel B, Feliu N, Vogt C, Abdelmonem AM, Parak WJ. Bridge over troubled waters: understanding the synthetic and biological identities of engineered nanomaterials. Wiley Interdiscip Rev Nanomed Nanobiotechnol. 2013;5(2):111–29.

Fytianos K, Drasler B, Blank F, von Garnier C, Seydoux E, Rodriguez-Lorenzo L, Petri-Fink A, Rothen-Rutishauser B. Current in vitro approaches to assess nanoparticle interactions with lung cells. Nanomedicine (Lond). 2016;11(18):2457–69.

Gajewski TF, Schreiber H, Fu YX. Innate and adaptive immune cells in the tumor microenvironment. Nat Immunol. 2013;14(10):1014–22.

Geiser M, Kreyling WG. Deposition and biokinetics of inhaled nanoparticles. Part Fibre Toxicol. 2010;7:2.

Hussain S, Vanoirbeek JA, Hoet PH. Interactions of nanomaterials with the immune system. Wiley Interdiscip Rev Nanomed Nanobiotechnol. 2012;4(2):169–83.

Hussell T, Bell TJ. Alveolar macrophages: plasticity in a tissue-specific context. Nat Rev Immunol. 2014;14(2):81–93.

Ihrie MD, Bonner JC. The toxicology of engineered nanomaterials in asthma. Curr Environ Health Rep. 2018;5(1):100–9.

Ihrie MD, Taylor-Just AJ, Walker NJ, Stout MD, Gupta A, Richey JS, Hayden BK, Baker GL, Sparrow BR, Duke KS, Bonner JC. Inhalation exposure to multi-walled carbon nanotubes alters the pulmonary allergic response of mice to house dust mite allergen. Inhal Toxicol. 2019;31(5):192–202.

Katwa P, Wang X, Urankar RN, Podila R, Hilderbrand SC, Fick RB, Rao AM, Ke PC, Wingard CJ, Brown JM. A carbon nanotube toxicity paradigm driven by mast cells and the IL-33/ST2 axis. Small. 2012;8(18):2904–12.

Konduru NV, Molina RM, Swami A, Damiani F, Pyrgiotakis G, Lin P, Andreozzi P, Donaghey TC, Demokritou P, Krol S, Kreyling W, Brain JD. Protein corona: implications for nanoparticle interactions with pulmonary cells. Part Fibre Toxicol. 2017;14(1):42.

Kreyling WG, Hirn S, Schleh C. Nanoparticles in the lung. Nat Biotechnol. 2010;28(12):1275–6.

Lambrecht BN, Hammad H. The immunology of asthma. Nat Immunol. 2015;16(1):45–56.

Li N, Xia T, Nel AE. The role of oxidative stress in ambient particulate matter-induced lung diseases and its implications in the toxicity of engineered nanoparticles. Free Radic Biol Med. 2008;44(9):1689–99.

Luster MI. A historical perspective of immunotoxicology. J Immunotoxicol. 2014;11:197–202.

Mangum JB, Turpin EA, Antao-Menezes A, Cesta MF, Bermudez E, Bonner JC. Single-walled carbon nanotube (SWCNT)-induced interstitial fibrosis in the lungs of rats is associated with increased levels of PDGF mRNA and the formation of unique intercellular carbon structures that bridge alveolar macrophages in situ. Part Fibre Toxicol. 2006;3:15.

Mitchell LA, Lauer FT, Burchiel SW, McDonald JD. Mechanisms for how inhaled multiwalled carbon nanotubes suppress systemic immune function in mice. Nat Nanotechnol. 2009;4(7):451–6.

Neagu M, Piperigkou Z, Karamanou K, Engin AB, Docea AO, Constantin C, Negrei C, Nikitovic D, Tsatsakis A. Protein bio-corona: critical issue in immune nanotoxicology. Arch Toxicol. 2017;91(3):1031–48.

Nel AE, Mädler L, Velegol D, Xia T, Hoek EM, Somasundaran P, Klaessig F, Castranova V, Thompson M. Understanding biophysicochemical interactions at the nano-bio interface. Nat Mater. 2009;8(7):543–57.

Nel AE, Nasser E, Godwin H, Avery D, Bahadori T, Bergeson L, Beryt E, Bonner JC, Boverhof D, Carter J, Castranova V, Deshazo JR, Hussain SM, Kane AB, Klaessig F, Kuempel E, Lafranconi M, Landsiedel R, Malloy T, Miller MB, Morris J, Moss K, Oberdorster G, Pinkerton K, Pleus RC, Shatkin JA, Thomas R, Tolaymat T, Wang A, Wong J. A multi-stakeholder perspective on the use of alternative test strategies for nanomaterial safety assessment. ACS Nano. 2013;7(8):6422–33.

Oberdorster G. Significance of particle parameters in the evaluation of exposure-dose-response relationships of inhaled particles. Inhal Toxicol. 1996;8 Suppl:73–89.

Pampaloni F, Florin EL. Microtubule architecture: inspiration for novel carbon nanotube-based biomimetic materials. Trends Biotechnol. 2008;26(6):302–10.

Polini A, Del Mercato LL, Barra A, Zhang YS, Calabi F, Gigli G. Towards the development of human immune-system-on-a-chip platforms. Drug Discov Today. 2019;24(2):517–25.

Sanpui P, Zheng X, Loeb JC, Bisesi JH Jr, Khan IA, Afrooz AR, Liu K, Badireddy AR, Wiesner MR, Ferguson PL, Saleh NB, Lednicky JA, Sabo-Attwood T. Single-walled carbon nanotubes increase pandemic influenza A H1N1 virus infectivity of lung epithelial cells. Part Fibre Toxicol. 2014;11:66.

Sargent LM, Shvedova AA, Hubbs AF, Salisbury JL, Benkovic SA, Kashon ML, et al. Induction of aneuploidy by single-walled carbon nanotubes. Environ Mol Mutagenesis. 2009;50(8):708–17.

Scoville DK, Botta D, Galdanes K, Schmuck SC, White CC, Stapleton PL, Bammler TK, MacDonald JW, Altemeier WA, Hernandez M, Kleeberger SR, Chen LC, Gordon T, Kavanagh TJ. Genetic determinants of susceptibility to silver nanoparticle-induced acute lung inflammation in mice. FASEB J. 2017;31(10):4600–11.

Septiadi D, Crippa F, Moore TL, Rothen-Rutishauser B, Petri-Fink A. Nanoparticle-cell interaction: a cell mechanics perspective. Adv Mater. 2018;30(19):e1704463.

Shvedova AA, Kisin ER, Murray AR, Kommineni C, Castranova V, Fadeel B, et al. Increased accumulation of neutrophils and decreased fibrosis in the lung of NADPH oxidase-deficient C57BL/6 mice exposed to carbon nanotubes. Toxicol Appl Pharmacol. 2008;231(2):235–40.

Stone V, Miller MR, Clift MJD, Elder A, Mills NL, Møller P, Schins RPF, Vogel U, Kreyling WG, Alstrup Jensen K, Kuhlbusch TAJ, Schwarze PE, Hoet P, Pietroiusti A, De Vizcaya-Ruiz A, Baeza-Squiban A, Teixeira JP, Tran CL, Cassee FR. Nanomaterials versus ambient ultrafine particles: an opportunity to exchange toxicology knowledge. Environ Health Perspect. 2017;125(10):106002.

Tkach AV, Shurin GV, Shurin MR, Kisin ER, Murray AR, Young SH, Star A, Fadeel B, Kagan VE, Shvedova AA. Direct effects of carbon nanotubes on dendritic cells induce immune suppression upon pulmonary exposure. ACS Nano. 2011;5:5755–62.

Thomas T, Bahadori T, Savage N, Thomas K. Moving toward exposure and risk evaluation of nanomaterials: challenges and future directions. Wiley Interdiscip Rev Nanomed Nanobiotechnol. 2009;1(4):426–33.

Thompson EA, Sayers BC, Glista-Baker EE, Shipkowski KA, Taylor AJ, Bonner JC. Innate immune responses to nanoparticle exposure in the lung. J Environ Immunol Toxicol. 2014;1(3):150–6.

van den Brule S, Ambroise J, Lecloux H, Levard C, Soulas R, De Temmerman PJ, Palmai-Pallag M, Marbaix E, Lison D. Dietary silver nanoparticles can disturb the gut microbiota in mice. Part Fibre Toxicol. 2016;13(1):38.

Xia T, Li N, Nel AE. Potential health impact of nanoparticles. Annu Rev Public Health. 2009;30:137–50.

Chapter 2
Macrophages: First Innate Immune Responders to Nanomaterials

Dorothy J. You, Ho Young Lee, and James C. Bonner

Abstract Macrophages are professional phagocytes that play key roles in immune surveillance and host defense against a variety of external stimuli, including engineered nanomaterials (ENMs). Physicochemical characteristics of ENMs such as size, shape, and surface charge are factors that determine macrophage recognition and uptake through various endocytotic mechanisms. Moreover, toll-like receptors (TLRs) and scavenger receptors (SRs) on the surface of macrophages facilitate binding, uptake, and intracellular signaling in response to specific types of ENMs. Biocorona formation further modulates the interaction of ENMs with cell-surface receptors. The immune response of macrophages to ENMs, including inflammasome activation and alternative polarization, plays important roles in immune-mediated diseases such as fibrosis, asthma, and cancer.

Keywords Macrophage · Nanomaterial · Toll-like receptors · Scavenger receptors · Reactive oxygen species · Inflammasome · Polarization

2.1 Introduction

Macrophages play critical roles in immune surveillance of pathogens and removal of inhaled particles and fibers. The immune response to engineered nanomaterials (ENMs) is similar in many ways to a variety of larger inhaled particles and fibers, although the nanoscale dimensions of ENMs allow for some unique interactions at the cell surface or with subcellular structures; termed the "*nano-bio interface*." The dynamic activity of activated macrophages can be appreciated in time-lapse photography under controlled cell culture conditions that reveals an aggressive, highly mobile amoeba-like phagocyte that has an impressive ability to engulf foreign

D. J. You · H. Y. Lee · J. C. Bonner (✉)
Toxicology Program, Department of Biological Sciences, North Carolina State University, Raleigh, NC, USA
e-mail: jcbonner@ncsu.edu

J. C. Bonner, J. M. Brown (eds.), *Interaction of Nanomaterials with the Immune System*, Molecular and Integrative Toxicology,
https://doi.org/10.1007/978-3-030-33962-3_2

particles or other injured cells. Macrophages are recruited to sites of ENM deposition by chemotactic factors released from lung epithelial cells at the deposition site. Upon arriving at sites of deposition, macrophages recognize various particles and fibers, including ENMs, in part through cell-surface receptors (discussed in detail below) that serve to tether particles to the cell membrane. In some cases, ENM interaction with cell-surface receptors activates signal transduction pathways prior to phagocytosis. After particle uptake, macrophages release a variety of factors (cytokines, proteases, and reactive oxygen species) that influence the local inflammatory reaction or systemic recruitment of other inflammatory cells. Macrophage-mediated clearance is facilitated via the *mucociliary escalator*; transport out of the lung via the unidirectional beating of cilia on the airway epithelium, which propels macrophages and their cargo out of the lungs in a fluid layer of mucosubstances. Alternatively, macrophages transport engulfed ENMs out of the lung via the lymphatic drainage system where they migrate to immune organs such as the thymus and spleen. Whether macrophages remain in the lung or migrate to other organs and tissues after phagocytosis, their cargo of nanoparticles (NPs) can persist in viable cells for months.

2.2 Macrophage Recognition and Uptake of ENMs

Particles and fibers are taken up by alveolar macrophages through endocytosis; a broad term that includes phagocytosis (typically applying to particles >5 microns), receptor-independent mechanisms such as macropinocytosis of micrometer-sized particles (typically outside of the nanoscale range >100 nm), clathrin-dependent endocytosis, and clathrin-independent endocytosis that may involve caveolin (Kuhn et al. 2014; Boraschi et al. 2017). Clathrin-dependent and clathrin-independent caveolin-dependent endocytosis are involved in the uptake of smaller nanoparticles (20–500 nm). For ENMs, recognition and uptake is determined by the physicochemical properties of the ENMs and coating of the ENM surface with biomolecules (i.e., biocorona). While endocytotic mechanisms such as phagocytosis and macropinocytosis generally apply to larger particles, ENMs with nanoscale dimensions <100 nm (e.g., graphene sheets or carbon nanotubes) combine via electrostatic charge to form micron-sized agglomerates that are recognized as larger particles and taken up via phagocytosis. A transmission electron microscopy image of a mouse lung macrophage containing MWCNT agglomerates after inhalation exposure is shown in Fig. 2.1. Because of the loose adherence of ENMs to one another in these agglomerates, there is extremely high surface area within an ENM agglomerate beyond what would be anticipated from a solid micron-sized particle of the same chemical composition. Greater surface area generates more ROS and toxicity.

The adhesion of ENMs to the cell surface is an important first step toward macrophage uptake. Physicochemical properties such as size, shape, rigidity, and surface charge determine the adhesion and uptake of ENMs to the macrophage cell

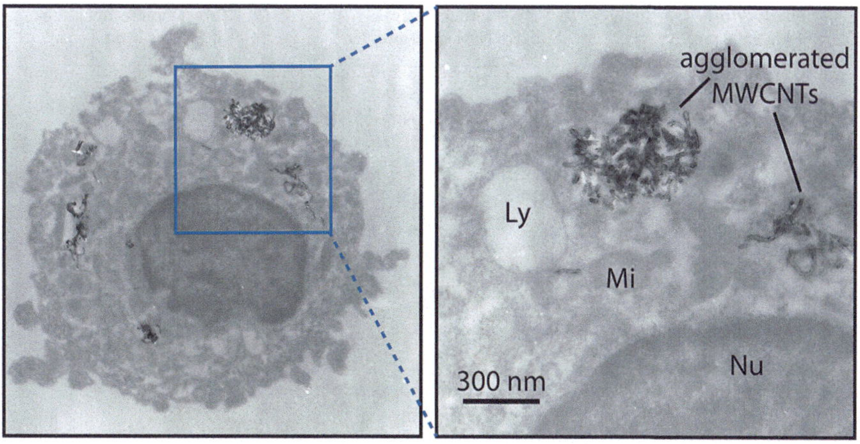

Fig. 2.1 Transmission electron microscopy of a macrophage in the lung tissue of a mouse after inhalation exposure to MWCNTs. Agglomerates of MWCNTs are indicated in relationship to subcellular structures: nucleus (Nu), mitochondria (Mi), and lysosome (Ly)

surface (Paul et al. 2013; Garapaty and Champion 2017). In biological fluids such as plasma or lung lining fluid, ENMs rapidly acquire a plethora of proteins on their surface, forming a "biocorona" (Monopoli et al. 2011). The coating process where plasma proteins form a biocorona on particles has been referred to in classical immunology as "opsonization" (Boraschi et al. 2017). Opsonization plays a role in the recognition of nanoparticles by circulating monocytes or tissue macrophages that express specific receptors for biomolecules found in the biocorona such as complement components, immunoglobulins, and apolipoproteins (Patra et al. 2016). Through opsonization and biocorona formation, ENMs are more readily recognized and bind to the surface of macrophages to stimulate an "eat me" signal that is required for mobilization of the actin cytoskeleton and the process of phagocytosis by professional phagocytes. Moreover, greater particle rigidity favors macrophage uptake over less rigid particles (Sosale et al. 2015). This greater propensity to engulf rigid ENMs can result in cytotoxic responses once particles are taken up in phagocytic vesicles and fuse with lysosomes, since long tube- or fiber-like ENMs cause damage to lysosomal membranes. For example, rod-like multiwalled carbon nanotubes (MWCNTs) cause frustrated phagocytosis in macrophages, which results in cell membrane damage and leakage of cellular constituents (Duke and Bonner 2018). The consequences of lysosomal membrane damage include assembly and activation of the inflammasome mechanism, which is discussed in detail below. In general, many of the principles that apply to fiber-like particles (e.g., asbestos, silica), such as frustrated phagocytosis of rigid fibers, also apply to fiber-like ENMs (Donaldson et al. 2006).

Macrophage receptors that bind specific types of ENMs and mediate intracellular uptake include Fc receptors, mannose receptors, toll-like receptors (TLRs), and scavenger receptors (SRs). Some of these cell-surface receptors are illustrated in Fig. 2.2.

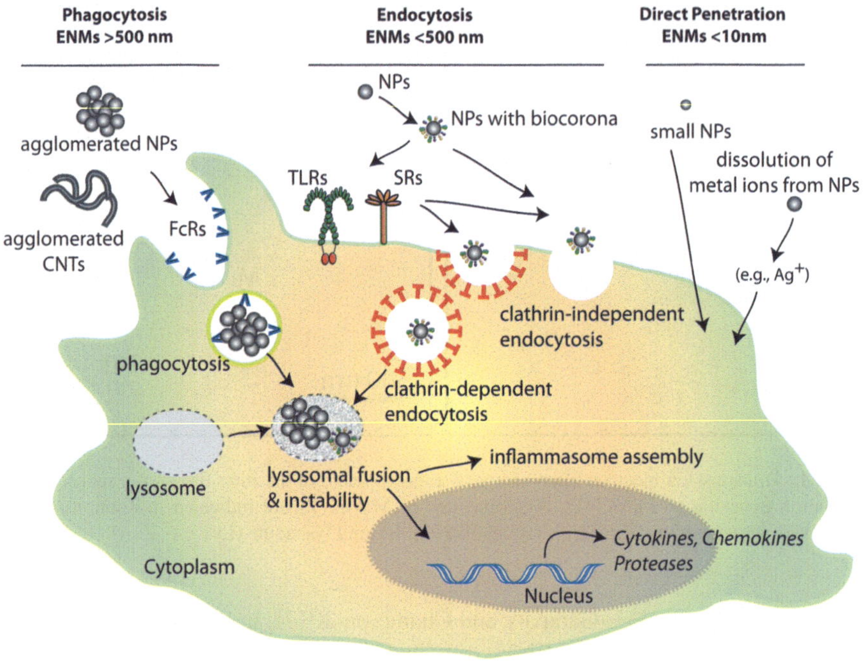

Fig. 2.2 Mechanisms of ENM recognition and uptake by macrophages. ENMs larger than 500 nm, including agglomerated nanoparticles (NPs) or carbon nanotubes (CNTs), are taken up by phagocytosis mediated by Fc receptors. ENMs smaller than 500 nm are taken up by clathrin-dependent or clathrin-independent endocytosis, and cell-surface toll-like receptors (TLRs) or scavenger receptors (SRs) could participate in this process. Smaller NPs or ions released from larger NPs could directly penetrate the cell membrane

Also, while some are nonsignaling receptors that tether ligands at the cell surface, others (e.g., TLRs) also mediate cell signaling events that result in immune- or inflammatory responses of macrophages to ENMs.

2.3 ENM Signaling via Toll-Like Receptors

Toll-like receptors (TLRs) are transmembrane proteins located on the plasma membrane or in the endosomes that are in the family of pattern recognition receptors (PRRs). PRRs recognize molecular structures called pathogen- or damage-associated molecular patterns (PAMPs or DAMPs) that are usually found in a variety of different microbes (Kuzmich et al. 2017). Among many different TLRs, TLR4 is known as a target for activation by gram-negative bacterial lipopolysaccharides (LPS) and plays a vital role toward generating the innate immune response to bacterial products (Kuzmich et al. 2017). LPS specifically binds to the myeloid differentiation factor 2 (MD-2) component of TLR4, initiating receptor dimerization and signal transduction (Kuzmich et al. 2017; Miller et al. 2005). TLR4 activation can lead to activation of mitogen-activated protein kinase (MAPK)/extracellular signal-regulated kinase (ERK), nuclear factor κB (NF-κB), and interferon regulatory factor (IRF) pathways, leading to the production of pro-inflammatory cytokines (e.g., IL-6, IL-8) (Kuzmich et al. 2017). These secreted cytokines mediate diverse biological responses in other resident pulmonary cells or stimulate the migration of circulating inflammatory cells (e.g., monocytes and neutrophils) to the sites of macrophage activation in the lung.

It is well known that many ENMs elicit a Th1 inflammatory response in the lungs of rodents that is characterized by neutrophilic infiltration (Thompson et al. 2014). Since LPS also stimulates a neutrophilic Th1-type of inflammatory response, it has been proposed that some ENMs mediate their pro-inflammatory effects via TLR4. While TLR4 is important in mediating the pro-inflammatory effects of LPS and plays a role in the inflammatory responses to some ENMs, the nature and amplitude of the macrophage response involving TLR4 is regulated by the integration of signals from other receptors that recognize microbial products, cytokines, and other soluble mediators (Wang et al. 2008). For example, there is cross talk between TLR4 signaling pathways and interferon signaling pathways that mediate M1 polarization of macrophages. This concept is illustrated in Fig. 2.3.

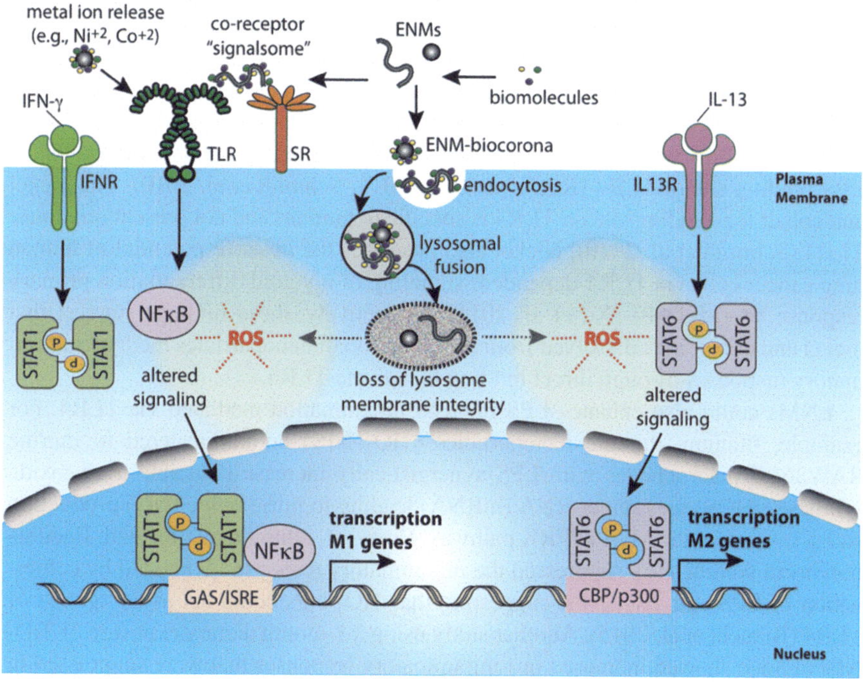

Fig. 2.3 Examples of molecular interactions at the "nano-bio interface" that influence macrophage activation and polarization. ENMs (e.g., nanoparticles or carbon nanotubes) form a biocorona of biomolecules that influences endocytosis or recognition by cell-surface toll-like receptors (TLRs) or scavenger receptors (SRs). Intracellular ENMs also generate ROS that could alter interferon receptor (IFNR) or IL-13 receptor (IL13R) signaling and macrophage polarization

In addition, reactive oxygen species (ROS) have been shown to account for pro-inflammatory potency of ENMs (Petrarca et al. 2015; Ma et al. 2016). ROS generation in the cytosol of macrophages disrupts intracellular signaling leading to membrane and DNA damage, but relatively low levels of ROS can also activate cell signaling intermediates, including transmembrane receptors, kinases, and transcription factors (Chan et al. 2017). ROS-induced activation of intracellular signaling pathways involving phosphorylation (e.g., receptor tyrosine kinases, MAP kinases, transcription factors) is mediated at least in part by deactivation of phosphatases that serve to dephosphorylate and inactivate signaling molecules (Samet and Tal 2010). Some components of ozone-induced oxidative stress, including neutrophilic inflammation in the lungs of mice, are mediated via TLR4, and TLR4-deficient mice lack ozone-induced neutrophilia (Gabehart et al. 2015). Therefore, oxidative stress activates TLR4 signaling, either directly or indirectly, and thereby affects downstream signaling molecules in this pathway.

There is limited information showing direct interaction of ENMs with TLR4. However, some metal nanoparticles undergo dissolution to release metal ions that elicit TLR4-dependent inflammatory responses. For example, nickel or cobalt ions bind to the MD2 component of TLR4 on human THP-1 monocyte-derived macrophages to promote NF-κB activation and cytokine production (Oblak et al. 2015; Potnis et al. 2013; Schmidt and Goebeler 2011). Human embryonic kidney (HEK) cells do not normally possess TLR4, yet when transfected with human TLR4 these cells are responsive to nickel- or cobalt-induced pro-inflammatory cytokine production. Moreover, imidazole groups H431, H456, and H458 of human TLR4 are required for nickel and cobalt ions to induce inflammatory responses (Schmidt et al. 2010). Furthermore, the nickel and cobalt ion-binding site is independent of LPS-binding site of TLR4 (Raghavan et al. 2012; Schmidt et al. 2010). The nickel and cobalt ion-binding site on TLR4 is specific to humans and not present on mouse TLR4 (Schmidt et al. 2010). Nickel also promotes the invasive potential of human lung cancer cells via TLR4-dependent signaling of myeloid differentiation primary response 88 (MyD88) (Xu et al. 2011). Collectively, these findings suggest that nickel and cobalt ions dissolved from nickel or cobalt nanoparticles mediate inflammatory responses through direct binding to activate TLR4.

ENMs could also enhance LPS-induced inflammation mediated via TLR4. For example, titanium dioxide nanoparticles (TiO_2 NPs) co-administered to murine RAW264.7 macrophages with LPS synergistically increase inducible nitric oxide synthase (iNOS) messenger RNA (mRNA) leading to nitric oxide (NO) production and activation of the MAPK/ERK pathway. However, inhibiting TLR4 with Bacillus polymyxa significantly suppressed the inflammatory responses generated by combination of LPS and TiO_2 NPs, suggesting that TiO_2 NPs could act as an agonist of TLR4 (Bianchi et al. 2015). Another study using a 1-month dietary exposure of TiO_2 NPs in mice showed increased pro-inflammatory responses that were suppressed in TLR4-deficient mice (Huang et al. 2017). TiO_2 nanoparticles also prime macrophages for enhanced inflammatory responses to LPS. For example, TiO_2 nanoparticles aggravated septic shock in mice in response to LPS, whereas TLR4-deficient mice and primary macrophages from these mice displayed a reduced pro-inflammatory response to TiO_2 nanoparticles (Huang et al. 2017). These studies show that TiO_2 nanoparticles enhance LPS-induced TLR4 activation in order to stimulate intracellular signaling, cytokine production, and consequent acute inflammation, although the precise mechanisms remain to be identified. Elucidating the mechanisms through which ENMs, such as TiO_2, enhance TLR4 signaling initiated by microbial agents is an important area for future research.

Lipid NPs, particularly those that are positively charged, also interact with TLR4 to promote a inflammatory response (Kedmi et al. 2010). DiC14-amidine is a synthetic cationic lipid nanoparticle used to transport nucleic acids in nanomedicine applications. This nanoparticle also acts as an agonist of TLR4. Point mutagenesis of TLR4 has shown that possible binding/docking site of DiC14-amidine would be different from both LPS and cobalt and nickel ions (Lonez et al. 2015). From this study, it was suggested that DiC14-amidine binds to the hydrophobic crevices of TLR4, and that DiC14-amidine can activate the TLR4 signaling pathway to generate immune responses similar to nickel and cobalt ions.

A recent study showed that endotoxin-free single-walled carbon nanotubes (SWCNTs) directly bind to TLRs and suggested that they act as nanoparticle-associated molecular patterns (NAMPs), like PAMPs, to activate the TLR–MyD88 signaling pathway when taken up by macrophages (Mukherjee et al. 2018). This study showed that either pristine or carboxylated SWCNTs can interact directly with TLR4 at different locations through both hydrophobic and electrostatic interactions. Molecular modeling with both pristine and carboxylated SWCNTs found two regions of TLR4 that bind to the tip and the sidewalls of SWCNTs. The first is a hydrophobic region of TLR4 from residues Ile 108 to Asn 265, while the second region is a loop around Ile 412 and Leu 434. Residues Arg 382, His 431, and His 458 are located at a close proximity enabling ion-pair interaction with the sidewalls of carboxylated SWCNTs only. Therefore, hydrophobic interaction with TLR4 seems to promote optimal binding to both pristine and carboxylated SWCNTs. Furthermore, in SWCNT-induced inflammation, NF-κB activation via TLR4 is critical for generating inflammatory responses and production of pro-inflammatory cytokines.

TLRs other than TLR4, including TLR2 and TLR6, also have been reported to play a role in mediating inflammatory responses from different ENM exposures. Interestingly, it has been reported that SWCNTs can activate both TLR2 and TLR4 for NF-kB activation to produce inflammatory responses in HEK293 cells transfected with TLR2 and TLR4 (Mukherjee et al. 2018). TLR2 can recognize a variety of microbes, such as gram-negative and gram-positive bacteria, fungi, lipoproteins, and mycoplasma, which are known to activate the NF-kB pathway (Yang et al. 1998). Furthermore, computational studies using X-ray structures showed that SWCNTs and C_{60} fullerenes have high binding scores for dimerized TLRs, including TLR1/2 and TLR2/6, suggesting that they may be recognized by TLRs (Turabekova et al. 2014). Therefore, different TLRs have the ability to recognize different types of ENMs and activate intracellular signaling for production of inflammatory mediators.

2.4 Role of Scavenger Receptors in ENM Recognition and Signaling

Scavenger receptors (SRs) encompass a large family of sequentially and structurally diverse transmembrane glycoproteins largely expressed by myeloid cells. First defined by Goldstein and Brown in 1979, SRs have since attracted much attention, leading to discoveries of variants that share similarity in the function of detecting and removing unwanted entities (Brown and Goldstein 1979). These variants, currently organized into nine different classes (A-I), consist of various combinations and permutations of structural and functional domains that enable recognition of a wide spectrum of targets as well as functional flexibility. Consequently, the broad binding specificity toward a variety of self- and non-self-ligands has implicated SRs in maintenance of homeostasis as well as innate immune responses (Brown and Goldstein 1979; Murphy et al. 2005; Canton et al. 2013).

Primarily, SRs have been well documented to behave as pattern recognition receptors (PRRs). They recognize and internalize danger-associated molecular patterns (DAMPs), mainly modified forms of low-density lipoproteins (LDL), as well as non-self, pathogen-associated molecular patterns (PAMPs) such as lipopolysaccharide (LPS), lipoteichoic acid (LTA), and other ligands associated with both gram-negative and -positive bacteria (Elomaa et al. 1995; van der Laan et al. 1999). More recent studies have provided evidence that suggest their ability to also recognize other endogenous host molecules such as nonmodified, native LDL, molecular chaperones, and extracellular matrix (ECM) proteins (Pluddemann et al. 2007; Pluddemann et al. 2009). Association with such a broad range of ligands has implicated SRs in a range of diverse signaling pathways involving gene transcription, cell polarity, actin polymerization, cell migration and adhesion, angiogenesis, production of pro-inflammatory cytokines, initiation of apoptosis, and various processes involved in the innate immune response. Detailed involvement of SRs in these cellular processes has been reviewed (Gough and Gordon 2000; Canton et al. 2013; Zani et al. 2015).

Although SRs have been known to engage multiple intracellular signaling pathways, the exact mechanisms involved in conveying the signals are unclear, as only a few members of the SR family display functional intracellular motifs or domains involved in signal transduction. Therefore, a fairly novel paradigm is that SRs function cooperatively as a component of *signalosomes*, hetero-multimeric signaling complexes consisting of multiple components that together activate intracellular signal transduction, and this mechanism has been introduced to explain the involvement of SRs in various physiological and pathological processes (Heit et al. 2013). However, this paradigm has been challenged by investigators attempting to investigate various aspects of SR, as conflicting findings have suggested that signalosome complexes tend to be highly influenced by cell-type as well as the context of the microenvironment in which the cell resides. For example, the association of SR-A1 with tumor necrosis factor receptor-associated factor 3 (TRAF3) was found to promote suppression of NF-κB leading to an overall anti-inflammatory response, while the association of SR-A1 with TLR4 in the presence of LPS activated the NF-κB pathway resulting in a pro-inflammatory response (Yu et al. 2012; Zong et al. 2018). In addition, fluorescently labeled high-mobility group box 1 (HMGB1), a DAMP that was found to internalize into both M1- and M2-type macrophages through class A scavenger receptor (SR-A) and TLR4, elicited inflammatory cytokine production only in M1-type macrophages. Inhibition of the binding of HMGB1 to the SR using maleylated bovine serum albumin not only reduced cytokine production but also reduced HMGB1–TLR4 interaction, suggesting that SR-A may function as a coreceptor to supplement HMGB1–TLR4 association and the resulting cellular signaling for cytokine production.

Aside from their interaction with conventional, biological ligands, the possibility of SR involvement in uptake and clearance of ENMs has gained much attention in recent years. Due to the promiscuity of ligands that can be recognized by a range of SRs, researchers have investigated SR as possible portal of entry through which ENMs can cross the cell membrane and/or initiate various cellular responses. With consideration that SRs structurally display either positively charged arginine (SR-A1 and MARCO), arginine and cysteine (MARCO), or lysine (SR-A2) clusters functioning as ligand-binding domains, there is wide speculation that negatively charged ENMs such as silica, TiO_2, superparamagnetic iron oxide (SPIO), asbestos, and MWCNTs may associate with these domains (Light and Wei 1977; Hamilton et al. 2000; Thakur et al. 2008). As mentioned above, SRs may function as a component of a signalosome in which SR may behave as a tethering receptor for other signaling receptors to interact with the tethered ENM to elicit a cellular response. This concept is illustrated in Fig. 2.3.

Currently, class A scavenger receptors (SR-A1, SR-A2, and macrophage receptor with collagenous structure (MARCO, also known as SR-A6)) have received much attention with regard to their interaction with ENMs. Other members of SR, such as class B scavenger receptor member 1 (SR-B1) and cluster of differentiation 36 (CD36, also known as SR-B2), have also been investigated to a lesser degree. One of the earliest studies into SR–particle interaction revealed that secreted domains on SR-A1 and SR-A2 were capable of binding to crocidolite asbestos (Resnick et al. 1993). In this study, SR inhibition via polyinosinic acid (poly(I)) and polyguanylic acid (poly(G)) significantly decreased such interaction and this supports earlier work on SR involvement in recognition of asbestos fibers (Holian and Scheule 1990). More recently, it was demonstrated that MARCO recognizes chrysotile asbestos and mediates macrophage polarization and development of pulmonary fibrosis (Murthy et al. 2015). MARCO-deficient macrophages generated via siRNA transfection displayed significantly less binding to chrysotile asbestos while mutation introduced to Arg^{432} and Arg^{434} in the extracellular domain V of MARCO showed decreased asbestos binding and decreased polarization of the macrophages toward a profibrotic, M2 phenotype. Both in vitro cell culture models and in vivo murine models investigating SR-A1 and MARCO have suggested that both members of SR-A can bind to silica and titanium particles (Hamilton Jr et al. 2006; Thakur et al. 2008). Additionally, a study involving human alveolar macrophages isolated from bronchoalveolar lavage (BAL) of healthy adults showed that antibodies against human MARCO resulted in decreased binding of titanium oxide and latex beads (Arredouani et al. 2005), thereby further supporting MARCO's involvement in ENM-cellular association. MARCO-transfected COS-7 cells recognized and mediated the uptake of ultrafine particles such as polystyrene nanoparticles with diameters ranging from 20 nm to 1 μm (Kanno et al. 2007). SR-A was also found to mediate cellular uptake of superparamagnetic iron oxide (SPIO) via the phagocytic pathway (Lunov et al. 2011; Raynal et al. 2004). SRs have also been shown to be involved in the uptake of carbon-based ENMs such as MWCNTs and graphene. Using a study design similar to the one involving MARCO-transfected COS-7 cells mentioned above, MARCO-transfected Chinese hamster ovary cells exhibited dendritic structures that tethered and engulfed MWCNTs, resulting in frustrated phagocytosis and cell death (Hirano et al. 2012). In the other work involving MWCNT exposure, the application of inhibitors of SR-A and pan-SR (2F8 and fucoidan, respectively) reduced the uptake of MWCNTs and MWCNT-induced apoptosis, suggesting a role for SR-A in the cellular association with MWCNTs (Gao et al. 2011; Wang et al. 2012). An investigation using bone marrow derived macrophages found that both SR-B1 and class B scavenger receptor member 2 (SR-B2) bind to silica and behave only as tethering receptors rather than internalizing the silica nanoparticles (Tsugita et al. 2017). In a separate study, SR-B1 was found to mediate the binding and uptake of silver nanoparticles (AgNPs) in macrophages, ultimately inducing apoptosis (Singh and Ramarao 2012). Furthermore, AgNP internalization and the subsequent apoptotic response were significantly diminished when a SR-B1 inhibitor was introduced.

Understanding the interaction of ENMs with biomolecules and cells at the nano-bio interface is a challenging task. The variables that are presented by the dynamic physicochemical properties of ENMs have made it extremely difficult to establish clear mechanistic information as to how ENMs interact with the structurally and functionally diverse family of SRs. Further complicating the issue is the interaction of ENMs with proteins, lipids, and carbohydrates in biological systems to form a biocorona that influences recognition of ENMs by cell-surface receptors. Recent findings show that when ENMs enter a biological system, they form a biocorona before being recognized by receptors present on the plasma membrane. Biocorona formation has been widely accepted to influence ENM–cell interaction and displays dynamic heterogeneity (Fleischer and Payne 2014a; Runa et al. 2018; Feiner-Gracia et al. 2017). This means that cell-surface receptors recognize elements of the biocorona rather than the ENM core, although ENMs could play a role in the presentation of biocorona components (Fleischer and Payne 2014b). Moreover, proteins in the biocorona may undergo structural alterations (e.g., surface denaturation) where they are no longer recognized as "self" by macrophages (Jayaram et al. 2017; Runa et al. 2018). Therefore, this raises the possibility that certain exposed motifs of the biocorona, coupled with versatile target recognition capability displayed by members of the SR family, promote (or possibly deter) ENM–cell interaction and modify downstream signaling and cellular responses. In addition to a few examples introduced above, coating silver nanoparticles (AgNPs) with albumin reduced their zeta potential and resulted in reduced uptake by macrophages (Shannahan et al. 2015). In addition, coating of AgNPs with high-density lipoproteins (HDL) resulted in decreased uptake and increased mRNA expression of interleukin 6 (IL-6) by endothelial cells, an observation that was diminished when SR-B1 was inhibited with SR-B1 inhibitor Blt-2 (Shannahan et al. 2015). Such findings collectively suggest that the formation of a biocorona may have a significant impact on the uptake process and that promiscuity in recognition of ligands by SRs may also elicit varied cellular responses.

It should be noted that the cellular response elicited by both TiO_2 and silica differs from that elicited by chrysotile asbestos, even though they have been found to bind to domain V of MARCO. One possible explanation for such observation is the different binding residues that are targeted by the ENMs. It has been previously documented that specific amino acids play key roles in the binding of these ENMs (Arredouani et al. 2005; Thakur et al. 2009). Moreover, it is possible that such differences can result in varied downstream effects.

Continued research into SRs has progressively expanded the SR superfamily in both number and their role in physiological and pathological processes. Such a structurally and functionally diverse family of receptors presents exciting new discoveries to come as well as a challenging road in elucidating mechanistic information that can be useful in unraveling the potential of SRs as therapeutic targets, including those involving nanoparticles. Moreover, SR's role in association with ENMs has become apparent and promises to be an important area of investigation in illuminating mechanisms involved in ENM-induced toxicology.

2.5 The Respiratory Burst: Innate Defense Mechanism with the Risk of Collateral Damage

The generation and release of reactive oxygen species (ROS) by macrophages, termed "respiratory burst," is an important innate host defense mechanism (Thomas 2017; Davies et al. 2019). The respiratory burst by macrophages, which likely evolved as a microbial killing mechanism, involves the conversion of oxygen to superoxide anion ($O_2^{\bullet-}$) that is primarily generated by the membrane nicotinamide adenine dinucleotide phosphate (NADPH) oxidase system (Mittal et al. 2014). $O_2^{\bullet-}$ is further converted to hydrogen peroxide (H_2O_2) by oxidation via the enzyme superoxide dismutase (SOD). H_2O_2 can be converted to highly reactive hydroxyl radical (•OH) via the iron-catalyzed Fenton reaction. During respiratory burst, these ROS are released by macrophages into the extracellular microenvironment, resulting not only in microbial killing but also in some collateral damage of host cells. Protective antioxidant enzymes, such as SOD, catalase, and the glutathione redox system, serve to convert ROS to water and molecular oxygen.

Phagocytosis of a variety of particles, fibers, and metals is known to stimulate the macrophage respiratory burst (Gwinn and Vallyathan 2006). In addition, specific types of ENMs, such as MWCNTs, cause loss of lysosomal membrane permeability and generation of ROS that has been implicated as a mechanism of inflammasome activation (discussed below in more detail). In addition to macrophage-derived ROS generated via NADPH or xanthine oxidases, foreign particles or fibers have the ability to generate ROS in the absence of cellular mechanisms via oxidation–reduction (i.e., redox) reactions. The oxidative potential of particles is generally attributed to transition metals (e.g., zinc, iron) that can induce both a respiratory burst in macrophages as well as extracellular redox reactions to form ROS.

ENMs are particularly potent generators of ROS compared to larger particles, due to their extremely high surface area per unit mass. A comparison of two different sizes of TiO_2 nanoparticles (20 nm versus 250 nm diameter) on lung inflammation in mice or rats demonstrated that smaller nanoparticles are more inflammatory than larger nanoparticles when normalized on a per mass basis (Oberdörster et al. 1994). This is due to a much greater surface area per unit mass and greater ROS-generating capacity at the nanoparticle surface (Oberdörster et al. 1992). In addition, the toxicity of smaller nanoparticles to macrophages is related to increased access to the distal lung compartments.

2.6 Inflammasome Activation

Fiber-like particles, including asbestos and silica, all exert at least part of their pro-inflammatory activity by activating inflammasomes; intracellular protein scaffolds assembled in response to lysosomal membrane damage that serve to cleave pro-IL-1β to mature IL-1β (Sayan and Mossman 2016). Cleavage is accomplished by the action of caspase-1 in the inflammasome scaffold. A number of studies have shown that MWCNTs stimulate inflammasome activation and consequent IL-1β secretion by macrophages (Palomäki et al. 2011; Meunier et al. 2012; Strowig et al. 2012; Sun et al. 2015). Inflammasome activation by MWCNTs and other high aspect ratio nanomaterials (e.g., nanofibers, TiO_2 nanobelts) is mediated, at least in part, by disruption of lysosomal membrane integrity, leading to the release of proteases and generation of ROS (Sun et al. 2015).

IL-1β plays important roles in acute inflammation. For example, IL-1β serves as a mediator of CXC chemokine production and subsequent neutrophil accumulation in the lungs; a hallmark of T-helper 1 (Th1) -mediated inflammation (Calkins et al. 2002; Miller et al. 2007; Biondo et al. 2014). Inflammasome activation and consequent IL-1β release have been proposed as a mechanism of pulmonary fibrosis. However, the role of the inflammasome and IL-1β in fibrosis remains controversial and its role as a profibrotic mediator may be dependent on the timing and duration of release. For example, IL-1β release by macrophages is an important innate immune response for recruiting neutrophils to the lung during acute lung inflammation initiated by LPS or other microbial products. Recruited neutrophils participate in microbial killing and the resolution of inflammation (Martinon et al. 2009).

Inflammasome activation occurs to a greater degree in classically activated macrophages (CAMs), which are important in microbial killing. In chronic pulmonary diseases such as asthma or fibrosis, macrophages are polarized to "alternatively activated macrophages" (AAMs), also referred to as "M2" or "M2 subtypes," in the presence of T-helper 2 (Th2) cytokines (e.g., IL-4, IL-13) or other specific stimuli and are thought to contribute to pathogenesis (Gordon and Martinez 2010). Inflammasome activation and IL-1β production are suppressed by the Th2 cytokines, IL-4 and IL-13, in human monocytic THP-1 cells in vitro and in the lungs of mice sensitized with house dust mite (HDM) allergen prior to MWCNT exposure by oropharyngeal aspiration (Shipkowski et al. 2015). The mechanism of inflammasome suppression by these Th2 cytokines was mediated via signal transducer and activator of transcription 6 (STAT6)-dependent suppression of pro-caspase-1, the precursor to caspase-1 that serves as a central inflammasome component to cleave pro-IL-1β to mature IL-1β. Interestingly, despite suppression of inflammasome activation by HDM allergen, MWCNT exposure exacerbated HDM allergen-induced airway fibrosis and reduced numbers of neutrophils in the lung, suggesting that inflammasome activation is not a mechanism of airway fibrosis but rather serves to resolve tissue injury by recruiting neutrophils.

Neutrophilic inflammation in the lung is a classic response to MWCNTs and has been linked to IL-1β production. This has been demonstrated in a study with IL-1 receptor knock-out (IL-1R KO) mice, which did not exhibit neutrophilic inflammation and yet developed pulmonary fibrosis to a greater degree than wild-type mice (Girtsman et al. 2014). Whether or not IL-1β is profibrogenic or antifibrogenic may depend on temporal expression, which in turn could determine the duration of neutrophilic inflammation in the lung.

2.7 Macrophage Polarization

Macrophages may develop distinct functional phenotypes that are determined by factors in the tissue microenvironment such as cytokines, which can lead to "alternative" transcriptional programming and gene expression (Martinez et al. 2008; Gordon and Martinez 2010). Alveolar macrophages isolated from the lungs of normal humans and treated ex vivo with IFN-γ develop a "classical" or "M1" phenotype, whereas the same alveolar macrophages treated with IL-4 or IL-13 develop an "alternative" or "M2" phenotype. The transcription factor STAT1 is an important intracellular signaling conduit that mediates M1 macrophage polarization in response to IFN-γ, while STAT6 is key to mediating M2 macrophage polarization in response to IL-4 or IL-13 (Gordon and Martinez 2010). Several M2-like subtypes have been recognized, indicating that macrophages likely display a range of different phenotypes.

Macrophage phenotypes are characterized by the expression of specific sets of genes. For example, M1 macrophages express genes encoding proteins such as TLR4, CXCL10, and CCL5, while M2 macrophages express genes encoding proteins such as mannose receptor (CD206), YM-1 (chitinase 3-like protein 3), and arginase (Hussell and Bell 2014). M1 macrophages are highly phagocytic and capable of a respiratory burst, both of which are key to innate host defense, whereas M2 macrophages are anti-inflammatory and associated with tissue remodeling and diseases such as fibrosis and cancer. It is most likely that healthy lungs predominantly have M1 macrophages since the IL-4 or IL-13 microenvironments that drive M2 polarization are mainly associated with asthma, pulmonary fibrosis, and exposure to cigarette smoke (Shaykhiev et al. 2009; Melgert et al. 2011; Cornwell et al. 2018).

ENMs have been reported to promote macrophage polarization to either an M1 or an M2 phenotype. Studies with human or murine macrophage cell lines or macrophages isolated from mice and exposed to ENMs ex vivo predominantly show M1 polarization, yet some have documented M2 polarization (Miao et al. 2017). For example, metal nanoparticles (Ag, TiO_2, ZnO) stimulate murine RAW264.7 macrophages in vitro to produce IL-6 and IL-1β via NF-κB signaling, which is consistent with M1 polarization (Giovanni et al. 2015). Another study showed that MWCNTs increased protein levels of iNOS, leading to increased NO production in murine RAW264.7 macrophages (Lee et al. 2012). Induction of iNOS is consistent with M1 macrophages (Miao et al. 2017). Graphene oxide stimulated a M1 phenotype in both murine and human macrophage cell lines (Ma et al. 2015). Some ENMs have been reported to promote M2 polarization. For example, gold nanorods polarized hepatic macrophages toward M1 or M2 activation, depending on the specific peptide sequence that was conjugated to the surface (Bartneck et al. 2012).

ENMs could influence macrophage polarization via indirect mechanisms. For example, ROS are generated by ENMs following uptake by macrophages and ROS are known to alter the phosphorylation patterns of cell signaling intermediates (e.g., kinases, transcription factors) by inhibiting phosphatase enzymes that are critical for dephosphorylation reactions. In a Th1 microenvironment where interferons activate STAT1 activation, or in a Th2 microenvironment where IL-4 and IL-13 activate STAT6, ROS generated intracellularly could alter the phosphorylation and activity of these transcription factors. This concept is illustrated in Fig. 2.2. While information on ENM-induced alteration of phosphorylation in macrophages is lacking, it has been reported that nickel nanoparticles prolong the phosphorylation of MAP kinases in rat mesothelial cells (Glista-Baker et al. 2012).

2.8 Consequences of ENMs in Impairment of the Sentinel Function of Macrophages

As discussed above, alveolar macrophages play important sentinel functions by engulfing and destroying foreign pathogens through phagocytosis and the respiratory burst. Alveolar macrophages are long-lived and have a turnover rate of less than 50% over a 1-year period (Maus et al. 2006; Janssen et al. 2011). Because of the relative longevity of macrophages, ENMs engulfed by these phagocytes could be bio-persistent within macrophages. In support of this concept, MWCNTs have been found within lung macrophages of mice for months after inhalation exposure (Ryman-Rasmussen et al. 2009).

ENMs have been reported to impair the protective sentinel functions of alveolar macrophages, such as phagocytosis and clearance of live bacteria. For example, mice exposed to SWCNTs have impaired clearance of the bacteria *Listeria monocytogenes* (Shvedova et al. 2008). Decreased bacterial clearance in mice pre-exposed to SWCNTs was associated with decreased phagocytosis of bacteria by macrophages and a decrease in nitric oxide production by these phagocytes. Helical carbon nanotubes delivered to the lungs of mice infected with *Pseudomonas aeruginosa* inhibited macrophage-mediated phagocytosis of the bacteria and yet primed the immune system for an enhanced neutrophilic inflammatory response to the bacteria (Walling et al. 2013).

Some unusual interactions with macrophages have also been observed with CNTs. For example, SWCNTs cause bridge-like structures between two or more macrophages (Mangum et al. 2006). It is intriguing to speculate that SWCNTs could interact directly with cytoskeletal filaments such as actin to form intercellular "bridge" connections. In any case, these structures would likely impede macrophage function. CNTs are also similar in size to the microtubules that form the mitotic spindle and this type of biomimicry could explain how these ENMs disrupt the mitotic spindle apparatus in lung epithelial cells (Sargent et al. 2009). Moreover, SWCNTs have approximately the same diameter (1–4 nm) as the DNA double helix and can physically intertwine with DNA (Pampaloni and Florin 2008). The nanoscale dimensions of ENMs offer numerous opportunities for interaction with cellular components of comparable size that could impact the innate immune responses of macrophages.

References

Arredouani MS, Palecanda A, Koziel H, Huang YC, Imrich A, Sulahian TH, Ning YY, Yang Z, Pikkarainen T, Sankala M, Vargas SO, Takeya M, Tryggvason K, Kobzik L. MARCO is the major binding receptor for unopsonized particles and bacteria on human alveolar macrophages. J Immunol. 2005;175(9):6058–64.

Bartneck M, Ritz T, Keul HA, Wambach M, Bornemann J, Gbureck U, Ehling J, Lammers T, Heymann F, Gassler N, Lüdde T, Trautwein C, Groll J, Tacke F. Peptide-functionalized gold nanorods increase liver injury in hepatitis. ACS Nano. 2012;6(10):8767–77.

Bianchi MG, Allegri M, Costa AL, et al. Titanium dioxide nanoparticles enhance macrophage activation by LPS through a TLR4-dependent intracellular pathway. Toxicol Res (Camb). 2015;4:385–98.

Biondo C, Mancuso G, Midiri A, Signorino G, Domina M, Lanza Cariccio V, Venza M, Venza I, Teti G, Beninati C. Essential role of interleukin-1 signaling in host defenses against group B streptococcus. MBio. 2014;5(5):e01428–14.

Boraschi D, Italiani P, Palomba R, Decuzzi P, Duschl A, Fadeel B, Moghimi SM. Nanoparticles and innate immunity: new perspectives on host defence. Semin Immunol. 2017;34:33–51.

Brown MS, Goldstein JL. Receptor-mediated endocytosis: insights from the lipoprotein receptor system. Proc Natl Acad Sci U S A. 1979;76(7):3330–7.

Calkins CM, Bensard DD, Shames BD, Pulido EJ, Abraham E, Fernandez N, Meng X, Dinarello CA, McIntyre RC Jr. IL-1 regulates in vivo C-X-C chemokine induction and neutrophil sequestration following endotoxemia. J Endotoxin Res. 2002;8(1):59–67.

Canton J, Neculai D, Grinstein S. Scavenger receptors in homeostasis and immunity. Nat Rev Immunol. 2013;13(9):621–34.

Chan TK, Tan WSD, Peh HY, Wong WSF. Aeroallergens induce reactive oxygen species production and DNA damage and dampen antioxidant responses in bronchial epithelial cells. J Immunol. 2017;199:39–47.

Cornwell WD, Kim V, Fan X, Vega ME, Ramsey FV, Criner GJ, Rogers TJ. Activation and polarization of circulating monocytes in severe chronic obstructive pulmonary disease. BMC Pulm Med. 2018;18(1):101.

Davies LC, Rice CM, McVicar DW, Weiss JM. Diversity and environmental adaptation of phagocytic cell metabolism. Journal of Leukocyte Biology. 2018;105(1):37–48

Donaldson K, Aitken R, Tran L, Stone V, Duffin R, Forrest G, Alexander A. Carbon nanotubes: a review of their properties in relation to pulmonary toxicology and workplace safety. Toxicol Sci. 2006;92(1):5–22.

Duke KS, Bonner JC. Mechanisms of carbon nanotube-induced pulmonary fibrosis: a physicochemical characteristic perspective. Wiley Interdiscip Rev Nanomed Nanobiotechnol. 2018;10(3):e1498.

Elomaa O, Kangas M, Sahlberg C, Tuukkanen J, Sormunen R, Liakka A, Thesleff I, Kraal G, Tryggvason K. Cloning of a novel bacteria-binding receptor structurally related to scavenger receptors and expressed in a subset of macrophages. Cell. 1995;80(4):603–9.

Feiner-Gracia N, Beck M, Pujals S, Tosi S, Mandal T, Buske C, Linden M, Albertazzi L. Super-resolution microscopy unveils dynamic heterogeneities in nanoparticle protein corona. Small. 2017;13(41):1701631.

Fleischer CC, Payne CK. Nanoparticle–cell interactions: molecular structure of the protein corona and cellular outcomes. Acc Chem Res. 2014a;47(8):2651–9.

Fleischer CC, Payne CK. Secondary structure of corona proteins determines the cell surface receptors used by nanoparticles. J Phys Chem B. 2014b;118(49):14017–26.

Gabehart K, Correll KA, Loader JE, White CW, Dakhama A. The lung response to ozone is determined by age and is partially dependent on Toll-like receptor 4. Respir Res. 2015;16:117.

Gao N, Zhang Q, Mu Q, Bai Y, Li L, Zhou H, Butch ER, Powell TB, Snyder SE, Jiang G, Yan B. Steering carbon nanotubes to scavenger receptor recognition by nanotube surface chemistry modification partially alleviates NFκB activation and reduces its immunotoxicity. ACS Nano. 2011;5(6):4581–91.

Garapaty A, Champion JA. Tunable particles alter macrophage uptake based on combinatorial effects of physical properties. Bioeng Transl Med. 2017;2(1):92–101.

Giovanni M, Yue J, Zhang L, Xie J, Ong CN, Leong DT. Pro-inflammatory responses of RAW264.7 macrophages when treated with ultralow concentrations of silver, titanium dioxide, and zinc oxide nanoparticles. J Hazard Mater. 2015;297:146–52.

Girtsman TA, Beamer CA, Wu N, Buford M, Holian A. IL-1R signaling is critical for regulation of multi-walled carbon nanotubes-induced acute lung inflammation in C57Bl/6 mice. Nanotoxicology. 2014;8(1):17–27.

Glista-Baker EE, Taylor AJ, Sayers BC, Thompson EA, Bonner JC. Nickel nanoparticles enhance platelet-derived growth factor-induced chemokine expression by mesothelial cells via prolonged mitogen-activated protein kinase activation. Am J Respir Cell Mol Biol. 2012;47(4):552–61.

Gordon S, Martinez FO. Alternative activation of macrophages: mechanism and functions. Immunity. 2010;32(5):593–604.

Gough PJ, Gordon S. The role of scavenger receptors in the innate immune system. Microbes Infect. 2000;2(3):305–11.

Gwinn MR, Vallyathan V. Respiratory burst: role in signal transduction in alveolar macrophages. J Toxicol Environ Health B Crit Rev. 2006;9(1):27–39.

Hamilton RF, de Villiers WJS, Holian A. Class A type II scavenger receptor mediates silica-induced apoptosis in Chinese hamster ovary cell line. Toxicol Appl Pharmacol. 2000;162(2):100–6.

Hamilton RF Jr, Thakur SA, Mayfair JK, Holian A. MARCO mediates silica uptake and toxicity in alveolar macrophages from C57BL/6 mice. J Biol Chem. 2006;281(45):34218–26.

Heit B, Kim H, Cosío G, Castaño D, Collins R, Lowell CA, Kain KC, Trimble WS, Grinstein S. Multimolecular signaling complexes enable Syk-mediated signaling of CD36 internalization. Dev Cell. 2013;24(4):372–83.

Hirano S, Fujitani Y, Furuyama A, Kanno S. Macrophage receptor with collagenous structure (MARCO) is a dynamic adhesive molecule that enhances uptake of carbon nanotubes by CHO-K1 cells. Toxicol Appl Pharmacol. 2012;259(1):96–103.

Holian A, Scheule RK. Alveolar macrophage biology. Hosp Pract (Off Ed). 1990;25(12):53–62.
Huang C, Sun M, Yang Y, Wang F, Ma X, Li J, Wang Y, Ding Q, Ying H, Song H, Wu Y, Jiang Y, Jia X, Ba Q, Wang H. Titanium dioxide nanoparticles prime a specific activation state of macrophages. Nanotoxicology. 2017;11(6):737–50.
Hussell T, Bell TJ. Alveolar macrophages: plasticity in a tissue-specific context. Nat Rev Immunol. 2014;14(2):81–93.
Janssen WJ, Barthel L, Muldrow A, Oberley-Deegan RE, Kearns MT, Jakubzick C, Henson PM. Fas determines differential fates of resident and recruited macrophages during resolution of acute lung injury. Am J Respir Crit Care Med. 2011;184(5):547–60.
Jayaram DT, Runa S, Kemp ML, Payne CK. Nanoparticle-induced oxidation of corona proteins initiates an oxidative stress response in cells. Nanoscale. 2017;9(22):7595–601.
Kanno S, Furuyama A, Hirano S. A murine scavenger receptor MARCO recognizes polystyrene nanoparticles. Toxicol Sci. 2007;97(2):398–406.
Kedmi R, Ben-Arie N, Peer D. The systemic toxicity of positively charged lipid nanoparticles and the role of Toll-like receptor 4 in immune activation. Biomaterials. 2010;31(26):6867–75.
Kuhn DA, Vanhecke D, Michen B, Blank F, Gehr P, Petri-Fink A, Rothen-Rutishauser B. Different endocytotic uptake mechanisms for nanoparticles in epithelial cells and macrophages. Beilstein J Nanotechnol. 2014;5:1625–36.
Kuzmich NN, Sivak KV, Chubarev VN, Porozov YB, Savateeva-Lyubimova TN, Peri F. TLR4 Signaling pathway modulators as potential therapeutics in inflammation and sepsis. Vaccines (Basel). 2017;5(4):34.
Lee JK, Sayers BC, Chun KS, Lao HC, Shipley-Phillips JK, Bonner JC, Langenbach R. Multi-walled carbon nanotubes induce COX-2 and iNOS expression via MAP kinase-dependent and -independent mechanisms in mouse RAW264.7 macrophages. Part Fibre Toxicol. 2012;9:14.
Light WG, Wei ET. Surface charge and asbestos toxicity. Nature. 1977;265(5594):537–9.
Lonez C, Irvine KL, Pizzuto M, et al. Critical residues involved in Toll-like receptor 4 activation by cationic lipid nanocarriers are not located at the lipopolysaccharide-binding interface. Cell Mol Life Sci. 2015;72(20):3971–82.
Lunov O, Zablotskii V, Syrovets T, Röcker C, Tron K, Nienhaus GU, Simmet T. Modeling receptor-mediated endocytosis of polymer-functionalized iron oxide nanoparticles by human macrophages. Biomaterials. 2011;32(2):547–55.
Ma J, Liu R, Wang X, Liu Q, Chen Y, Valle RP, Zuo YY, Xia T, Liu S. Crucial role of lateral size for graphene oxide in activating macrophages and stimulating pro-inflammatory responses in cells and animals. ACS Nano. 2015;9(10):10498–515.
Ma J, Li R, Qu G, et al. Carbon nanotubes stimulate synovial inflammation through inducing systemic pro-inflammatory cytokines. Nanoscale. 2016;8:18070–86.
Mangum JB, Turpin EA, Antao-Menezes A, Cesta MF, Bermudez E, Bonner JC. Single-walled carbon nanotube (SWCNT)-induced interstitial fibrosis in the lungs of rats is associated with increased levels of PDGF mRNA and the formation of unique intercellular carbon structures that bridge alveolar macrophages in situ. Part Fibre Toxicol. 2006;3:15.
Martinez FO, Sica A, Mantovani A, Locati M. Macrophage activation and polarization. Front Biosci. 2008;13:453–61.
Martinon F, Mayor A, Tschopp J. The inflammasomes: guardians of the body. Annu Rev Immunol. 2009;27:229–65.
Maus UA, Janzen S, Wall G, Srivastava M, Blackwell TS, Christman JW, Seeger W, Welte T, Lohmeyer J. Resident alveolar macrophages are replaced by recruited monocytes in response to endotoxin-induced lung inflammation. Am J Respir Cell Mol Biol. 2006;35(2):227–35.
Meunier E, Coste A, Olagnier D, et al. Double-walled carbon nanotubes trigger IL-1β release in human monocytes through Nlrp3 inflammasome activation. Nanomedicine. 2012;8(6):987–95.
Melgert BN, ten Hacken NH, Rutgers B, Timens W, Postma DS, Hylkema MN. More alternative activation of macrophages in lungs of asthmatic patients. Journal of Allergy and Clinical Immunology. 2011;127(3):831–833
Miao X, Leng X, Zhang Q. The current state of nanoparticle-induced macrophage polarization and reprogramming research. Int J Mol Sci. 2017;18(2):pii: E336.

Miller SI, Ernst RK, Bader MW. LPS, TLR4 and infectious disease diversity. Nat Rev Microbiol. 2005;3:36–46.

Miller LS, Pietras EM, Uricchio LH, Hirano K, Rao S, Lin H, O'Connell RM, Iwakura Y, Cheung AL, Cheng G, Modlin RL. Inflammasome-mediated production of IL-1beta is required for neutrophil recruitment against Staphylococcus aureus in vivo. J Immunol. 2007;179(10):6933–42.

Mittal M, Siddiqui MR, Tran K, Reddy SP, Malik AB. Reactive oxygen species in inflammation and tissue injury. Antioxid Redox Signal. 2014;20(7):1126–67.

Monopoli MP, Walczyk D, Campbell A, Elia G, Lynch I, Bombelli FB, Dawson KA. Physical-chemical aspects of protein corona: relevance to in vitro and in vivo biological impacts of nanoparticles. J Am Chem Soc. 2011;133(8):2525–34.

Mukherjee SP, Bondarenko O, Kohonen P, et al. Macrophage sensing of single-walled carbon nanotubes via Toll-like receptors. Sci Rep. 2018;8:1115.

Murphy JE, Tedbury PR, Homer-Vanniasinkam S, Walker JH, Ponnambalam S. Biochemistry and cell biology of mammalian scavenger receptors. Atherosclerosis. 2005;182(1):1–15.

Murthy S, Larson-Casey JL, Ryan AJ, He C, Kobzik L, Carter AB. Alternative activation of macrophages and pulmonary fibrosis are modulated by scavenger receptor, macrophage receptor with collagenous structure. FASEB J. 2015;29(8):3527–36.

Oberdörster G, Ferin J, Gelein R, Soderholm SC, Finkelstein J. Role of the alveolar macrophage in lung injury: studies with ultrafine particles. Environ Health Perspect. 1992;97:193–9.

Oberdörster G, Ferin J, Lehnert BE. Correlation between particle size, in vivo particle persistence, and lung injury. Environ Health Perspect. 1994;102(Suppl 5):173–9.

Oblak A, Pohar J, Jerala R. MD-2 determinants of nickel and cobalt-mediated activation of human TLR4. PLoS One. 2015;10(3):e0120583.

Palomäki J, Välimäki E, Sund J, Vippola M, Clausen PA, Jensen KA, Savolainen K, Matikainen S, Alenius H. Long, needle-like carbon nanotubes and asbestos activate the NLRP3 inflammasome through a similar mechanism. ACS Nano. 2011;5(9):6861–70.

Pampaloni F, Florin EL. Microtubule architecture: inspiration for novel carbon nanotube-based biomimetic materials. Trends Biotechnol. 2008;26(6):302–10.

Patra A, Ding T, Engudar G, Wang Y, Dykas MM, Liedberg B, Kah JC, Venkatesan T, Drum CL. Component-specific analysis of plasma protein corona formation on gold nanoparticles using multiplexed surface plasmon resonance. Small. 2016;12(9):1174–82.

Paul D, Achouri S, Yoon YZ, Herre J, Bryant CE, Cicuta P. Phagocytosis dynamics depends on target shape. Biophys J. 2013;105(5):1143–50.

Petrarca C, Clemente E, Amato V, Pedata P, Sabbioni E, Bernardini G, Iavicoli I, Cortese S, Niu Q, Otsuki T, Paganelli R, Di Gioacchino M. Engineered metal based nanoparticles and innate immunity. Clin Mol Allergy. 2015;13:13.

Pluddemann A, Neyen C, Gordon S. Macrophage scavenger receptors and host-derived ligands. Methods. 2007;43(3):207–17.

Pluddemann A, Mukhopadhyay S, Sankala M, Savino S, Pizza M, Rappuoli R, Tryggvason K, Gordon S. SR-A, MARCO and TLRs differentially recognise selected surface proteins from Neisseria meningitidis: an example of fine specificity in microbial ligand recognition by innate immune receptors. J Innate Immun. 2009;1(2):153–63.

Potnis PA, Dutta DK, Wood SC. Toll-like receptor 4 signaling pathway mediates proinflammatory immune response to cobalt-alloy particles. Cell Immunol. 2013;282:53–65.

Raghavan B, Martin SF, Esser PR, Goebeler M, Schmidt M. Metal allergens nickel and cobalt facilitate TLR4 homodimerization independently of MD2. EMBO Rep. 2012;13(12):1109–15.

Raynal I, Prigent P, Peyramaure S, Najid A, Rebuzzi C, Corot C. Macrophage endocytosis of superparamagnetic iron oxide nanoparticles: mechanisms and comparison of ferumoxides and ferumoxtran-10. Investig Radiol. 2004;39(1):56–63.

Resnick D, Freedman NJ, Xu S, Krieger M. Secreted extracellular domains of macrophage scavenger receptors form elongated trimers which specifically bind crocidolite asbestos. J Biol Chem. 1993;268(5):3538–45.

Runa S, Hussey M, Payne CK. Nanoparticle-cell interactions: relevance for public health. J Phys Chem B. 2018;122(3):1009–16.

Ryman-Rasmussen JP, Cesta MF, Brody AR, Shipley-Phillips JK, Everitt JI, Tewksbury EW, Moss OR, Wong BA, Dodd DE, Andersen ME, Bonner JC. Inhaled carbon nanotubes reach the subpleural tissue in mice. Nat Nanotechnol. 2009;4(11):747–51.

Samet JM, Tal TL. Toxicological disruption of signaling homeostasis: tyrosine phosphatases as targets. Annu Rev Pharmacol Toxicol. 2010;50:215–35.

Sargent LM, Shvedova AA, Hubbs AF, Salisbury JL, Benkovic SA, Kashon ML, et al. Induction of aneuploidy by single-walled carbon nanotubes. Environ Mol Mutagen. 2009;50(8):708–17.

Sayan M, Mossman BT. The NLRP3 inflammasome in pathogenic particle and fibre-associated lung inflammation and diseases. Part Fibre Toxicol. 2016;13(1):51.

Schmidt M, Goebeler M. Nickel allergies: paying the Toll for innate immunity. J Mol Med. 2011;89:961–70.

Schmidt M, Raghavan B, Müller V, Vogl T. Crucial role for human Toll-like receptor 4 in the development of contact allergy to nickel. Nat Immunol. 2010;11:814–9.

Shannahan JH, Fritz KS, Raghavendra AJ, Podila R, Persaud I, Brown JM. Formation of a protein corona on silver nanoparticles mediates cellular toxicity via scavenger receptors. Toxicol Sci. 2015;143(1):136–46.

Shaykhiev R, Krause A, Salit J, Strulovici-Barel Y, Harvey BG, O'Connor TP, Crystal RG. Smoking-dependent reprogramming of alveolar macrophage polarization: implication for pathogenesis of chronic obstructive pulmonary disease. J Immunol. 2009;183(4):2867–83.

Shipkowski KA, Taylor AJ, Thompson EA, Glista-Baker EE, Sayers BC, Messenger ZJ, Bauer RN, Jaspers I, Bonner JC. An allergic lung microenvironment suppresses carbon nanotube-induced inflammasome activation via STAT6-dependent inhibition of caspase-1. PLoS One. 2015;10(6):e0128888.

Shvedova AA, Fabisiak JP, Kisin ER, Murray AR, Roberts JR, Tyurina YY, Antonini JM, Feng WH, Kommineni C, Reynolds J, Barchowsky A, Castranova V, Kagan VE. Sequential exposure to carbon nanotubes and bacteria enhances pulmonary inflammation. Am J Respir Cell Mol Biol. 2008;38(5):579–90.

Singh RP, Ramarao P. Cellular uptake, intracellular trafficking and cytotoxicity of silver nanoparticles. Toxicol Lett. 2012;213(2):249–59.

Sosale NG, Spinler KR, Alvey C, Discher DE. Macrophage engulfment of a cell or nanoparticle is regulated by unavoidable opsonization, a species-specific 'Marker of Self' CD47, and target physical properties. Curr Opin Immunol. 2015;35:107–12.

Strowig T, Henao-Mejia J, Elinav E, et al. Inflammasomes in health and disease. Nature. 2012; 481(7381):278–86.

Sun B, Wang X, Ji Z, Wang M, Liao YP, Chang CH, Li R, Zhang H, Nel AE, Xia T. NADPH oxidase-dependent NLRP3 inflammasome activation and its important role in lung fibrosis by multiwalled carbon nanotubes. Small. 2015;11(17):2087–97.

Thakur SA, Hamilton RF Jr, Holian A. Role of scavenger receptor a family in lung inflammation from exposure to environmental particles. J Immunotoxicol. 2008;5(2):151–7.

Thakur SA, Hamilton R, Pikkarainen T, Holian A. Differential binding of inorganic particles to MARCO. Toxicological Sciences. Toxicol Sci. 2009;107(1):238–246

Thomas DC. The phagocyte respiratory burst: historical perspectives and recent advances. Immunol Lett. 2017;192:88–96.

Thompson EA, Sayers BC, Glista-Baker EE, Sipkowski KA, Taylor AJ, Bonner JC. Innate immune responses to nanoparticle exposure in the lung. J Environ Immunol Toxicol. 2014;1(3):150–6.

Tsugita M, Morimoto N, Tashiro M, Kinoshita K, Nakayama M. SR-B1 is a silica receptor that mediates canonical inflammasome activation. Cell Rep. 2017;18(5):1298–311.

Turabekova M, Rasulev B, Theodore M, Jackman J, Leszcynska D, Leszczynski J. Immunotoxicity of nanoparticles: computational study suggests that CNTs and C60 fullerenes might be recognized as pathogens by Toll-like receptors. Nanoscale. 2014;6:3488.

van der Laan LJ, Döpp EA, Haworth R, Pikkarainen T, Kangas M, Elomaa O, Dijkstra CD, Gordon S, Tryggvason K, Kraal G. Regulation and functional involvement of macrophage scavenger receptor MARCO in clearance of bacteria in vivo. J Immunol. 1999;162(2):939–47.

Walling BE, Kuang Z, Hao Y, Estrada D, Wood JD, Lian F, Miller LA, Shah AB, Jeffries JL, Haasch RT, Lyding JW, Pop E, Lau GW. Helical carbon nanotubes enhance the early immune response and inhibit macrophage-mediated phagocytosis of Pseudomonas aeruginosa. PLoS One. 2013;8(11):e80283.

Wang L, Tassiulas I, Park-Min KH, Reid AC, Gil-Henn H, Schlessinger J, Baron R, Zhang JJ, Ivashkiv LB. 'Tuning' of type I interferon-induced Jak-STAT1 signaling by calcium-dependent kinases in macrophages. Nat Immunol. 2008;9(2):186–93.

Wang X, Guo J, Chen T, Nie H, Wang H, Zang J, Cui X, Jia G. Multi-walled carbon nanotubes induce apoptosis via mitochondrial pathway and scavenger receptor. Toxicol In Vitro. 2012;26(6):799–806.

Xu Z, Ren T, Xiao C, Li H, Wu T. Nickel promotes the invasive potential of human lung cancer cells via TLR4/MyD88 signaling. Toxicology. 2011;285:25–30.

Yang RB, Mark MR, Gray A, et al. Toll-like receptor-2 mediates lipopolysaccharide-induced cellular signaling. Nature. 1998;395:284–8.

Yu H, Ha T, Liu L, Wang X, Gao M, Kelley J, Kao R, Williams D, Li C. Scavenger receptor A (SR-A) is required for LPS-induced TLR4 mediated NF-κB activation in macrophages. Biochim Biophys Acta. 2012;1823(7):1192–8.

Zani IA, Stephen SL, Mughal NA, Russell D, Homer-Vanniasinkam S, Wheatcroft SB, Ponnambalam S. Scavenger receptor structure and function in health and disease. Cell. 2015;4(2):178–201.

Zong G, Zhu Y, Zhang Y, Wang Y, Bai H, Yang Q, Ben J, Zhang H, Li X, Zhu X, Chen Q. SR-A1 suppresses colon inflammation and tumorigenesis through negative regulation of NF-κB signaling. Biochem Pharmacol. 2018;154:335–43.

Chapter 3
Nanomaterials and Neutrophils

Sandeep Keshavan and Bengt Fadeel

Abstract Neutrophils are key components of the innate arm of the immune system and represent the front line of host defence against intruding pathogens. However, neutrophils can also cause damage to the host. Understanding nanomaterial interactions with neutrophils is of critical importance. In this chapter, we discuss how nanomaterials may recruit or activate neutrophils and how neutrophils, in turn, may target and digest certain types of nanomaterials. We also reflect on lessons learned from previous studies of host responses towards implantable biomaterials.

Keywords Biomaterial · Degradation · Inflammasome · Inflammation · Nanomaterial · Neutrophil

The immune system is traditionally divided into two branches: innate immunity and adaptive immunity. However, this is, of course, purely academic – in reality, there is only one immune system that comprises multiple players including many different cell types and soluble mediators. The immune system is better viewed as a network of cells and organs that acts as a guardian against infection or other foreign intrusion. Indeed, the strict dichotomization of immune cells into innate and adaptive, respectively, does not capture the remarkable cross-talk that occurs in the immune system, and cells such as the neutrophil that were previously thought of as 'primitive' are known to interact with and instruct cells of the so-called adaptive immune system (Nathan 2006; Bardoel et al. 2014). Furthermore, the view that only the adaptive immune system is endowed with immunological memory (i.e., the occurrence of a more potent response on exposure to an antigen encountered previously) has also been challenged in recent years. Hence, the innate immune system can

S. Keshavan · B. Fadeel (✉)
Division of Molecular Toxicology, Institute of Environmental Medicine, Karolinska Institutet, Stockholm, Sweden
e-mail: bengt.fadeel@ki.se

J. C. Bonner, J. M. Brown (eds.), *Interaction of Nanomaterials with the Immune System*, Molecular and Integrative Toxicology,
https://doi.org/10.1007/978-3-030-33962-3_3

mount resistance to reinfection, a phenomenon referred to as 'trained' immunity or innate immune memory (Farber et al. 2016; Netea et al. 2016).

Understanding the interactions of engineered nanomaterials with the immune system is of considerable relevance both in toxicology and in medicine (Farrera and Fadeel 2015; Boraschi et al. 2017). However, there is a preponderance of publications on nanomaterials and macrophages, while less attention is devoted to neutrophils, despite the fact that neutrophils are key players in inflammation. Here, we will focus on the reciprocal interactions of various nano(bio)materials with neutrophils, the most abundant white blood cell in the peripheral blood.

3.1 Inflammation: Introducing the Main Players

Inflammation is a complex biological response involving soluble factors and cells that can arise in a tissue in response to harmful stimuli, including pathogens, toxicants, or dead cells. The process normally leads to recovery and to healing. However, inflammation can also lead to persistent tissue damage and may even promote neoplastic transformation (Nathan 2002; Medzhitov 2008). Indeed, the distinction between *acute* and *chronic* inflammation is important. Inflammation is fundamentally a normal, protective physiological response to injury or infection. However, if inflammatory responses are persistent due to an exaggerated or dysregulated response, including a failure of resolution, a pathological response can emerge (Johnston et al. 2018). This is important to bear in mind not least in nanotoxicology where 'inflammation' is often used as a proxy for hazard, but this is clearly an oversimplification as it is the persistent inflammation that does not resolve that is the problem, not inflammation *per se*.

The main cell types of the innate arm of the immune system can be categorized into (a) phagocytic cells, such as macrophages and neutrophils, and (b) non-phagocytic cells, including mast cells, eosinophils and basophils. Neutrophils are the most abundant type of white blood cell, and they have traditionally been thought of as 'microbe killers', but research in the last decade or more has revealed that these cells may also inform and shape adaptive immune responses (Nathan 2006). Thus, in addition to their 'traditional' role in host defence, neutrophils are also implicated in many diseases (Bardoel et al. 2014). Paradoxically, in spite of their abundance, less is known about neutrophils in comparison to other immune cells; it is also worth noting that much of what we do know is derived from neutrophils that have been isolated from peripheral blood, yet the main functions of neutrophils are exerted in tissues and not in the blood. To further compound the situation, neutrophils are short-lived and terminally differentiated, meaning that they do not divide and cannot be genetically modified. Moreover, as pointed out by Bardoel et al. (2014), mice are an imperfect model of the human condition albeit one that is widely used in immunology. For instance, the balance of lymphocytes and neutrophils is remarkably different between the two species: human blood is neutrophil rich (50–70% neutrophils, 30–50% lymphocytes), whereas mouse blood is dominated by lymphocytes (75–90% lymphocytes, 10–25% neutrophils) (Mestas and Hughes 2004). On the other hand, valuable lessons may be gleaned from the study of patients with mutations in genes encoding for proteins important for neutrophil function, such as chronic granulomatous disease (CGD) patients who suffer from life-threatening infections as a result of mutations in one of the components of the NADPH oxidase (Thomas 2018). Other so-called innate immune cells, including monocytes/macrophages and mast cells, are discussed in other chapters in this volume, and we shall, therefore, focus mainly on the neutrophil. However, it is certainly worth noting that other innate immune cells which have been studied mainly in relation to their roles in pathology (e.g., asthma, allergy and autoimmune diseases) are increasingly being recognized for their roles as protectors of human health and not only as mediators of disease (Dudeck et al. 2019).

Neutrophils or polymorphonuclear granulocytes are short-lived granular leukocytes possessing a nucleus with three or more lobes (hence, polymorphonuclear when viewed under a microscope). Neutrophils are produced in the bone marrow and differentiate into mature neutrophils equipped with cytoplasmic granules that are released into the peripheral circulation, poised to attack bacteria or other invaders. Granules, the hallmark of all granulocytes (eosinophils, basophils and neutrophils), are filled with proteins that can kill microbes, but may also cause tissue damage (Borregaard 2010). Neutrophil granules are classified into three subsets based on the presence of characteristic proteins: primary or azurophil granules [containing myeloperoxidase (MPO)], secondary or specific granules (lactoferrin), and tertiary or gelatinase granules (gelatinase) (Borregaard 2010). Neutrophils may enter tissues by crossing the vascular wall, and they destroy invading pathogens through the release of microbicidal agents stored in granules or through proteolytic and oxidative destruction of ingested pathogens. Neutrophils can also capture and destroy pathogens extracellularly through the release of neutrophil extracellular traps (NETs) (see below). Dying neutrophils are normally removed by macrophages, though evidence for neutrophil 'cannibalism' has been presented (in mice), and this may serve as a back-up when macrophage clearance is insufficient (Rydell-Törmänen et al. 2006). However, as has already been alluded to, the traditional view of neutrophils as short-lived effector cells with limited functional capacity is clearly obsolete, and a growing number of studies have shown that neutrophils may act as regulators of innate and adaptive networks (reviewed in: Scapini and Cassatella 2014). Indeed, once recruited into inflamed tissues, the neutrophil engages in complex interactions with other immune cells, including macrophages, natural killer cells, dendritic cells, and B and T lymphocytes, and neutrophils may in this way orchestrate not only the inflammatory response but also other immune responses, tissue remodelling and so on (Scapini and Cassatella 2014). Figure 3.1 illustrates how neutrophils interface with adaptive immune cells (Nauseef and Borregaard 2014).

Neutrophil heterogeneity was postulated several decades ago (Gallin 1984), but this subject did not receive much attention at the time. In more recent years, the existence of different subsets of neutrophils or neutrophils with relevant phenotypic differences has gained traction (Kruger et al. 2015; Silvestre-Roig et al. 2016). Neutrophil heterogeneity may be partly explained by the differential activation of

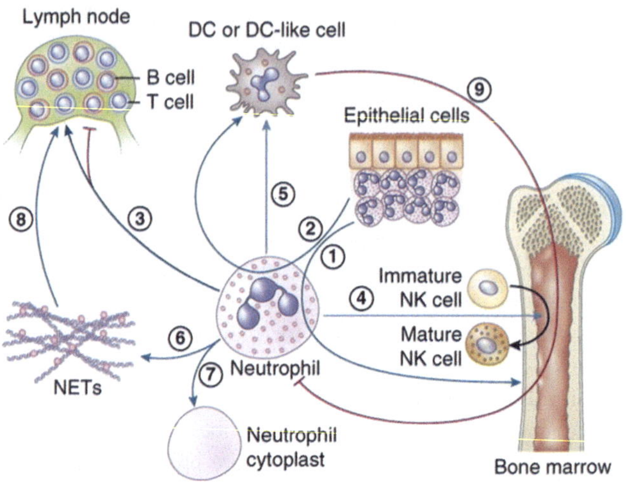

Fig. 3.1 Neutrophils interface with elements of adaptive immunity. In addition to capturing and killing bacteria and fungi (the 'traditional' role of these cells), neutrophils also ferry antigens from the epithelium to the bone marrow (1) or to dendritic cells (DCs) or DC-like cells (2); they inhibit proliferation of T cells in lymph nodes, and they stimulate maturation of antibody-producing B cells (3). Moreover, neutrophils are essential for the generation of natural killer (NK) cells (4) and can transform into functional DC-like cells that maintain essential neutrophil antigens (5). Neutrophils can also release NETs (6) and transform into anuclear cytoplasts (7). NETs, in turn, can stimulate autoimmunity (8). Reciprocally, the adaptive immune system can modulate neutrophil activity: thus, DCs can inhibit the generation and function of neutrophils (9). (Reprinted from: Nauseef and Borregaard (2014), with permission from Springer Nature)

neutrophil subpopulations. Indeed, activation of neutrophil at the site of inflammation may result in exposure of distinct surface markers and/or synthesis of specific cytokines, and such features may be used to categorize neutrophils (Silvestre-Roig et al. 2016). However, whether or not these differences in phenotypes correspond to distinct subsets of neutrophils remains to be elucidated (Hong 2017). Nevertheless, the implications for in vitro studies are obvious: one cannot assume that neutrophils will display homogenous responses when assayed as a bulk population. Neutrophils may display primed, activated and exhausted phenotypes (Fig. 3.2). Priming refers to a process of augmentation of neutrophil functions in response to specific stimuli, such as pathogen-associated molecular patterns (PAMPs) and damage-associated molecular patterns (DAMPs) released from dying cells. Primed neutrophils show enhanced ROS generation, granule release and NET formation. On the other hand, when neutrophils are excessively stimulated, they are thought to undergo exhaustion and show diminished granule and NET release in response to activating stimuli. ROS generation is also greatly decreased in these cells (Hong 2017). The possibility of artefactual activation of neutrophils during ex vivo isolation is something that should be noted as this may influence subsequent cellular assays. For instance, the choice of anticoagulant influences the number of neutrophils isolated by the gradient density method, the intracellular Ca^{2+} level and the sensibility to activators of the respiratory burst (Freitas et al. 2008).

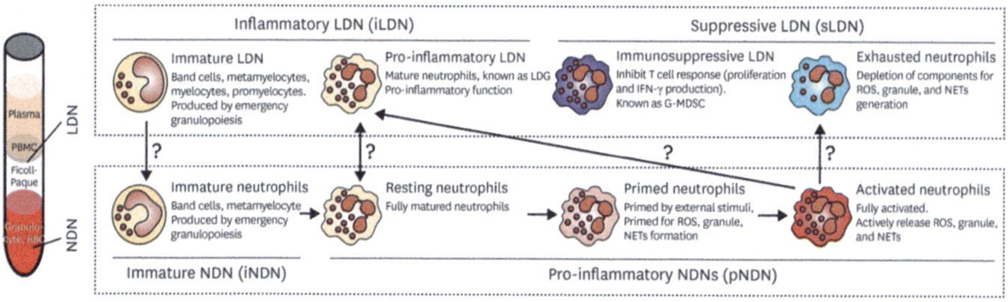

Fig. 3.2 Neutrophils: homogenous no more. Schematic diagram depicting different phenotypes of circulating neutrophils. Neutrophils are usually obtained from blood by density gradient centrifugation using Ficoll™, a highly branched, hydrophilic polysaccharide, and based on their weight, neutrophils are divided into either 'low-density' neutrophils (LDNs) or 'normal-density' neutrophils (NDNs). NDNs include immature neutrophils, resting neutrophils, primed neutrophils and activated neutrophils. LDNs include immature LDNs, pro-inflammatory LDNs, immunosuppressive LDNs and so-called exhausted neutrophils (refer to main text). Immature LDNs and pro-inflammatory LDNs are further classified as inflammatory LDNs, whereas immunosuppressive LDNs and exhausted neutrophils are classified as suppressive LDNs. However, while this illustration depicts a classification of neutrophils, it remains unclear whether this corresponds to distinct neutrophil subsets. (The figure is reprinted under a Creative Commons Attribution Non-Commercial License from: Hong (2017))

3.2 Neutrophil Traps: A Necessary Nuisance?

Brinkmann et al. reported almost 15 years ago that neutrophils can kill pathogens extracellularly by releasing so-called neutrophil extracellular traps or NETs (Brinkmann et al. 2004). NETs comprise a backbone of nuclear chromatin decorated with antimicrobial proteins. NETs can be triggered by different stimuli, such as phorbol-12-myristate-13-acetate (PMA), lipopolysaccharide (LPS) and calcium ionophores, as well as by various bacterial and fungal pathogens (Kenny et al. 2017). There are at least two mechanisms of NET formation: oxidant dependent and oxidant independent (Douda et al. 2015). Neutrophils from CGD patients show decreased NET release in response to PMA (Fuchs et al. 2007; van der Linden et al. 2017), and the restoration of NADPH oxidase function and NET formation was shown to cure refractory aspergillosis in a patient with X-linked CGD (Bianchi et al. 2009). In addition to the 'classical' or most commonly studied form of NADPH oxidase-dependent NETs which consist of nuclear chromatin (Brinkmann et al. 2004), some studies have shown that neutrophils can under certain conditions release NETs comprised of mitochondrial DNA (Yousefi et al. 2009). Moreover, NETs enriched in oxidized mitochondrial DNA were found to contribute to autoimmune disease in a mouse model of systemic lupus erythematosus (SLE) (Lood et al. 2016).

NET formation is frequently viewed as a specialized form of neutrophil cell death that is distinct from apoptosis and necrosis (Fuchs et al. 2007), and this cell death has been dubbed 'NETosis'. However, not all NET formation leads to cell death, and not all neutrophil death is coupled with NET formation. This has led to some confusion in the literature as the term 'NETosis' is commonly equated with NET formation, and some authors also deploy the term 'vital NETosis' (Clark et al. 2007; Yipp et al. 2012), which is certainly a misnomer; in our opinion, it is advisable to simply refer to neutrophil formation of NETs with or without attendant cell death. In fact, the terms suicidal and vital NETosis are controversial (Yang et al. 2016). On the other hand, it is clear that neutrophil stimulation with PMA leads to a non-apoptotic form of cell death (Takei et al. 1996; Fadeel et al. 1998). Interestingly, very recent data suggest some commonalities between NET formation and necroptosis (Sollberger et al. 2018).

Neutrophils release extracellular traps in response to a wide variety of bacteria (Brinkmann et al. 2004; Urban et al. 2006). It is worth noting that neutrophils have the ability to modulate their antimicrobial strategy according to the size of the pathogen (Branzk et al. 2014). Hence, neutrophils selectively released NETs in response to large pathogens such as *Candida albicans* hyphae and aggregates of *Mycobacterium bovis*, but not in response to small yeast or single bacteria. The authors speculated that such size-tailored responses may enable the clearance of large microbes while minimizing pathology related to excessive formation of NETs when microbes are small enough to be phagocytosed (Branzk et al. 2014). However, this does not explain why other studies have shown vigorous formation of NETs in response to bacteria such as *E. coli* that are small enough to be phagocytosed by neutrophils. Nevertheless, while NETs are thought to play a role during infection by allowing neutrophils to capture and kill pathogens extracellularly, mounting evidence suggests that uncontrolled or excessive production of NETs, or defective degradation or removal of NETs, is related to the exacerbation of inflammation and to the development of several diseases (Papayannopoulos 2018). Hakkim et al. (2010) reported that impairment of DNaseI-mediated NET degradation is associated with SLE. Excessive formation of NETs, on the other hand, could clog blood vessels and provide a scaffold for thrombus formation (Fuchs et al. 2010), while a recent study has shown that both DNase1 and DNase1-like 3 are capable of degrading NETs in circulation (Jiménez-Alcázar et al. 2017). Notably, metastatic breast cancer cells can 'hijack' neutrophils and trigger the formation of metastasis-supporting NETs in the absence of infection (Park et al. 2016). Treatment with DNaseI-coated nanoparticles markedly reduced lung metastases in mice. This suggests that the triggering of NETs by cancer cells is a potential therapeutic target (Park et al. 2016).

Recently, it was found that NETs are generated at the site of immunization and drive the activity of aluminium adjuvant (alum), thereby enhancing adjuvant-induced adaptive immune responses (Stephen et al. 2017). Hence, the very early inflammatory response to alum drives neutrophil recruitment and generates free DNA through NET formation. This enhances antigen-specific T-cell expansion and B-cell differentiation leading to an increased, class-switched antibody response. These data show that the liberation of DNA *via* NET formation plays an important role in adjuvant activity and support the view that neutrophils orchestrate adaptive immune responses. Thus, it transpires that the net balance between the formation and degradation of NETs is important.

3.3 Effects of Nanomaterials on Neutrophils

Neutrophils are key players in inflammation, and numerous studies have shown that engineered nanomaterials may elicit acute and/or chronic inflammation (Boraschi et al. 2017; Leso et al. 2018). Therefore, in order to evaluate the potential toxicity of an engineered nanomaterial or nanobiomaterial (i.e., a nanomaterial designed with biomedical applications in mind), it is of vital importance to understand its interactions with neutrophils. However, despite numerous studies showing tissue infiltration of neutrophils upon exposure to nanoparticles, it can be argued that neutrophils are a somewhat neglected cell in nanotoxicology as there are relatively few studies on direct interactions of nanomaterials with these cells while there is a considerable number of papers on nanomaterials and macrophages. Nevertheless, neutrophils are normally the first responders in an inflammatory reaction, while macrophages arrive in the second wave of inflammation and serve mainly to remove cell debris and to promote tissue healing (Underhill and Goodridge 2012). Similarly, it is worth noting that macrophages are not the only cells that are involved in the clearance of nanoparticles from the blood; in fact, a recent study showed that neutrophils also play a major role in nanoparticle clearance, at least in certain mouse strains (Jones et al. 2013), suggesting that neutrophils ought to be considered in the context of nanomedicine. In the following, a few examples of particle–neutrophil interactions are provided.

Girard and co-workers have published a series of papers in which various metal and metal oxide nanoparticles including nanoparticles of titanium dioxide, zinc oxide and silver were shown to activate neutrophils and/or to inhibit neutrophil apoptosis (Gonçalves et al. 2010; Babin et al. 2013; Goncalves and Girard 2014; Babin et al. 2015). In contrast, gold nanoparticles were found to activate or accelerate neutrophil apoptosis (Noël et al. 2016). Needless to say, careful attention to endotoxin contamination of the tested particles is required, as neutrophils respond to lipopolysaccharides (LPS) (Gonçalves and Girard 2013). Other authors have shown that silver nanoparticles differently affect distinct subpopulations of neutrophils (Fraser et al. 2018). Notably, in a recent in vivo study, interactions between injected nanoparticles and circulating neutrophils were documented which could drive particle clearance, but could also alter neutrophil responses in a mouse model of acute lung injury (Fromen et al. 2017). Again, these results suggest that neutrophils play an important role both in toxicology and in medicine. Remarkably, though the attachment of poly(ethylene glycol) (PEG) onto the surface of nanoparticles is commonly through to prevent particle opsonization and macrophage uptake, a recent study suggested, instead, that neutrophils preferentially internalized PEGylated particles (i.e., polystyrene microspheres), specifically in the presence of human plasma (Kelley et al. 2018).

Bisso et al. (2018) recently conducted an in-depth study of nanomaterial interactions with human neutrophils focusing on biomaterials, that is, polymeric and liposomal particles ranging in size from 20 nm to 5 μm. The authors found that nanoparticles are readily internalized by neutrophils ex vivo in the absence of serum proteins, and that the internalization was size-dependent insofar as a significant increase in uptake of the 200 nm particles was observed over particles <100 nm in diameter. Interestingly, gold nanoparticles ranging from 5 to 20 nm were not taken up at all. The inclusion of albumin in the cell culture medium prevented uptake of polystyrene particles and reduced the uptake of liposomal nanoparticles, but enhanced neutrophil uptake of poly(lactic-*co*-glycolic acid) (PLGA) particles at all particle sizes tested. Notably, particle-laden neutrophils (i.e., 1 μg/mL of polystyrene particles or 5 μg/mL of liposomes) were found to undergo normal degranulation upon stimulation with conventional agonists, as evidenced by the release of the specific granule protein, neutrophil gelatinase-associated lipocalin. Overall, these results are important as they provide an initial step towards clinically relevant particle design that takes neutrophil behaviour into account (Bisso et al. 2018).

Carbon-based nanomaterials including carbon nanotubes (CNTs) are widely studied in nanotoxicology and nanomedicine (Bhattacharya et al. 2016). Vlasova et al. (2016) reported that adsorbed plasma proteins influenced neutrophil responses caused by PEGylated, single-walled CNTs (SWCNTs). Specifically, the coating of SWCNTs with IgG resulted in neutrophil activation, as determined by degranulation and release of MPO. Similar results were obtained for carboxylated (non-PEGylated) SWCNTs (Lu et al. 2018). In a recent study conducted in the Horizon2020 project BIORIMA ('biomaterial risk management'), the toxicity of three multi-walled CNTs (MWCNTs) with varying physicochemical properties was evaluated in neutrophils versus macrophages. While macrophages were susceptible only to the long and thin or 'fibre-like' MWCNTs, neutrophil cell viability was affected by all three CNTs, and the mechanism of cell death differed between the two cell types (Keshavan et al., manuscript in preparation). Both neutrophils and macrophages were maintained in the presence of 10% serum. Thus, while macrophages are a widely used model in nanotoxicology, and certainly a relevant one as these cells are involved in innate immune responses, neutrophils should not be neglected.

Muñoz et al. (2016) reported that nanodiamonds cause plasma membrane damage and signs of lysosomal instability in neutrophils and found that these nanoparticles triggered the formation of NETs, though it is noted that very high concentrations of nanodiamonds were applied (200 μg/mL). The authors found that larger particles were relatively inert. In contrast, in a more recent study, low concentrations of large (micrometer-sized) graphene oxide (GO) sheets were found to trigger NETs in primary human neutrophils more effectively than small (nano-sized) GO (Mukherjee et al. 2018a). Interestingly, the large GO sheets initiated the oxidation of cholesterol in the plasma membrane of the neutrophil as evidenced by time-of-flight secondary ion mass spectrometry (ToF-SIMS), and this was found to set into motion a series of intracellular events leading to the formation of NETs; furthermore, the release of NETs was reduced by the vitamin E analogue, Trolox, a potent lipid antioxidant (Mukherjee et al. 2018a). In a companion paper, small and large GO sheets (endotoxin-free) were shown to undergo MPO-dependent biodegradation in NETs purified from activated human neutrophils (Mukherjee et al. 2018b). Neutrophils can also enzymatically 'digest' several forms of CNTs (discussed below). Thus, it appears that neutrophils are capable of handling at least some carbon-based nanomaterials leading to the destruction of the offending agents (Farrera and Fadeel 2015).

3.4 Inflammasomes: Double-Edged Swords?

Inflammation plays a critical role in host defence against pathogens. However, inflammation may also occur under sterile conditions as a result of a variety of physical, chemical, or metabolic insults, and in response to silica, asbestos, and other environmental agents (Leso et al. 2018). Nanomaterials (with or without a corona of proteins or other biomolecules) may be considered as a particular case of danger signals that are able to trigger sterile inflammatory responses. Indeed, we have postulated that engineered nanomaterials may present 'nanomaterial-associated molecular patterns' or NAMPs to cells of the immune system (Fadeel 2012).

How do nanomaterials and other exogenous substances trigger inflammation? Studies published in the past decade have revealed that the so-called inflammasome, so named by Tschopp and co-workers in their seminal publication 15 years ago (Martinon et al. 2002), is a key signalling hub that regulates innate immunity (Broz and Dixit 2016; Mangan et al. 2018). The inflammasomes are multi-protein complexes that activate the protease, caspase-1, leading to the secretion of IL-1β and are assembled by the protein pyrin or by members of the NOD-like receptor (NLR) and pyrin and HIN domain-containing (PYHIN) protein families, in response to diverse pathogen- and host-derived danger signals (Broz and Dixit 2016). NLRP3 (also known as cryopyrin) stands out as it responds to a surprisingly diverse array of different stimuli including crystalline and particulate matter, such as uric acid crystals, silica, asbestos, and alum, as well as to pathogens (Martinon et al. 2006; Dostert et al. 2008; Cassel et al. 2008; Hornung et al. 2008). NLRP3 is also a sensor of various carbon-based and other nanomaterials (Sun et al. 2013). Indeed, not only long and rigid (needle-like) CNTs (Palomäki et al. 2011) but also ultra-thin graphene oxide (GO) sheets are able to trigger NRLP3-dependent IL-1β secretion in human macrophages (Mukherjee et al. 2018c). However, less is known regarding inflammasome activation in neutrophils. In fact, neutrophils, often viewed as targets of IL-1β, are capable of activating the NLRC4 inflammasome in response to bacterial challenge. However, unlike macrophages, neutrophils do not undergo cell death (pyroptosis) upon inflammasome activation (Chen et al. 2014). Further studies are needed to understand if and how nanomaterials trigger neutrophil inflammasome activation and/or cell death. It is also important to keep in mind that while inflammasome activation is a key component in the host defence against pathogens and other xenobiotic compounds, and may also play an important role in the immunostimulatory actions of alum and of other adjuvants, excessive inflammasome activation is linked to severe, autoinflammatory disorders. Understanding how nanomaterials trigger the inflammasome(s) is, therefore, of great importance – both in nanotoxicology and in medicine.

3.5 Lessons from Studies of Biomaterials

Implanted biomedical devices are commonplace, and there is a wealth of literature on the biocompatibility and biodegradability of such devices; hence, the nanotoxicology and nanomedicine communities would do well to take note of these results. Therefore, in this section, we focus on implantable biomaterials and on their interactions with the innate immune system.

The so-called foreign body reaction, centred on the recruitment of macrophages and on macrophage fusion to form foreign body giant cells in response to objects that are too large to engulf, has been well-described in the context of (implantable) biomaterials (Anderson et al. 2008). The role of other cell types, however, should not be neglected. Several studies have shown that neutrophil–biomaterial interactions may result in neutrophil activation. Cohen et al. (2013) cultured monocytes in the presence of neutrophil-conditioned medium on poly(ethylene glycol) (PEG) hydrogels and noted distinct differences in adhesion, viability and cytokine secretion in such monocytes when compared with monocytes cultured alone. The same authors investigated the release of MPO, a primary granule marker, and matrix metalloproteinase-9 (MMP-9), a marker of tertiary granules, from neutrophils cultured on various biomaterial surfaces, and they observed that neutrophils cultured on PEG hydrogels released more MPO than cells cultured on polydimethylsiloxane (PDMS) or tissue culture polystyrene. The release of MMP-9, however, was similar for all the biomaterials tested (Cohen et al. 2014). Furthermore, PEG-containing hydrogels promoted the release of antibacterial α-defensins (Lieberthal et al. 2015). Hydroxyapatite (HA) is widely used as a bone substitute or for the coating of metal prostheses. Velard et al. (2009) reported that HA is capable of activating neutrophils in vitro and suggested that this could potentially contribute to the onset of implant-associated inflammation. In a subsequent study, the authors showed that exposure to zinc-substituted HA resulted in decreased levels of the pro-inflammatory mediators, IL-8 and MMP-9 (Velard et al. 2010).

However, it is important to recognize that *inflammation* is not a detrimental response, but it is, in fact, involved in the orchestration of salutary immune signalling to promote tissue healing. Therefore, one should not seek to prevent (acute) inflammation at every cost. Instead, careful design of nano(bio)materials in order to control immune reactions or even to harness and exploit the innate and adaptive arms of the immune system is a highly desirable goal (Vasconcelos et al. 2019). The ultimate aim may not be immune 'stealth', but rather a form of peaceful co-existence – the immune system is not the enemy, though one should strive to avoid protracted (chronic) responses. Vegas et al. (2016) used a combinatorial approach for covalent chemical modification to generate a large library of variants of the hydrogel biomaterial, alginate, and identified a unique hydrogel surface that caused the recruitment of fewer macrophages and neutrophils and substantially reduced foreign body reactions in rodents and non-human primate (up to 6 months). The same authors reported in a more recent study that macrophages are indispensable for the fibrotic response to biomaterials while neutrophils are not, even though both cell types were present on the surface of the implanted alginate microspheres

(500 μm in diameter) (Doloff et al. 2017). Furthermore, colony-stimulating factor-1 receptor (CSF1R) was significantly increased following implantation of multiple biomaterial classes (ceramic, polymer and hydrogel), and its inhibition resulted in a complete absence of fibrosis, presumably as a result of reprogramming of the macrophages from a pro-fibrotic to a pro-resolving phenotype (Doloff et al. 2017). It will be of interest to determine whether similar reactions occur in response to other elongated or flat biomaterials as opposed to such spherical particles.

The adsorption of proteins on biomaterial surfaces is known to influence subsequent immune responses. For instance, Tang and Eaton (1993) reported 25 years ago that spontaneous adsorption of fibrinogen appears to initiate the acute inflammatory response to an implanted polymer. Notably, hypofibrinogenemic mice were unable to mount an inflammatory response to implanted polyester terephthalate film (PET) unless the material was coated with fibrinogen or the animals were injected with fibrinogen before implantation. In a subsequent study, the authors showed that fibrinogen promoted inflammatory responses through interactions with CD11b/CD18 (Mac-1) as cellular accumulation on the implants was abrogated by administration of recombinant neutrophil inhibitory factor which blocks interactions with CD11b (Tang et al. 1996). Furthermore, the adsorption and denaturation of fibrinogen on the biomaterial were shown to display a cryptic epitope that interacts with Mac-1 and is normally hidden in the soluble form of fibrinogen (Hu et al. 2001). Ten years later, Deng et al. (2011) reported that negatively charged poly(acrylic acid)-conjugated gold nanoparticles bind to and induce unfolding of fibrinogen, which promotes interaction with the integrin receptor, Mac-1 and TNF-α secretion. Hence, the discovery of the bio-corona on nanoparticles is in many ways a re-discovery of the well-known phenomenon of opsonization of biomaterials and pathogens.

Recent studies have shown that a variety of bioimplants including spherical microcapsules and electrospun fibres elicit the formation of NETs in vitro as well as in vivo (Vitkov et al. 2015; Jhunjhunwala et al. 2015; Fetz et al. 2017). However, further studies are needed to understand whether the generation of NETs interferes with implant performance and whether this acute reaction to implanted materials also would translate into a chronic inflammatory (aka foreign body) reaction and/or fibrosis. Certain pathogens have the ability to circumvent NETs through the secretion of DNases (Buchanan et al. 2006; Berends et al. 2010), and it would be interesting to see whether implants could be protected from NETs using a similar strategy (Jhunjhunwala 2018). Sperling et al. (2017) have suggested that NET formation may contribute to the thrombogenicity of clinically applied hydrophobic materials and showed that disintegration of surface-bound NETs by DNase treatment resulted in significantly lower pro-coagulant effects. As already mentioned, large GO sheets also trigger NETs (Mukherjee et al. 2018a), a process which involves lipid oxidation in the plasma membrane; while this does not exclude a role for phagocytosis, the latter results are in accordance with a previous study showing that the adjuvant, alum, triggers responses in dendritic cells by altering membrane lipid structures (Flach et al. 2011). These data suggest that not all immune signalling is receptor mediated and imply that the plasma membrane may act as a sensor for solid structures. Moreover, not all innate immune responses depend on phagocytosis of the offending agent or nano(bio)material.

3.6 Effect of Neutrophils on Nanomaterials

Neutrophils are remarkably resourceful when it comes to the killing of microorganisms. The NADPH oxidase generates reactive oxygen species that are instrumental for the killing of ingested pathogens. Furthermore, degranulation resulting in the release of MPO is also an important aspect of the microbicidal function of neutrophils (Klebanoff et al. 2013). In addition to its role in antimicrobial defence, MPO is also reported to be involved in the degradation of CNTs, and the clearance of CNTs from the lungs was markedly less efficient in MPO-deficient mice when compared to wild-type mice (Shvedova et al. 2012) (Fig. 3.3). Neutrophil-mediated, MPO-dependent destruction of SWCNTs was first described by Kagan et al. (2010). Subsequently, MPO-dependent degradation of PEGylated SWCNTs was reported (Bhattacharya et al. 2014), and this was suggested to occur in a two-step process whereby the polymers were first removed from the CNTs by neutrophil elastase followed by the degradation of the CNTs themselves by MPO. Furthermore, SWCNTs were shown to undergo acellular, MPO-dependent degradation in NETs purified from activated neutrophils (Farrera et al. 2014). More recent studies conducted in

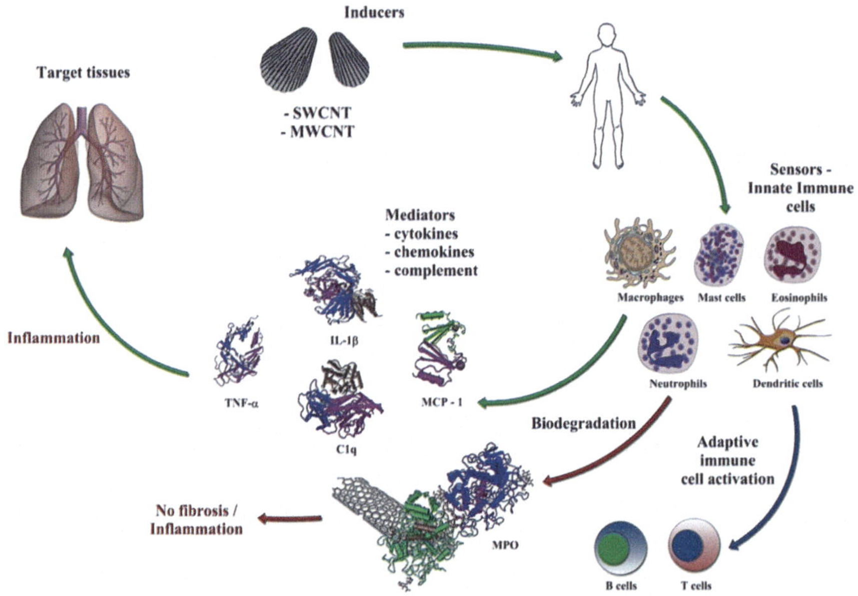

Fig. 3.3 Nanotubes and inflammation: reciprocal interactions. Exposure to *inducers* (exemplified here by single- and multi-walled carbon nanotubes, CNTs) initiates a multifactorial response from the *sensors* (i.e., innate immune cells). These cells produce soluble *mediators* (cytokines, chemokines, complement factors and others), leading to inflammation in target tissues. On the other hand, recent studies have shown that inflammatory cells are capable of biodegradation of CNTs through expression of myeloperoxidase (MPO) and related enzymes, leading to a reduced inflammatory and fibrotic response in mouse lungs. (Reprinted from: Bhattacharya et al. (2013), with permission from Elsevier)

the Graphene Flagship project have revealed that graphene oxide (GO) is also susceptible to biodegradation in the presence of ex vivo activated neutrophils, and prompt biodegradation of GO was also observed in purified NETs (Mukherjee et al. 2018b). Even graphene can undergo neutrophil-mediated degradation though the process is considerably slower when compared to GO (Kurapati et al. 2018), indicating that graphene is not necessarily biopersistent (Fadeel et al. 2018). The latter studies were conducted ex vivo using human neutrophils. Notably, a recent in vivo study has shown that GO elicits an inflammatory reaction in mice in response to implantation consistent with a foreign body reaction; moreover, a reduction in the degree of oxidation resulted in faster immune cell infiltration, uptake and clearance following both subcutaneous and peritoneal implantation (Sydlik et al. 2015). Tuning the properties of graphene-based materials to achieve optimal performance in a biomedical setting while maintaining an acceptable degree of biocompatibility and biodegradability (in cases when this is desired) remains an important challenge.

3.7 Concluding Remarks

Neutrophils are key cells of the innate arm of the immune system and play important roles in host defence against pathogens, but paradoxically, they are also involved in numerous pathological conditions characterized by chronic inflammation and/or by dysrepair and malfunction of cells and tissues following an inflammatory episode. Indeed, it has been suggested that the pathological potential of inflammation is unprecedented for a physiological process (Medzhitov 2008). Nano(bio)materials may be viewed as a special class of sterile or non-infectious danger signals, and emerging studies have shown that such materials can modulate and activate neutrophils and other immune cells. Moreover, activated neutrophils may also 'attack' and digest certain carbon-based nanomaterials as if they were pathogens (Mukherjee et al. 2017). Deciphering the reciprocal interactions between nano(bio)materials and neutrophils is critically important for the development of safe materials for biomedical applications.

Acknowledgements Supported by the European Commission's H2020 programme through BIORIMA (grant agreement no. 760928) and the Graphene Flagship (grant agreement no. 785219), and the Swedish Research Council, and Swedish Research Council for Health, Working Life and Welfare.

References

Anderson JM, Rodriguez A, Chang DT. Foreign body reaction to biomaterials. Semin Immunol. 2008;20(2):86–100.

Babin K, Antoine F, Goncalves DM, Girard D. TiO_2, CeO2 and ZnO nanoparticles and modulation of the degranulation process in human neutrophils. Toxicol Lett. 2013;221(1):57–63.

Babin K, Goncalves DM, Girard D. Nanoparticles enhance the ability of human neutrophils to exert phagocytosis by a Syk-dependent mechanism. Biochim Biophys Acta. 2015;1850(11):2276–82.
Bardoel BW, Kenny EF, Sollberger G, Zychlinsky A. The balancing act of neutrophils. Cell Host Microbe. 2014;15(5):526–36.
Berends ET, Horswill AR, Haste NM, Monestier M, Nizet V, von Köckritz-Blickwede M. Nuclease expression by *Staphylococcus aureus* facilitates escape from neutrophil extracellular traps. J Innate Immun. 2010;2(6):576–86.
Bhattacharya K, Andón FT, El-Sayed R, Fadeel B. Mechanisms of carbon nanotube-induced toxicity: focus on pulmonary inflammation. Adv Drug Deliv Rev. 2013;65(15):2087–97.
Bhattacharya K, Sacchetti C, El-Sayed R, Fornara A, Kotchey GP, Gaugler JA, Star A, Bottini M, Fadeel B. Enzymatic 'stripping' and degradation of PEGylated carbon nanotubes. Nanoscale. 2014;6(24):14686–90.
Bhattacharya K, Mukherjee SP, Gallud A, Burkert SC, Bistarelli S, Bellucci S, Bottini M, Star A, Fadeel B. Biological interactions of carbon-based nanomaterials: from coronation to degradation. Nanomedicine. 2016;12(2):333–51.
Bianchi M, Hakkim A, Brinkmann V, Siler U, Seger RA, Zychlinsky A, Reichenbach J. Restoration of NET formation by gene therapy in CGD controls aspergillosis. Blood. 2009;114(13):2619–22.
Bisso PW, Gaglione S, Guimaraes PP, Mitchell MJ, Langer R. Nanomaterial interactions with human neutrophils. ACS Biomater Sci Eng. 2018;4:4255–65.
Boraschi D, Italiani P, Palomba R, Decuzzi P, Duschl A, Fadeel B, Moghimi SM. Nanoparticles and innate immunity: new perspectives on host defense. Semin Immunol. 2017;34:33–51.
Borregaard N. Neutrophils, from marrow to microbes. Immunity. 2010;33(5):657–70.
Branzk N, Lubojemska A, Hardison SE, Wang Q, Gutierrez MG, Brown GD, Papayannopoulos V. Neutrophils sense microbe size and selectively release neutrophil extracellular traps in response to large pathogens. Nat Immunol. 2014;15(11):1017–25.
Brinkmann V, Reichard U, Goosmann C, Fauler B, Uhlemann Y, Weiss DS, Weinrauch Y, Zychlinsky A. Neutrophil extracellular traps kill bacteria. Science. 2004;303(5663):1532–5.
Broz P, Dixit VM. Inflammasomes: mechanism of assembly, regulation and signalling. Nat Rev Immunol. 2016;16(7):407–20.
Buchanan JT, Simpson AJ, Aziz RK, Liu GY, Kristian SA, Kotb M, Feramisco J, Nizet V. DNase expression allows the pathogen group A *Streptococcus* to escape killing in neutrophil extracellular traps. Curr Biol. 2006;16(4):396–400.
Cassel SL, Eisenbarth SC, Iyer SS, Sadler JJ, Colegio OR, Tephly LA, Carter AB, Rothman PB, Flavell RA, Sutterwala FS. The Nalp3 inflammasome is essential for the development of silicosis. Proc Natl Acad Sci U S A. 2008;105(26):9035–40.
Chen KW, Groß CJ, Sotomayor FV, Stacey KJ, Tschopp J, Sweet MJ, Schroder K. The neutrophil NLRC4 inflammasome selectively promotes IL-1β maturation without pyroptosis during acute *Salmonella* challenge. Cell Rep. 2014;8(2):570–82.
Clark SR, Ma AC, Tavener SA, McDonald B, Goodarzi Z, Kelly MM, Patel KD, Chakrabarti S, McAvoy E, Sinclair GD, Keys EM, Allen-Vercoe E, Devinney R, Doig CJ, Green FH, Kubes P. Platelet TLR4 activates neutrophil extracellular traps to ensnare bacteria in septic blood. Nat Med. 2007;13(4):463–9.
Cohen HC, Joyce EJ, Kao WJ. Biomaterials selectively modulate interactions between human blood-derived polymorphonuclear leukocytes and monocytes. Am J Pathol. 2013;182(6):2180–90.
Cohen HC, Lieberthal TJ, Kao WJ. Poly(ethylene glycol)-containing hydrogels promote the release of primary granules from human blood-derived polymorphonuclear leukocytes. J Biomed Mater Res A. 2014;102(12):4252–61.
Deng ZJ, Liang M, Monteiro M, Toth I, Minchin RF. Nanoparticle-induced unfolding of fibrinogen promotes Mac-1 receptor activation and inflammation. Nat Nanotechnol. 2011;6(1):39–44.
Doloff JC, Veiseh O, Vegas AJ, Tam HH, Farah S, Ma M, Li J, Bader A, Chiu A, Sadraei A, Aresta-Dasilva S, Griffin M, Jhunjhunwala S, Webber M, Siebert S, Tang K, Chen M, Langan E, Dholokia N, Thakrar R, Qi M, Oberholzer J, Greiner DL, Langer R, Anderson DG. Colony stimulating factor-1 receptor is a central component of the foreign body response to biomaterial implants in rodents and non-human primates. Nat Mater. 2017;16(6):671–80.

Dostert C, Pétrilli V, Van Bruggen R, Steele C, Mossman BT, Tschopp J. Innate immune activation through Nalp3 inflammasome sensing of asbestos and silica. Science. 2008;320(5876):674–7.

Douda DN, Khan MA, Grasemann H, Palaniyar N. SK3 channel and mitochondrial ROS mediate NADPH oxidase-independent NETosis induced by calcium influx. Proc Natl Acad Sci U S A. 2015;112(9):2817–22.

Dudeck A, Köberle M, Goldmann O, Meyer N, Dudeck J, Lemmens S, Rohde M, Roldán NG, Dietze-Schwonberg K, Orinska Z, Medina E, Hendrix S, Metz M, Zenclussen AC, von Stebut E, Biedermann T. Mast cells as protectors of health. J Allergy Clin Immunol. 2019;144:S4–S18.

Fadeel B. Clear and present danger? Engineered nanoparticles and the immune system. Swiss Med Wkly. 2012;142:w13609.

Fadeel B, Åhlin A, Henter J-I, Orrenius S, Hampton MB. Involvement of caspases in neutrophil apoptosis: regulation by reactive oxygen species. Blood. 1998;92(12):4808–18.

Fadeel B, Bussy C, Merino S, Vázquez E, Flahaut E, Mouchet F, Evariste L, Gauthier L, Koivisto AJ, Vogel U, Martín C, Delogu LG, Buerki-Thurnherr T, Wick P, Beloin-Saint-Pierre D, Hischier R, Pelin M, Candotto Carniel F, Tretiach M, Cesca F, Benfenati F, Scaini D, Ballerini L, Kostarelos K, Prato M, Bianco A. Safety assessment of graphene-based materials: focus on human health and the environment. ACS Nano. 2018;12(11):10582–620.

Farber DL, Netea MG, Radbruch A, Rajewsky K, Zinkernagel RM. Immunological memory: lessons from the past and a look to the future. Nat Rev Immunol. 2016;16(2):124–8.

Farrera C, Fadeel B. It takes two to tango: understanding the interactions between engineered nanomaterials and the immune system. Eur J Pharm Biopharm. 2015;95(Pt A):3–12.

Farrera C, Bhattacharya K, Lazzaretto B, Andón FT, Hultenby K, Kotchey GP, Star A, Fadeel B. Extracellular entrapment and degradation of single-walled carbon nanotubes. Nanoscale. 2014;6(12):6974–83.

Fetz AE, Neeli I, Rodriguez IA, Radic MZ, Bowlin GL. Electrospun template architecture and composition regulate neutrophil NETosis *in vitro* and *in vivo*. Tissue Eng Part A. 2017;23(19–20):1054–63.

Flach TL, Ng G, Hari A, Desrosiers MD, Zhang P, Ward SM, Seamone ME, Vilaysane A, Mucsi AD, Fong Y, Prenner E, Ling CC, Tschopp J, Muruve DA, Amrein MW, Shi Y. Alum interaction with dendritic cell membrane lipids is essential for its adjuvanticity. Nat Med. 2011;17:479–87.

Fraser JA, Kemp S, Young L, Ross M, Prach M, Hutchison GR, Malone E. Silver nanoparticles promote the emergence of heterogeneic human neutrophil sub-populations. Sci Rep. 2018;8(1):7506.

Freitas M, Porto G, Lima JL, Fernandes E. Isolation and activation of human neutrophils *in vitro*. The importance of the anticoagulant used during blood collection. Clin Biochem. 2008;41(7–8):570–5.

Fromen CA, Kelley WJ, Fish MB, Adili R, Noble J, Hoenerhoff MJ, Holinstat M, Eniola-Adefeso O. Neutrophil-particle interactions in blood circulation drive particle clearance and alter neutrophil responses in acute inflammation. ACS Nano. 2017;11(11):10797–807.

Fuchs TA, Abed U, Goosmann C, Hurwitz R, Schulze I, Wahn V, Weinrauch Y, Brinkmann V, Zychlinsky A. Novel cell death program leads to neutrophil extracellular traps. J Cell Biol. 2007;176(2):231–41.

Fuchs TA, Brill A, Duerschmied D, Schatzberg D, Monestier M, Myers DD Jr, Wrobleski SK, Wakefield TW, Hartwig JH, Wagner DD. Extracellular DNA traps promote thrombosis. Proc Natl Acad Sci U S A. 2010;107(36):15880–5.

Gallin JI. Human neutrophil heterogeneity exists, but is it meaningful? Blood. 1984;63(5):977–83.

Gonçalves DM, Girard D. Evidence that polyhydroxylated C_{60} fullerenes (fullerenols) amplify the effect of lipopolysaccharides to induce rapid leukocyte infiltration *in vivo*. Chem Res Toxicol. 2013;26(12):1884–92.

Goncalves DM, Girard D. Zinc oxide nanoparticles delay human neutrophil apoptosis by a *de novo* protein synthesis-dependent and reactive oxygen species-independent mechanism. Toxicol In Vitro. 2014;28(5):926–31.

Gonçalves DM, Chiasson S, Girard D. Activation of human neutrophils by titanium dioxide (TiO_2) nanoparticles. Toxicol In Vitro. 2010;24(3):1002–8.
Hakkim A, Fürnrohr BG, Amann K, Laube B, Abed UA, Brinkmann V, Herrmann M, Voll RE, Zychlinsky A. Impairment of neutrophil extracellular trap degradation is associated with lupus nephritis. Proc Natl Acad Sci U S A. 2010;107(21):9813–8.
Hong CW. Current understanding in neutrophil differentiation and heterogeneity. Immune Netw. 2017;17(5):298–306.
Hornung V, Bauernfeind F, Halle A, Samstad EO, Kono H, Rock KL, Fitzgerald KA, Latz E. Silica crystals and aluminum salts activate the NALP3 inflammasome through phagosomal destabilization. Nat Immunol. 2008;9(8):847–56.
Hu WJ, Eaton JW, Ugarova TP, Tang L. Molecular basis of biomaterial-mediated foreign body reactions. Blood. 2001;98(4):1231–8.
Jhunjhunwala S. Neutrophils at the biological-material interface. ACS Biomater Sci Eng. 2018;4(4):1128–36.
Jhunjhunwala S, Aresta-DaSilva S, Tang K, Alvarez D, Webber MJ, Tang BC, Lavin DM, Veiseh O, Doloff JC, Bose S, Vegas A, Ma M, Sahay G, Chiu A, Bader A, Langan E, Siebert S, Li J, Greiner DL, Newburger PE, von Andrian UH, Langer R, Anderson DG. Neutrophil responses to sterile implant materials. PLoS One. 2015;10(9):e0137550.
Jiménez-Alcázar M, Rangaswamy C, Panda R, Bitterling J, Simsek YJ, Long AT, Bilyy R, Krenn V, Renné C, Renné T, Kluge S, Panzer U, Mizuta R, Mannherz HG, Kitamura D, Herrmann M, Napirei M, Fuchs TA. Host DNases prevent vascular occlusion by neutrophil extracellular traps. Science. 2017;358(6367):1202–6.
Johnston HJ, Verdon R, Gillies S, Brown DM, Fernandes TF, Henry TB, Rossi AG, Tran L, Tucker C, Tyler CR, Stone V. Adoption of *in vitro* systems and zebrafish embryos as alternative models for reducing rodent use in assessments of immunological and oxidative stress responses to nanomaterials. Crit Rev Toxicol. 2018;48(3):252–71.
Jones SW, Roberts RA, Robbins GR, Perry JL, Kai MP, Chen K, Bo T, Napier ME, Ting JP, Desimone JM, Bear JE. Nanoparticle clearance is governed by Th1/Th2 immunity and strain background. J Clin Invest. 2013;123(7):3061–73.
Kagan VE, Konduru NV, Feng W, Allen BL, Conroy J, Volkov Y, Vlasova II, Belikova NA, Yanamala N, Kapralov A, Tyurina YY, Shi J, Kisin ER, Murray AR, Franks J, Stolz D, Gou P, Klein-Seetharaman J, Fadeel B, Star A, Shvedova AA. Carbon nanotubes degraded by neutrophil myeloperoxidase induce less pulmonary inflammation. Nat Nanotechnol. 2010;5(5):354–9.
Kelley WJ, Fromen CA, Lopez-Cazares G, Eniola-Adefeso O. PEGylation of model drug carriers enhances phagocytosis by primary human neutrophils. Acta Biomater. 2018;79:283–93.
Kenny EF, Herzig A, Krüger R, Muth A, Mondal S, Thompson PR, Brinkmann V, Bernuth HV, Zychlinsky A. Diverse stimuli engage different neutrophil extracellular trap pathways. Elife. 2017;6:pii: e24437.
Klebanoff SJ, Kettle AJ, Rosen H, Winterbourn CC, Nauseef WM. Myeloperoxidase: a front-line defender against phagocytosed microorganisms. J Leukoc Biol. 2013;93(2):185–98.
Kruger P, Saffarzadeh M, Weber AN, Rieber N, Radsak M, von Bernuth H, Benarafa C, Roos D, Skokowa J, Hartl D. Neutrophils: between host defence, immune modulation, and tissue injury. PLoS Pathog. 2015;11(3):e1004651.
Kurapati R, Mukherjee SP, Martín C, Bepete G, Vázquez E, Pénicaud A, Fadeel B, Bianco A. Degradation of single-layer and few-layer graphene by neutrophil myeloperoxidase. Angew Chem Int Ed Engl. 2018;57(36):11722–7.
Leso V, Fontana L, Iavicoli I. Nanomaterial exposure and sterile inflammatory reactions. Toxicol Appl Pharmacol. 2018;355:80–92.
Lieberthal TJ, Cohen HC, Kao WJ. Poly(ethylene glycol)-containing hydrogels modulate α-defensin release from polymorphonuclear leukocytes and monocyte recruitment. J Biomed Mater Res A. 2015;103(12):3772–80.
Lood C, Blanco LP, Purmalek MM, Carmona-Rivera C, De Ravin SS, Smith CK, Malech HL, Ledbetter JA, Elkon KB, Kaplan MJ. Neutrophil extracellular traps enriched in oxi-

dized mitochondrial DNA are interferogenic and contribute to lupus-like disease. Nat Med. 2016;22(2):146–53.

Lu N, Sui Y, Tian R, Peng YY. Adsorption of plasma proteins on single-walled carbon nanotubes reduced cytotoxicity and modulated neutrophil activation. Chem Res Toxicol. 2018;31(10):1061–8.

Mangan MSJ, Olhava EJ, Roush WR, Seidel HM, Glick GD, Latz E. Targeting the NLRP3 inflammasome in inflammatory diseases. Nat Rev Drug Discov. 2018;17(8):588–606.

Martinon F, Burns K, Tschopp J. The inflammasome: a molecular platform triggering activation of inflammatory caspases and processing of proIL-beta. Mol Cell. 2002;10(2):417–26.

Martinon F, Pétrilli V, Mayor A, Tardivel A, Tschopp J. Gout-associated uric acid crystals activate the NALP3 inflammasome. Nature. 2006;440(7081):237–41.

Medzhitov R. Origin and physiological roles of inflammation. Nature. 2008;454(7203):428–35.

Mestas J, Hughes CC. Of mice and not men: differences between mouse and human immunology. J Immunol. 2004;172(5):2731–8.

Mukherjee SP, Bottini M, Fadeel B. Graphene and the immune system: a romance of many dimensions. Front Immunol. 2017;8:673.

Mukherjee SP, Lazzaretto B, Hultenby K, Newman L, Rodrigues AF, Lozano N, Kostarelos K, Malmberg P, Fadeel B. Graphene oxide elicits membrane lipid changes and neutrophil extracellular trap formation. Chem. 2018a;4:334–58.

Mukherjee SP, Gliga AR, Lazzaretto B, Brandner B, Fielden M, Vogt C, Newman L, Rodrigues AF, Shao W, Fournier PM, Toprak MS, Star A, Kostarelos K, Bhattacharya K, Fadeel B. Graphene oxide is degraded by neutrophils and the degradation products are non-genotoxic. Nanoscale. 2018b;10(3):1180–8.

Mukherjee SP, Kostarelos K, Fadeel B. Cytokine profiling of primary human macrophages exposed to endotoxin-free graphene oxide: size-independent NLRP3 inflammasome activation. Adv Healthc Mater. 2018c;7(4):1700815.

Muñoz LE, Bilyy R, Biermann MH, Kienhöfer D, Maueröder C, Hahn J, Brauner JM, Weidner D, Chen J, Scharin-Mehlmann M, Janko C, Friedrich RP, Mielenz D, Dumych T, Lootsik MD, Schauer C, Schett G, Hoffmann M, Zhao Y, Herrmann M. Nanoparticles size-dependently initiate self-limiting NETosis-driven inflammation. Proc Natl Acad Sci U S A. 2016;113(40):E5856–65.

Nathan C. Points of control in inflammation. Nature. 2002;420(6917):846–52.

Nathan C. Neutrophils and immunity: challenges and opportunities. Nat Rev Immunol. 2006;6(3):173–82.

Nauseef WM, Borregaard N. Neutrophils at work. Nat Immunol. 2014;15(7):602–11.

Netea MG, Joosten LA, Latz E, Mills KH, Natoli G, Stunnenberg HG, O'Neill LA, Xavier RJ. Trained immunity: a program of innate immune memory in health and disease. Science. 2016;352(6284):aaf1098.

Noël C, Simard JC, Girard D. Gold nanoparticles induce apoptosis, endoplasmic reticulum stress events and cleavage of cytoskeletal proteins in human neutrophils. Toxicol In Vitro. 2016;31:12–22.

Palomäki J, Välimäki E, Sund J, Vippola M, Clausen PA, Jensen KA, Savolainen K, Matikainen S, Alenius H. Long, needle-like carbon nanotubes and asbestos activate the NLRP3 inflammasome through a similar mechanism. ACS Nano. 2011;5(9):6861–70.

Papayannopoulos V. Neutrophil extracellular traps in immunity and disease. Nat Rev Immunol. 2018;18(2):134–47.

Park J, Wysocki RW, Amoozgar Z, Maiorino L, Fein MR, Jorns J, Schott AF, Kinugasa-Katayama Y, Lee Y, Won NH, Nakasone ES, Hearn SA, Küttner V, Qiu J, Almeida AS, Perurena N, Kessenbrock K, Goldberg MS, Egeblad M. Cancer cells induce metastasis-supporting neutrophil extracellular DNA traps. Sci Transl Med. 2016;8(361):361ra138.

Rydell-Törmänen K, Uller L, Erjefält JS. Neutrophil cannibalism--a back up when the macrophage clearance system is insufficient. Respir Res. 2006;7:143.

Scapini P, Cassatella MA. Social networking of human neutrophils within the immune system. Blood. 2014;124(5):710–9.

Shvedova AA, Kapralov AA, Feng WH, Kisin ER, Murray AR, Mercer RR, St Croix CM, Lang MA, Watkins SC, Konduru NV, Allen BL, Conroy J, Kotchey GP, Mohamed BM, Meade AD, Volkov Y, Star A, Fadeel B, Kagan VE. Impaired clearance and enhanced pulmonary inflammatory/fibrotic response to carbon nanotubes in myeloperoxidase-deficient mice. PLoS One. 2012;7(3):e30923.

Silvestre-Roig C, Hidalgo A, Soehnlein O. Neutrophil heterogeneity: implications for homeostasis and pathogenesis. Blood. 2016;127(18):2173–81.

Sollberger G, Choidas A, Burn GL, Habenberger P, Di Lucrezia R, Kordes S, Menninger S, Eickhoff J, Nussbaumer P, Klebl B, Krüger R, Herzig A, Zychlinsky A. Gasdermin D plays a vital role in the generation of neutrophil extracellular traps. Sci Immunol. 2018;3(26):eaar6689.

Sperling C, Fischer M, Maitz MF, Werner C. Neutrophil extracellular trap formation upon exposure of hydrophobic materials to human whole blood causes thrombogenic reactions. Biomater Sci. 2017;5(10):1998–2008.

Stephen J, Scales HE, Benson RA, Erben D, Garside P, Brewer JM. Neutrophil swarming and extracellular trap formation play a significant role in Alum adjuvant activity. NPJ Vaccines. 2017;2(1):1.

Sun B, Wang X, Ji Z, Li R, Xia T. NLRP3 inflammasome activation induced by engineered nanomaterials. Small. 2013;9(9–10):1595–607.

Sydlik SA, Jhunjhunwala S, Webber MJ, Anderson DG, Langer R. *In vivo* compatibility of graphene oxide with differing oxidation states. ACS Nano. 2015;9(4):3866–74.

Takei H, Araki A, Watanabe H, Ichinose A, Sendo F. Rapid killing of human neutrophils by the potent activator phorbol 12-myristate 13-acetate (PMA) accompanied by changes different from typical apoptosis or necrosis. J Leukoc Biol. 1996;59(2):229–40.

Tang L, Eaton JW. Fibrin(ogen) mediates acute inflammatory responses to biomaterials. J Exp Med. 1993;178(6):2147–56.

Tang L, Ugarova TP, Plow EF, Eaton JW. Molecular determinants of acute inflammatory responses to biomaterials. J Clin Invest. 1996;97(5):1329–34.

Thomas DC. How the phagocyte NADPH oxidase regulates innate immunity. Free Radic Biol Med. 2018;125:44–52.

Underhill DM, Goodridge HS. Information processing during phagocytosis. Nat Rev Immunol. 2012;12(7):492–502.

Urban CF, Reichard U, Brinkmann V, Zychlinsky A. Neutrophil extracellular traps capture and kill *Candida albicans* yeast and hyphal forms. Cell Microbiol. 2006;8(4):668–76.

van der Linden M, Westerlaken GHA, van der Vlist M, van Montfrans J, Meyaard L. Differential signalling and kinetics of neutrophil extracellular trap release revealed by quantitative live imaging. Sci Rep. 2017;7(1):6529.

Vasconcelos DP, Águas AP, Barbosa MA, Pelegrín P, Barbosa JN. The inflammasome in host response to biomaterials: bridging inflammation and tissue regeneration. Acta Biomater. 2019;83:1–12.

Vegas AJ, Veiseh O, Doloff JC, Ma M, Tam HH, Bratlie K, Li J, Bader AR, Langan E, Olejnik K, Fenton P, Kang JW, Hollister-Locke J, Bochenek MA, Chiu A, Siebert S, Tang K, Jhunjhunwala S, Aresta-Dasilva S, Dholakia N, Thakrar R, Vietti T, Chen M, Cohen J, Siniakowicz K, Qi M, McGarrigle J, Graham AC, Lyle S, Harlan DM, Greiner DL, Oberholzer J, Weir GC, Langer R, Anderson DG. Combinatorial hydrogel library enables identification of materials that mitigate the foreign body response in primates. Nat Biotechnol. 2016;34(3):345–52.

Velard F, Laurent-Maquin D, Guillaume C, Bouthors S, Jallot E, Nedelec JM, Belaaouaj A, Laquerriere P. Polymorphonuclear neutrophil response to hydroxyapatite particles, implication in acute inflammatory reaction. Acta Biomater. 2009;5(5):1708–15.

Velard F, Laurent-Maquin D, Braux J, Guillaume C, Bouthors S, Jallot E, Nedelec JM, Belaaouaj A, Laquerriere P. The effect of zinc on hydroxyapatite-mediated activation of human polymorphonuclear neutrophils and bone implant-associated acute inflammation. Biomaterials. 2010;31(8):2001–9.

Vitkov L, Krautgartner WD, Obermayer A, Stoiber W, Hannig M, Klappacher M, Hartl D. The initial inflammatory response to bioactive implants is characterized by NETosis. PLoS One. 2015;10(3):e0121359.

Vlasova II, Mikhalchik EV, Barinov NA, Kostevich VA, Smolina NV, Klinov DV, Sokolov AV. Adsorbed plasma proteins modulate the effects of single-walled carbon nanotubes on neutrophils in blood. Nanomedicine. 2016;12(6):1615–25.

Yang H, Biermann MH, Brauner JM, Liu Y, Zhao Y, Herrmann M. New insights into neutrophil extracellular traps: mechanisms of formation and role in inflammation. Front Immunol. 2016;7:302.

Yipp BG, Petri B, Salina D, Jenne CN, Scott BN, Zbytnuik LD, Pittman K, Asaduzzaman M, Wu K, Meijndert HC, Malawista SE, de Boisfleury Chevance A, Zhang K, Conly J, Kubes P. Infection-induced NETosis is a dynamic process involving neutrophil multitasking *in vivo*. Nat Med. 2012;18(9):1386–93.

Yousefi S, Mihalache C, Kozlowski E, Schmid I, Simon HU. Viable neutrophils release mitochondrial DNA to form neutrophil extracellular traps. Cell Death Differ. 2009;16(11):1438–44.

Chapter 4
Mast Cells and Nanomaterials

Ryan P. Mendoza and Jared M. Brown

Abstract Mast cells are an essential cell type within the immune system, playing a prominent role in a variety of cell functions—both adaptive and innate immune system as well as others. This central role throughout many systems means that any inappropriate modulation of mast cell activation can, and does, contribute to many pathologies, including lung disease, autoimmune disorders, and cancer as examples. There are many exogenous compounds that can lead to activation of mast cells and potentially cause irregular function, one of which are engineered nanomaterials (ENMs). The human population is ubiquitously exposed to a range of ENM at a high rate through exposure mechanisms associated with high mast cell number, making these nanoparticles suspect when considering mast cell dysfunction. Research looking into ENM-induced mast cell pathologies is underwhelming; however; current research on the topic shows a clear ability for ENM to change mast cell activity and must be further explored.

Keywords Mast cells · Engineered nanomaterials · Immune system · Toxicity

4.1 Introduction

Mast cells were identified by histologist Paul Eulich in 1878 by the presence of large preformed metachromatic-stained granules present within the cell (Saito et al. 2013). Since then, many general unique functions of mast cells have been identified, including elimination of helminths, defense against venomous bites, and allergic response (Mukai et al. 2016; Shimokawa et al. 2017). However, current research is starting to demonstrate that this cell type has a much wider range of involvement in the immune system, acting as an essential effector cell in both innate and adaptive

R. P. Mendoza · J. M. Brown (✉)
Department of Pharmaceutical Sciences, Skaggs School of Pharmacy,
University of Colorado, Anschutz Medical Campus, Aurora, CO, USA
e-mail: jared.brown@cuanschutz.edu

J. C. Bonner, J. M. Brown (eds.), *Interaction of Nanomaterials with the Immune System*, Molecular and Integrative Toxicology,
https://doi.org/10.1007/978-3-030-33962-3_4

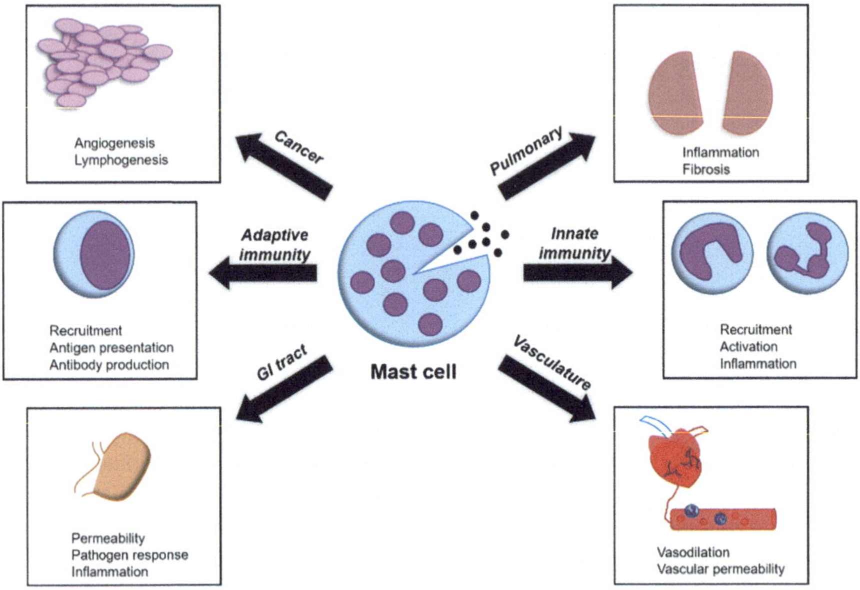

Fig. 4.1 Mast cells are major effector cells that majorly influence several other organs and system in the human body. Above is a schematic of a mast cell being activated (early-phase degranulation and/or late-phase activation) and examples of several systems throughout the body that could be affected by this activation

immunity as well as a first responder to foreign invaders. This central role in the immune system as well as its ability to identify and respond to many compounds implicates mast cells as an important regulator of immunotoxicity.

Mast cells' role in the immune system gives them the capacity to influence various pathophysiological outcomes, and inappropriate activation can manifest as pathologies across a wide range of systems in the body (Fig. 4.1). Mast cells are major effector cells in the cardiovascular system and can contribute to lung fibrosis and pulmonary inflammation as well as modulating vascular reactivity (Brown et al. 2007; Grashoff et al. 1997). Mediators released from mast cells have been correlated with angiogenesis and lymphomagenesis, which are important for tumor development in cancer (Varricchi et al. 2017). Mast cells also reside throughout the nervous system, including high numbers in the brain, and have been known to regulate nervous system communication and neurotransmitter release (Hendriksen et al. 2017). Gastrointestinal (GI) tract function is highly dependent on mast cell activity. For example, mast cells act as first responders in GI response to ingested pathogens. Inappropriate mast cell activation is related to gut pathologies including Crohn's disease and ulcerative colitis (Brenner et al. 2014; Stasikowska-Kanicka et al. 2012). Therefore, any foreign substance, such as engineered nanomaterials (ENMs), that can affect mast cell function has potential to disrupt proper homeostasis of essential physiological systems. This chapter explores mast cells' multilevel biological response to a variety of possibly damaging agents as well as the potential toxicity that arises from inappropriate mast cell activation. Of central focus will be the mast cell's ability to release preformed granules (i.e., early-phase degranulation) and the de novo synthesis and release of compounds, including a range of cytokines, chemokines, and inflammatory mediators (i.e., late-phase activation). This chapter reviews the biology of mast cells, their role in the immune system, and potential activation of this important cell type by ENMs. Since mast cells are found in higher numbers around areas of the body associated with ENM exposure (e.g., skin, lung, gastrointestinal tract), it is of utmost importance for toxicologists to explore how ENMs can either (1) activate mast cells directly or (2) modulate levels of mast cell response to other compounds such as allergens.

4.2 Mast Cell

4.2.1 Origins and Characterization

Mast cells are major effector cells within the innate and adaptive immune system, playing a pivotal role in several functions including inflammation and immune cell activation and recruitment. Mast cells originate in the bone marrow as $CD34^{+}$/$CD117^{+}$ hematopoietic stem cells that develop to circulate as multipotent progenitor (MMP)/common myeloid progenitor (CMP)/granulocyte-monocyte progenitor (GMP) cells in the vasculature. They leave the vasculature to enter the tissue as committed mast cell progenitor cells (MCp) and fully differentiate in the tissue as resident granulocytes (Dahlin and Hallgren 2015). The organs with the highest density of mast cells tend to be mucosal/connective tissue organs with high exposure to the external environment, including skin, lung, and gastrointestinal tract (Chernetsova et al. 2016; Gschwandtner et al. 2017; Ravindran et al. 2018). This allows the cells to interact with foreign pathogens as first responders to further recruit and activate other appropriate immune cells depending on the pathogen. However, mast cells are present in most organs throughout the body, including the heart, brain, and lymph nodes (Dong et al. 2017; Ingason et al. 2019; Wang et al. 1998). They are identified in the tissue by the presence of surface receptors c-kit and FcεR1 (the high-affinity IgE receptor).

Mast cells contain preformed granules filled with a variety of mediators, including cytokines, chemokines, proteases, and inflammatory mediators. The release of these granule contents constitutes *early-phase degranulation* and includes important mediators, such as histamine, proteases, heparin, and serotonin (de Souza Jr. et al. 2012; Ringvall et al. 2008; Samoszuk et al. 2005). Mast cells also have the ability to synthesize and release mediators over a longer time period (8–24 hours), which constitutes *late-phase activation*. This late-phase response consists of the cellular release of cytokines, growth factors, and importantly, lipid mediators including arachidonic acid metabolites such as prostaglandins (PG) and leukotrienes (LT) (Abe et al. 2000; Iwanaga et al. 2014). Mast cells have a complex role in the immune system; for example, they can release cytokines that can induce Th1 immunity, such as IL-1β and IL-6, or induce Th2 immunity, such as IL-2 and IL-4, depending on the source of activation (Ebenezer et al. 2018; Kandere-Grzybowska et al. 2003; Nakae et al. 2007). Also, several of the proteases contained in mast cell granules are mast cell specific (e.g., chymase and tryptase), providing them with unique functions within the immune system (Abraham et al. 2007). There is strong evidence demonstrating the importance of mast cell involvement in regulating the immune system and playing a role in cellular metabolism, proliferation, and cytotoxicity (Misto et al. 2019; Ningyan et al. 2015; Ozdemir 2011). Importantly, mast cells have two generalized pathways of activation: IgE mediated and non-IgE mediated.

4.2.2 Early-Phase Activation

There are a diverse range of intrinsic and extrinsic compounds that induce early-phase mast cell activation (i.e., degranulation). When these compounds interact with mast cells, degranulation occurs within seconds to minutes. Degranulation of mast cells is heterogenous—not all mast cells release all their granules and not all granules contain the same mediators (Puri and Roche 2008). This inconsistency poses a complex issue when analyzing the levels of degranulation as well as deciphering the systemic effects of mast cells.

The most studied mechanism of mast cell degranulation is classical IgE degranulation and occurs through a very well-described pathway in the cell following exposure to an allergen (Fig. 4.2). When an allergen is first introduced into the body, it is taken up by antigen-presenting cells (APC), often dendritic cells, and, along with helper T cells, is presented to a B cell. This will activate B cells to produce IgE antibodies with complimentary determining regions specific to that allergen. Mast cells can be involved in this process as nonprofessional APCs and/or by releasing mediators that skew antibody production toward IgE over other classes, to be discussed later. The IgE antibodies bind to high-affinity IgE receptors (FcεR1) present on the surface of mast cells. This constitutes the primary sensitization step of the mechanism. Upon second exposure, the allergen is reintroduced and crosslinks the surface bound IgE on the cell surface; this activates Syk and Lyn proteins on the basal side of the cell membrane, attached to FcεR1 through the transmembrane subunits beta and gamma (Xiao et al. 2005). The signal will propagate through the cell, activating phospholipase C (PLCγ) and phosphoinositide 3-kinase (PI3K) cascades that results in $Ca2^{+}$-dependent release of granules (Chen et al. 2017; Slodka et al. 2013). IgE-mediated degranulation is ATP dependent and causes an increase in glycolysis in the cell, demonstrated by an increase in lactic acid release (Caslin et al. 2018).

While IgE-driven mast cell activation is the most described mechanism of degranulation, it has been identified that many compounds can activate mast cells in a non-FcεR1 fashion without requiring sensitization. For example, Substance P can lead to mast cell degranulation through association with Neurokinin-1 surface receptors and cause release of histamine containing granules, leading to physiological responses similar to IgE mediated (Abraham et al. 2007). Another well-known inducer of mast cell degranulation is compound 48/80, which acts through

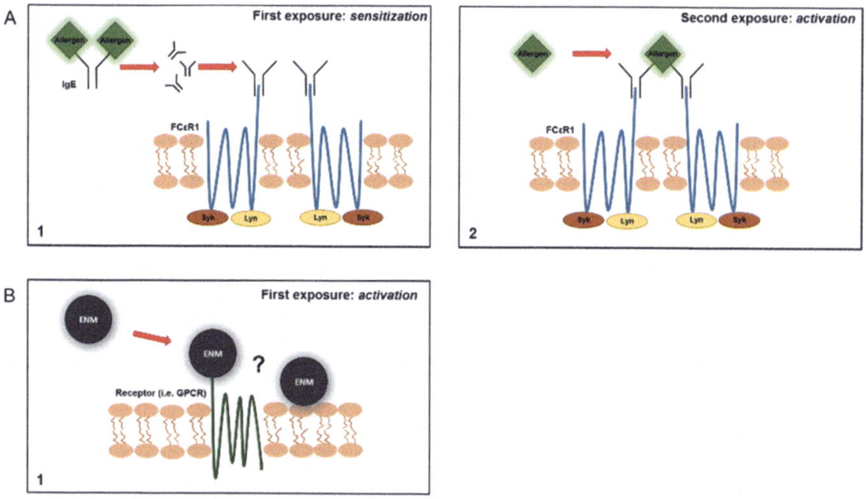

Fig. 4.2 Schematic of (**a**) IgE pathway and (**b**) non-IgE pathway of mast cell activation. (**a**) IgE pathway requires two exposures to be activated: an initial sensitization step (1) where IgE antibodies are produced against a specific allergen introduced to the body and attach to the FcεE1 receptor on mast cells. Then, a secondary activation step (2) when the allergen is reintroduced and causes crosslinking of IgE antibodies attached to mast cells leading to activation of Lyn and Syk proteins on the basal side of the membrane to initiate IgE pathway. (**b**) Non-IgE pathway (ENM used as an example) only requires one exposure for activation (1) and occurs through mostly undiscovered receptor interactions

G-protein-coupled receptors (GPCR), specifically the human analog Mas-related G-protein-coupled receptor X1 and X2 (MrgprX1/X2), mouse analog MrgprB2, and rat analog MrgprB3 (Irman-Florjanc and Erjavec 1983; Kashem et al. 2011; Zylka et al. 2003). Complement can induce mast cell degranulation through the C3a receptor, which is another GPCR (Nilsson et al. 1996). There appears to be an important role for G-protein-coupled receptors to propagate non-IgE signals into the cell. Several ENMs have been shown to lead to degranulation of mast cells through an unknown non-IgE-mediated pathway, potentially associated with a G-protein-coupled receptor (Johnson et al. 2017). This ability to cause mast cell degranulation raises questions to the safety of widespread ENM implementation. Some of these activating compounds lead to early-phase degranulation only, some lead to late-phase activation, and some, including IgE-mediated activation, will lead to both.

4.2.3 *Late-Phase Activation*

Along with releasing contents of prefilled granules through degranulation, mast cells have the ability to synthesize and release a different set of mediators over a longer period of time (e.g., 8–24 hours). Among these newly synthesized mediators are additional inflammatory-related mediators, including cytokines, arachidonic acid metabolites, and growth factors. Similar to degranulation, there are several mediators that can lead to late-phase activation through associations with a variety of surface receptors including FcεR1 and Toll-like receptors (TLRs).

IgE-mediated mast cell activation will lead to both degranulation and late-phase activation through diverging cell signaling pathways. With late-phase activation, phosphorylation of Syk will result in PLCγ phosphorylation, as with early-phase degranulation, but this will further activate Ras and PI3K. The Ras protein is important for initiating phosphorylation of the mitogen-activated protein kinase (MAP kinase) cascade that ends in the production of transcription factors leading to synthesis of cytokines and inflammatory mediators (Graham et al. 1998). MAP kinase phosphorylation will also produce phospholipase 2 (PLA_2), leading to metabolism of arachidonic acid to form lipid mediators known as eicosanoids (Zhang et al. 1997). This group includes leukotrienes (LT) and prostaglandins (PG), as examples, which are known to mediate inflammation, cell growth, and have autocrine function that can further affect mast cell activity (Iwanaga et al. 2014; Kondeti et al. 2016; Serra-Pages et al. 2015).

In addition to mast cell activation via FcεR1, there are number of other non-IgE surface receptors on mast cells that contribute to late-phase activation either through initiation or amplification of mast cell responses. For example, lipopolysaccharide (LPS) can interact with Toll-like receptors (TLRs) on the cell surface, specifically TLR4 which can recognize gram-negative bacteria, to lead to late activation and production of IL-6 and IL-1β (Kirshenbaum et al. 2008). Certain cytokines have also been shown to activate mast cells. For example, IL-33 secreted from epithelial cells and macrophages will induce late activation in mast cells (Salamon et al. 2017). In this instance, mast cell activation leads to synthesis and secretion of pro-Th2 mediators, and this process is involved in several pathologies (Kaur et al. 2015). As mentioned, mast cells are important effector cells in the immune system. Not only do they act to recruit and activate innate immune cells, they play a prominent role in adaptive immunity.

4.2.4 *Mast Cell Role in Adaptive Immunity*

Although mast cells are primarily innate immune system cells and act within that realm, they can interact with several adaptive immune cells to modulate their function as well. One example of mast cells acting in the adaptive immune system is their ability to be antigen-presenting cells (APC). When primed with IFNγ, mast cells will display MHCII, essential for antigen presenting, as well as necessary costimulatory molecules including CD40 (Love et al. 1996). Although they are not professional antigen-presenting cells, under certain conditions they can uptake pathogens, degrade them, and relocate to lymph nodes to present epitopes to T cells leading to activation. Mast cells also have the ability to release necessary costimulatory cytokines for antigen presentation including IL-4 and IL-13, that latter of which promotes IgE antibody production from B cells (Pawankar et al. 1997). Mast cells, however, are not professional antigen-presenting cells, and this phenomenon only occurs under certain conditions.

Mast cells also release mediators during both degranulation and late-phase activation that can affect the adaptive immune system. An inflammatory mediator present both in preformed granules and produced strongly during late activation is TNFα. The quick release nature of mast cells makes them likely the first immune cell to release this cytokine in response to foreign pathogens. This cytokine is mostly associated with an acute inflammatory effect through activation of NF-kB and MAP kinase pathways in order to quickly recruit neutrophils and eosinophils to site of infection (Yang et al. 2017). However, mast cell TNFα also has the ability to relocate dendritic cells to lymph nodes (Suto et al. 2006). Research has also shown that mast cell-specific TNFα acts on T cell to affect proliferation and maturity by increasing expression of OX40 as well as changing their pattern of interaction during production of antibodies, potentially leading to higher production of IgE class antibodies (Nakae et al. 2005). This was shown to be a mast cell-specific TNFα effect, not TNFα from other cell sources.

A major mediator released during degranulation is histamine, and while it is largely associated with inflammatory response and allergies, it can also have a strong regulatory effect on different adaptive immune cells. Histamine released from mast cells has been shown to inhibit the suppressive function of regulatory T cells, leading to increased production of CD4+ and CD8+ T cells (Forward et al. 2009). Histamine can activate endothelial cells which will lead to recruitment of an assortment of innate immune cells and T lymphocytes (Ashina et al. 2015). Histamine is also associated with dendritic cell maturation, further implicating mast cells in the antigen presentation process (Mazzoni et al. 2001). Mast cells have several mechanisms through which they can regulate adaptive immune system activity in both degranulation and late-phase activation responses.

It is clear that mast cells have an important influence on T-cell activity in the adaptive immune system. As mentioned above, mast cells can directly activate T cells, present epitopes as an APC, and relocate T cells to the lymph. They can also release molecules that control helper T cells and Tregs. Mast cells are responsive to IL-33, especially in the lungs where it is produced and released by epithelial cells. Mast cells respond to this stimulus by producing and secreting IL-2, which will promote Treg production and activity. These Tregs can release IL-10, which mast cells can also produce, that ultimately suppresses germinal centers and T follicular helper cell function (Lin et al. 2019; Morita et al. 2015). This homeostasis is essential for proper T-cell activity and, therefore, proper B cell activation and antibody production. Modulating Treg and helper cell population is another mechanism through which mast cells can potentially control antibody production.

4.2.5 Mast Cell Role in Disease

Since mast cells have such a central role in response to pathogens, including both the innate and adaptive immune systems, it is not surprising that this cell type is implicated in an array of diseases. In organs containing a high number of mast cells, they can play a role in inflammatory pathologies. For example, mast cells are quite abundant in the lung and are one of the major driving cell types for allergic asthma (Amin et al. 2005; Jin et al. 2017). They are strong responders to IL-33 produced by epithelial cells, a key inflammatory mediator leading to asthmatic airway constriction, to be discussed later. Mast cells are also abundant in the skin and correlated strongly with atopic dermatitis, among other inflammatory skin diseases, through the production of IL-17 (Katayama et al. 1992). Mast cells are present in high density throughout the GI tract, which has linked mast cell-specific cytokines with Crohn's disease and ulcerative colitis (Raithel et al. 1995; Stasikowska-Kanicka et al. 2012). Since mast cell mediator release is often unique and mast cell specific, it is straightforward to correlate the cell type with diseases that occur strictly from those mediators.

Mast cells are also implicated in the mechanism of several autoimmune diseases including multiple sclerosis and Type 1 diabetes. For example, in multiple sclerosis (MS), there are increased numbers of mast cells found at the sites of inflammatory demyelination that characterizes the disease (Toms et al. 1990). Also, mast cell-deficient MS mouse models do not develop the disease in contrast to wild-type mice containing sufficient numbers of mast cells (Secor et al. 2000). There are high levels of mast cell-specific granule contents present in MS patients including histamine and tryptase (Christy et al. 2013). In regard to Type 1 diabetes, mast cells can localize to the pancreas near pancreatic islets and an increased number of mast cells are found in prediabetic mouse models, suggesting their involvement in this pathology (Martino et al. 2015). Also, an arachidonic acid metabolite synthesized and released from mast cells, leukotriene B4 (LTB4), is increased in mice with diabetes compared to wild-type mice, providing a potential cellular mechanism of action (Talahalli et al. 2010). Mast cells' role in the adaptive immune system implicates them as a possible cell that can initiate or progress several different autoimmune and inflammatory diseases.

Mast cells are major effector cells in the immune system milieu and, therefore, have potential as important regulators of the tumor immune microenvironment. While mast cells have demonstrated a role in tumorigenesis and cancer, the outcome of their activity varies between tumors. Tumor cells release several chemoattractant molecules that can cause mast cell migration to that site, including stem cell factor (SCF), histamine, and osteopontin (Giannou et al. 2015; Godot et al. 2007; Huang et al. 2008). However, their function can be both protumorigenic in some solid tumors, as with pancreatic, bladder, and gastric cancer, or antitumorigenic in some tumors, as with breast cancer (Ammendola et al. 2013; Chang et al. 2011; Dabiri et al. 2004; Rao et al. 2016). Mast cells can also release vascular endothelial growth factors (VEGF) which can promote angiogenesis and lymphogenesis, two mechanisms that strongly encourage tumor growth (Detoraki et al. 2009; Grutzkau et al. 1998). They can release CXCL8/IL-8 which upregulates epithelial-to-mesenchymal transition, a hallmark of cancer (Visciano et al. 2015). But mast cells also release heparin and prostaglandin D2 (PGD2), which have been shown to inhibit tumor growth, as well as TNFα which is cytotoxic to tumor cells (Iwanaga et al. 2014; Samoszuk et al. 2005; Shikotra et al. 2016). While their role in cancer is uncertain, it is clear that mast cells have an important regulatory role in tumorigenesis and in a variety of disease across several organs.

4.3 ENM Interaction with Mast Cells

Current in vitro and in vivo research and case studies have indicated that engineered nanomaterials (ENMs) are capable of activating mast cells through an unknown non-IgE pathway. Since mast cells are present at tissue sites prone to environmental exposures, they are a prime candidate in ENM-mediated toxicity. In most cases, only preliminary toxicity screens have been performed with much of the mechanistic work unexplored. However, it has been elucidated that a range of ENM exposure can lead to mast cell activation either: (1) independently through a non-IgE pathway resulting in degranulation and/or de novo synthesis of late activation mediators, or (2) modulate IgE-mediated activation of mast cells (Fig. 4.3). The following section will focus on literature that pertains to ENM's ability to interact with mast cells including both in vitro and in vivo experiments.

4.3.1 *In Vitro Evidence*

Metal-based ENMs comprise a range of material compositions with the most commonly implemented being silver, titanium, and zinc nanoparticles. Silver nanoparticles (AgNP) are often employed because of their naturally antimicrobial activity,

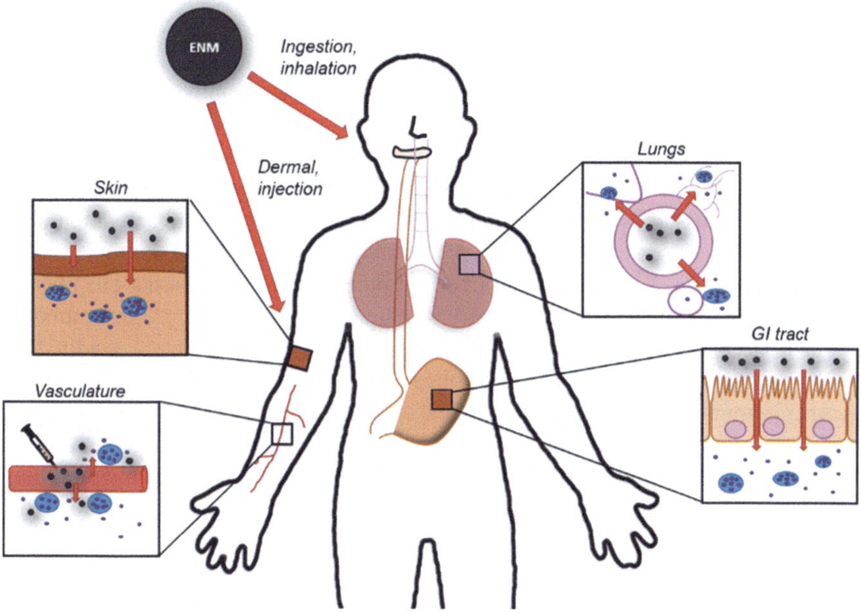

Fig. 4.3 Mast cells are at high concentration in tissues exposed to the external environment, making them a prime target for ENM-mediated toxicity. ENM can be ingested, inhaled, injected, or absorbed through dermal contact. Above is a schematic of potential routes of exposure to organs where ENM could most likely interact with mast cells to modulate their function

whereas titanium oxide (TiO_2) and zinc oxide (ZnO) ENMs are commonly used as UV protectants in sunscreens, as examples. Johnson et al. explored mast cell response to over 20 different metallic nanomaterials and showed that there was a wide range of mast cell early-phase activation between ENM, measured by release of beta-hexosaminidase (a protease present in mast cell granules). The strongest response was to 20-nm AgNP, but several different ENM, including copper oxide (CuO) and silica oxide (SiO2) nanomaterials, induced activation but to a lesser extent (Johnson et al. 2017). The intensity of response varied based on physicochemical properties; for example, AgNP of lower hydrodynamic diameter (e.g., 5 nm) was more likely to lead to activation. Interestingly, there were ENMs that resulted in inhibition of degranulation, such as iron oxide (Fe_2O_3) and TiO_2, providing potential for therapeutic uses. Additional studies on TiO_2 nanomaterials revealed contrasting results. For example, the above study showed reduced mast cell degranulation, while another study showed that exposure to this ENM induces increased release of histamine from mast cells (Chen et al. 2012). Despite this, it is clear that ENM of certain physicochemical properties induces mast cell degranulation. Further, it has been demonstrated that mast cell response to AgNP induces a robust global protein tyrosine phosphorylation as well as phosphorylation of PLCγ and PIK3 (Alsaleh et al. 2016). It is unknown how these signaling mechanisms are initiated ultimately leading to degranulation. However, studies to date strongly suggest that ENM activation of mast cells is non-IgE mediated and possibly occurs through a novel mechanism.

Further, there is in vitro evidence that ENMs not only affect early-phase degranulation, they can independently induce mast cell late-phase activation. The release of late-phase mediators, including inflammatory markers and eicosanoids, strongly controls mast cell regulation of other immune cells, especially of the adaptive immune system. Exposure of isolated mast cells to cerium dioxide (CeO_2) nanomaterials increased late-phase production of osteopontin, TNFα, and the arachidonic acid metabolite prostaglandin D2 (PGD2) (Wingard et al. 2011). TNFα has been previously discussed as having a central role in the recruitment and activation of T cells, as well as the process of antigen presentation to B cells. Osteopontin can suppress the activity of IL-10 and can shift development of T cells toward a Th1 phenotype as well as changing B-cell proliferation and modulating IgE antibody production (Kurokawa et al. 2009; Murugaiyan et al. 2008). Prostaglandin D2, which is mast cell specific, has been implicated in recruitment of Th2 T cells as well as other immune cells such as basophils and eosinophils (Jandl et al. 2016; Monneret et al. 2001). It can also act as a potent vasodilator which can drastically affect chemotaxis and cell activation (Ellis et al. 1979). Although results from investigating ENM ability to cause late activation are mostly negative, it is important to continue to analyze this potential toxicity.

While the above nanomaterials have been shown to induce degranulation or late activation independently through a non-IgE pathway, it is also important to recognize ENM's ability to manipulate the IgE response. Exposure to zinc oxide (ZnO) particles has been shown to inhibit IgE-mediated early-phase release of beta-hexosaminidase and histamine from mast cells (Yamaki and Yoshino 2009).

Furthermore, when those same ZnO particles are doped with aluminum, they were shown to further inhibit MAPK and NF-kB activation more than bare ZnO, leading to reduced late-phase synthesis and release of inflammatory cytokines IL-6 and TNFα (Kim et al. 2016). This has led to consideration of implementing this ENM as a therapeutic against allergic response; however, there are confounding results when analyzing mast cell activation with this material in vivo. Although the majority of ENMs produced and studied are metallic based, a number of organic-based nanomaterials have also been shown to inhibit mast cell response to IgE activation. Fullerene, a hollow spherical carbon particle (e.g., buckyball), demonstrated an ability to downregulate protein activation necessary for IgE-mediated degranulation, including phosphorylation of Syk (Ryan et al. 2007). Preexposure to poly(lactic-co-glycolic acid) (PLGA) nanoparticles was shown to decrease histamine release from IgE-mediated early-phase degranulation by acting as an antagonist of the SNARE pathways, which is essential for proper histamine release (Tahara et al. 2012). While ENM effects on mast cells in vitro have not been extensively investigated, there is a body of evidence that shows ENM of certain physicochemical properties can be reactive.

4.3.2 *In Vivo Evidence*

Along with in vitro research indicating potentially toxic side effects of ENMs on mast cells, there has been some in vivo evidence to support activation of mast cells following ENM exposure. One of the most commonly used animal models to explore the contribution of mast cells to pathology is the mast cell-deficient mouse model, $Kit^{W\text{-}sh}$. These mice have a spontaneous inversion mutation that occurs in the regulatory elements upstream of the gene for c-kit receptor, which is required for the proper maturation and proliferation of mast cells, thereby preventing mast cell development in vivo. By comparing this mast cell-deficient model to wild-type mice, a true role of mast cells can be determined. Furthermore, primary mast cells can be reintroduced back into $Kit^{W\text{-}sh}$, allowing further confirmation of mast cell's involvement in different pathologies and systemic effects.

Many experiments exploring toxicity of ENM exposure to mast cells focus on pulmonary and cardiovascular pathologies that can occur. For example, mice exposed to CeO_2 ENM exhibit cardiovascular and pulmonary dysfunction that is completely abrogated in the mast cell-deficient mouse model (Wingard et al. 2011). Specifically, it has been recognized that inhalation of carbon nanotubes (CNT) can lead to lung fibrosis and other pathologies including pulmonary inflammation (Duke et al. 2017; Fujita et al. 2016). It was shown that these issues do not occur in the mast cell-deficient mouse (Katwa et al. 2012). Further, IL-33 is shown to be required for the development of these issues and, as previously mentioned, mast cells are a major responder to IL-33 produced from epithelial cells within the lungs, hinting that these cells have a role in this response. Another study explored how shape of carbon nanotubes would affect the ENM's ability to induce asthma in the lung.

They found that rod-like carbon nanotubes, but not tangled carbon nanotubes, will lead to activation of allergic response in the lung and induce a shift toward Th2 immunity which is linked to allergic asthma (Rydman et al. 2014). The authors found a significant contribution from mast cells to this allergy-related response.

ENM toxicity related to mast cells has been explored in organs other than the lungs as well. It was shown that there was a significant increase of mast cells in the liver of rats exposed to silica nanoparticles (Zhuravskii et al. 2016). Liver fibrosis developed in these rats, providing a possible link between mast cells and this disease. Similarly, a significant increase in mast cells infiltrating the liver was observed after exposure of mice to superparamagnetic iron oxide nanoparticles, but there was no fibrosis observed in this organ (Sabareeswaran et al. 2016). Exposure to ZnO nanoparticles showed different results in vivo than in vitro. As mentioned above, this ENM is being considered as a therapeutic for mast cell activation because of its success in reducing IgE-mediated activation in cell model studies. However, in Balb/c mice, exposure to ZnO leads to an increase of mast cells in the lung and spleen (Roy et al. 2014). Although there was a decrease in late-phase production and release of IL-10 and TNFα, there was an increase in other mast cell mediators, including IL-2, IL-4, and IL-17. This points to a shift from Th1 immune response toward an allergic-type Th2 immune response. The development of Th2 immunity can affect the development of autoimmune diseases and other adaptive immune system-based diseases.

As previously mentioned, ENM is implicated in modulating IgE-mediated effects and pathologies arising from them. When a mouse model prone to develop atopic dermatitis from IgE activation was preexposed to 5 nm AgNP before IgE pathway activation, there was an increase in the severity of the disease (Kang et al. 2017). Results revealed a significant increase in total mast cell number and total circulating IgE levels in the preexposed group compared to IgE pathway activation alone. Certain lipid-based nanoparticles (LNPs) were also shown to cause hypersensitivity to complement-mediated mast cell activation in different animal models including dogs and pigs. LNPs at a low dose were also shown to directly cause mast cell activation similar to complement-based activation, displaying both independent effects and modulation of non-IgE pathways (Szebeni et al. 2007). It is clear that mast cell-mediated ENM toxicity can negatively affect several organ systems and lead to unwanted pathologies. This further implicates a potential key role for these important effector cells in the toxicity of engineered nanomaterials both in cell and animal models.

4.4 Conclusions

Mast cells are most commonly linked with allergic response because they are the main histamine producing cells in the body. However, they have also been shown to be major effector cells in the innate immune system and are centrally important for appropriate immune cell recruitment and activation. They have also displayed a regulatory role in adaptive immune system affecting T-cell proliferation and development as well as antigen presentation to B cells and antibody production. It is because of this central role in the immune system that mast cells are implicated in the initiation and progression of several diseases, such as diabetes mellitus and multiple sclerosis, as well other pathologies such as cardiovascular and pulmonary dysfunction.

Although the mechanism is often understudied or unknown, there is evidence that a range of ENMs can lead to inappropriate activation of mast cells. Cell models have shown that ENM can directly lead to mast cell degranulation leading to allergic response and inflammation. ENMs have also been shown to modulate late activation, changing the synthesis and release of many important mediators, as well as manipulate sensitivity to other pathways of mast cell activation including IgE mediated. Further studies in animal models show a more direct correlation of inappropriate mast cell activation with pathologies arising from ENM exposure. While most research has been done exploring pulmonary diseases because of the abundance of mast cells, it has been shown that mast cells are responsible for ENM-mediated damage in other organs such as the liver and skin. It is clear from the body of evidence that ENMs are reactive with mast cells and this modulation of mast cell activity can lead to larger systemic damage that goes much further than allergic response.

References

Abe M, Kurosawa M, Ishikawa O, Miyachi Y. Effect of mast cell-derived mediators and mast cell-related neutral proteases on human dermal fibroblast proliferation and type I collagen production. J Allergy Clin Immunol. 2000;106:S78–84.

Abraham D, Oster H, Huber M, Leitges M. The expression pattern of three mast cell specific proteases during mouse development. Mol Immunol. 2007;44:732–40.

Alsaleh NB, Persaud I, Brown JM. Silver nanoparticle-directed mast cell degranulation is mediated through calcium and PI3K signaling independent of the high affinity IgE receptor. PLoS One. 2016;11:e0167366.

Amin K, Janson C, Boman G, Venge P. The extracellular deposition of mast cell products is increased in hypertrophic airways smooth muscles in allergic asthma but not in nonallergic asthma. Allergy. 2005;60:1241–7.

Ammendola M, Sacco R, Donato G, Zuccala V, Russo E, Luposella M, Vescio G, Rizzuto A, Patruno R, De Sarro G, et al. Mast cell positivity to tryptase correlates with metastatic lymph nodes in gastrointestinal cancer patients treated surgically. Oncology. 2013;85:111–6.

Ashina K, Tsubosaka Y, Nakamura T, Omori K, Kobayashi K, Hori M, Ozaki H, Murata T. Histamine induces vascular hyperpermeability by increasing blood flow and endothelial barrier disruption in vivo. PLoS One. 2015;10:e0132367.

Brenner SA, Zacheja S, Schaffer M, Feilhauer K, Bischoff SC, Lorentz A. Soluble CD14 is essential for lipopolysaccharide-dependent activation of human intestinal mast cells from macroscopically normal as well as Crohn's disease tissue. Immunology. 2014;143:174–83.

Brown JM, Swindle EJ, Kushnir-Sukhov NM, Holian A, Metcalfe DD. Silica-directed mast cell activation is enhanced by scavenger receptors. Am J Respir Cell Mol Biol. 2007;36:43–52.

Caslin HL, Taruselli MT, Haque T, Pondicherry N, Baldwin EA, Barnstein BO, Ryan JJ. Inhibiting glycolysis and ATP production attenuates IL-33-mediated mast cell function and peritonitis. Front Immunol. 2018;9:3026.

Chang DZ, Ma Y, Ji B, Wang H, Deng D, Liu Y, Abbruzzese JL, Liu YJ, Logsdon CD, Hwu P. Mast cells in tumor microenvironment promotes the in vivo growth of pancreatic ductal adenocarcinoma. Clin Cancer Res. 2011;17:7015–23.

Chen EY, Garnica M, Wang YC, Mintz AJ, Chen CS, Chin WC. A mixture of anatase and rutile TiO(2) nanoparticles induces histamine secretion in mast cells. Part Fibre Toxicol. 2012;9:2.

Chen YC, Chang YC, Chang HA, Lin YS, Tsao CW, Shen MR, Chiu WT. Differential Ca(2+) mobilization and mast cell degranulation by FcepsilonRI- and GPCR-mediated signaling. Cell Calcium. 2017;67:31–9.

Chernetsova E, Sullivan K, de Nanassy J, Barkey J, Mack D, Nasr A, El Demellawy D. Histologic analysis of eosinophils and mast cells of the gastrointestinal tract in healthy Canadian children. Hum Pathol. 2016;54:55–63.

Christy AL, Walker ME, Hessner MJ, Brown MA. Mast cell activation and neutrophil recruitment promotes early and robust inflammation in the meninges in EAE. J Autoimmun. 2013;42:50–61.

Dabiri S, Huntsman D, Makretsov N, Cheang M, Gilks B, Bajdik C, Gelmon K, Chia S, Hayes M. The presence of stromal mast cells identifies a subset of invasive breast cancers with a favorable prognosis. Mod Pathol. 2004;17:690–5.

Dahlin JS, Hallgren J. Mast cell progenitors: origin, development and migration to tissues. Mol Immunol. 2015;63:9–17.

de Souza DA Jr, Toso VD, Campos MR, Lara VS, Oliver C, Jamur MC. Expression of mast cell proteases correlates with mast cell maturation and angiogenesis during tumor progression. PLoS One. 2012;7:e40790.

Detoraki A, Staiano RI, Granata F, Giannattasio G, Prevete N, de Paulis A, Ribatti D, Genovese A, Triggiani M, Marone G. Vascular endothelial growth factors synthesized by human lung mast cells exert angiogenic effects. J Allergy Clin Immunol. 2009;123:1142–9, 1149 e1141–1145.

Dong H, Zhang X, Wang Y, Zhou X, Qian Y, Zhang S. Suppression of brain mast cells degranulation inhibits microglial activation and central nervous system inflammation. Mol Neurobiol. 2017;54:997–1007.

Duke KS, Taylor-Just AJ, Ihrie MD, Shipkowski KA, Thompson EA, Dandley EC, Parsons GN, Bonner JC. STAT1-dependent and -independent pulmonary allergic and fibrogenic responses in mice after exposure to tangled versus rod-like multi-walled carbon nanotubes. Part Fibre Toxicol. 2017;14:26.

Ebenezer AJ, Prasad K, Rajan S, Thangam EB. Silencing of H4R inhibits the production of IL-1beta through SAPK/JNK signaling in human mast cells. J Recept Signal Transduct Res. 2018;38:204–12.

Ellis EF, Wei EP, Kontos HA. Vasodilation of cat cerebral arterioles by prostaglandins D2, E2, G2, and I2. Am J Phys. 1979;237:H381–5.

Forward NA, Furlong SJ, Yang Y, Lin TJ, Hoskin DW. Mast cells down-regulate CD4+CD25+ T regulatory cell suppressor function via histamine H1 receptor interaction. J Immunol. 2009;183:3014–22.

Fujita K, Fukuda M, Endoh S, Maru J, Kato H, Nakamura A, Shinohara N, Uchino K, Honda K. Pulmonary and pleural inflammation after intratracheal instillation of short single-walled and multi-walled carbon nanotubes. Toxicol Lett. 2016;257:23–37.

Giannou AD, Marazioti A, Spella M, Kanellakis NI, Apostolopoulou H, Psallidas I, Prijovich ZM, Vreka M, Zazara DE, Lilis I, et al. Mast cells mediate malignant pleural effusion formation. J Clin Invest. 2015;125:2317–34.

Godot V, Arock M, Garcia G, Capel F, Flys C, Dy M, Emilie D, Humbert M. H4 histamine receptor mediates optimal migration of mast cell precursors to CXCL12. J Allergy Clin Immunol. 2007;120:827–34.

Graham TE, Pfeiffer JR, Lee RJ, Kusewitt DF, Martinez AM, Foutz T, Wilson BS, Oliver JM. MEK and ERK activation in ras-disabled RBL-2H3 mast cells and novel roles for geranylgeranylated and farnesylated proteins in fc epsilonRI-mediated signaling. J Immunol. 1998;161:6733–44.

Grashoff WF, Sont JK, Sterk PJ, Hiemstra PS, de Boer WI, Stolk J, Han J, van Krieken JM. Chronic obstructive pulmonary disease: role of bronchiolar mast cells and macrophages. Am J Pathol. 1997;151:1785–90.

Grutzkau A, Kruger-Krasagakes S, Baumeister H, Schwarz C, Kogel H, Welker P, Lippert U, Henz BM, Moller A. Synthesis, storage, and release of vascular endothelial growth factor/vascular permeability factor (VEGF/VPF) by human mast cells: implications for the biological significance of VEGF206. Mol Biol Cell. 1998;9:875–84.

Gschwandtner M, Paulitschke V, Mildner M, Brunner PM, Hacker S, Eisenwort G, Sperr WR, Valent P, Gerner C, Tschachler E. Proteome analysis identifies L1CAM/CD171 and DPP4/CD26 as novel markers of human skin mast cells. Allergy. 2017;72:85–97.

Hendriksen E, van Bergeijk D, Oosting RS, Redegeld FA. Mast cells in neuroinflammation and brain disorders. Neurosci Biobehav Rev. 2017;79:119–33.

Huang B, Lei Z, Zhang GM, Li D, Song C, Li B, Liu Y, Yuan Y, Unkeless J, Xiong H, et al. SCF-mediated mast cell infiltration and activation exacerbate the inflammation and immunosuppression in tumor microenvironment. Blood. 2008;112:1269–79.

Ingason AB, Mechmet F, Atacho DAM, Steingrimsson E, Petersen PH. Distribution of mast cells within the mouse heart and its dependency on Mitf. Mol Immunol. 2019;105:9–15.

Irman-Florjanc T, Erjavec F. Compound 48/80 and substance P induced release of histamine and serotonin from rat peritoneal mast cells. Agents Actions. 1983;13:138–41.

Iwanaga K, Nakamura T, Maeda S, Aritake K, Hori M, Urade Y, Ozaki H, Murata T. Mast cell-derived prostaglandin D2 inhibits colitis and colitis-associated colon cancer in mice. Cancer Res. 2014;74:3011–9.

Jandl K, Stacher E, Balint Z, Sturm EM, Maric J, Peinhaupt M, Luschnig P, Aringer I, Fauland A, Konya V, et al. Activated prostaglandin D2 receptors on macrophages enhance neutrophil recruitment into the lung. J Allergy Clin Immunol. 2016;137:833–43.

Jin C, Shelburne CP, Li G, Potts EN, Riebe KJ, Sempowski GD, Foster WM, Abraham SN. Particulate allergens potentiate allergic asthma in mice through sustained IgE-mediated mast cell activation. J Clin Invest. 2017;127:3913.

Johnson MM, Mendoza R, Raghavendra AJ, Podila R, Brown JM. Contribution of engineered nanomaterials physicochemical properties to mast cell degranulation. Sci Rep. 2017;7:43570.

Kandere-Grzybowska K, Letourneau R, Kempuraj D, Donelan J, Poplawski S, Boucher W, Athanassiou A, Theoharides TC. IL-1 induces vesicular secretion of IL-6 without degranulation from human mast cells. J Immunol. 2003;171:4830–6.

Kang H, Kim S, Lee KH, Jin S, Kim SH, Lee K, Jeon H, Song YG, Lee SW, Seo J, et al. 5 nm silver nanoparticles amplify clinical features of atopic dermatitis in mice by activating mast cells. Small. 2017;13.

Kashem SW, Subramanian H, Collington SJ, Magotti P, Lambris JD, Ali H. G protein coupled receptor specificity for C3a and compound 48/80-induced degranulation in human mast cells: roles of Mas-related genes MrgX1 and MrgX2. Eur J Pharmacol. 2011;668:299–304.

Katayama I, Yokozeki H, Nishioka K. Mast-cell-derived mediators induce epidermal cell proliferation: clue for lichenified skin lesion formation in atopic dermatitis. Int Arch Allergy Immunol. 1992;98:410–4.

Katwa P, Wang X, Urankar RN, Podila R, Hilderbrand SC, Fick RB, Rao AM, Ke PC, Wingard CJ, Brown JM. A carbon nanotube toxicity paradigm driven by mast cells and the IL-(3)(3)/ST(2) axis. Small. 2012;8:2904–12.

Kaur D, Gomez E, Doe C, Berair R, Woodman L, Saunders R, Hollins F, Rose FR, Amrani Y, May R, et al. IL-33 drives airway hyper-responsiveness through IL-13-mediated mast cell: airway smooth muscle crosstalk. Allergy. 2015;70:556–67.

Kim MH, Seo JH, Kim HM, Jeong HJ. Aluminum-doped zinc oxide nanoparticles attenuate the TSLP levels via suppressing caspase-1 in activated mast cells. J Biomater Appl. 2016;30:1407–16.

Kirshenbaum AS, Swindle E, Kulka M, Wu Y, Metcalfe DD. Effect of lipopolysaccharide (LPS) and peptidoglycan (PGN) on human mast cell numbers, cytokine production, and protease composition. BMC Immunol. 2008;9:45.

Kondeti V, Al-Azzam N, Duah E, Thodeti CK, Boyce JA, Paruchuri S. Leukotriene D4 and prostaglandin E2 signals synergize and potentiate vascular inflammation in a mast cell-dependent manner through cysteinyl leukotriene receptor 1 and E-prostanoid receptor 3. J Allergy Clin Immunol. 2016;137:289–98.

Kurokawa M, Konno S, Takahashi A, Plunkett B, Rittling SR, Matsui Y, Kon S, Morimoto J, Uede T, Matsukura S, et al. Regulatory role of DC-derived osteopontin in systemic allergen sensitization. Eur J Immunol. 2009;39:3323–30.

Lin X, Wang X, Xiao F, Ma K, Liu L, Wang X, Xu D, Wang F, Shi X, Liu D, et al.. IL-10-producing regulatory B cells restrain the T follicular helper cell response in primary Sjogren's syndrome. Cell Mol Immunol. 2019; https://doi.org/10.1038/s41423-019-0227-z. [Epub ahead of print]

Love KS, Lakshmanan RR, Butterfield JH, Fox CC. IFN-gamma-stimulated enhancement of MHC class II antigen expression by the human mast cell line HMC-1. Cell Immunol. 1996;170:85–90.

Martino L, Masini M, Bugliani M, Marselli L, Suleiman M, Boggi U, Nogueira TC, Filipponi F, Occhipinti M, Campani D, et al. Mast cells infiltrate pancreatic islets in human type 1 diabetes. Diabetologia. 2015;58:2554–62.

Mazzoni A, Young HA, Spitzer JH, Visintin A, Segal DM. Histamine regulates cytokine production in maturing dendritic cells, resulting in altered T cell polarization. J Clin Invest. 2001;108:1865–73.

Misto A, Provensi G, Vozella V, Passani MB, Piomelli D. Mast cell-derived histamine regulates liver ketogenesis via oleoylethanolamide signaling. Cell Metab. 2019;29:91–102 e105.

Monneret G, Gravel S, Diamond M, Rokach J, Powell WS. Prostaglandin D2 is a potent chemoattractant for human eosinophils that acts via a novel DP receptor. Blood. 2001;98:1942–8.

Morita H, Arae K, Unno H, Miyauchi K, Toyama S, Nambu A, Oboki K, Ohno T, Motomura K, Matsuda A, et al. An interleukin-33-mast cell-interleukin-2 axis suppresses papain-induced allergic inflammation by promoting regulatory T cell numbers. Immunity. 2015;43:175–86.

Mukai K, Tsai M, Starkl P, Marichal T, Galli SJ. IgE and mast cells in host defense against parasites and venoms. Semin Immunopathol. 2016;38:581–603.

Murugaiyan G, Mittal A, Weiner HL. Increased osteopontin expression in dendritic cells amplifies IL-17 production by CD4+ T cells in experimental autoimmune encephalomyelitis and in multiple sclerosis. J Immunol. 2008;181:7480–8.

Nakae S, Ho LH, Yu M, Monteforte R, Iikura M, Suto H, Galli SJ. Mast cell-derived TNF contributes to airway hyperreactivity, inflammation, and TH2 cytokine production in an asthma model in mice. J Allergy Clin Immunol. 2007;120:48–55.

Nakae S, Suto H, Kakurai M, Sedgwick JD, Tsai M, Galli SJ. Mast cells enhance T cell activation: importance of mast cell-derived TNF. Proc Natl Acad Sci U S A. 2005;102:6467–72.

Nilsson G, Johnell M, Hammer CH, Tiffany HL, Nilsson K, Metcalfe DD, Siegbahn A, Murphy PM. C3a and C5a are chemotaxins for human mast cells and act through distinct receptors via a pertussis toxin-sensitive signal transduction pathway. J Immunol. 1996;157:1693–8.

Ningyan G, Xu Y, Hongfei S, Jingjing C, Min C. The role of macrophage migration inhibitory factor in mast cell-stimulated fibroblast proliferation and collagen production. PLoS One. 2015;10:e0122482.

Ozdemir O. Flow cytometric mast cell-mediated cytotoxicity assay: a three-color flow cytometric approach using monoclonal antibody staining with annexin V/propidium iodide co-labeling to assess human mast cell-mediated cytotoxicity by fluorosphere-adjusted counts. J Immunol Methods. 2011;365:166–73.

Pawankar R, Okuda M, Yssel H, Okumura K, Ra C. Nasal mast cells in perennial allergic rhinitics exhibit increased expression of the Fc epsilonRI, CD40L, IL-4, and IL-13, and can induce IgE synthesis in B cells. J Clin Invest. 1997;99:1492–9.

Puri N, Roche PA. Mast cells possess distinct secretory granule subsets whose exocytosis is regulated by different SNARE isoforms. Proc Natl Acad Sci U S A. 2008;105:2580–5.

Raithel M, Matek M, Baenkler HW, Jorde W, Hahn EG. Mucosal histamine content and histamine secretion in Crohn's disease, ulcerative colitis and allergic enteropathy. Int Arch Allergy Immunol. 1995;108:127–33.

Rao Q, Chen Y, Yeh CR, Ding J, Li L, Chang C, Yeh S. Recruited mast cells in the tumor microenvironment enhance bladder cancer metastasis via modulation of ERbeta/CCL2/CCR2 EMT/MMP9 signals. Oncotarget. 2016;7:7842–55.

Ravindran A, Ronnberg E, Dahlin JS, Mazzurana L, Safholm J, Orre AC, Al-Ameri M, Peachell P, Adner M, Dahlen SE, et al. An optimized protocol for the isolation and functional analysis of human lung mast cells. Front Immunol. 2018;9:2193.

Ringvall M, Ronnberg E, Wernersson S, Duelli A, Henningsson F, Abrink M, Garcia-Faroldi G, Fajardo I, Pejler G. Serotonin and histamine storage in mast cell secretory granules is dependent on serglycin proteoglycan. J Allergy Clin Immunol. 2008;121:1020–6.

Roy R, Kumar S, Verma AK, Sharma A, Chaudhari BP, Tripathi A, Das M, Dwivedi PD. Zinc oxide nanoparticles provide an adjuvant effect to ovalbumin via a Th2 response in Balb/c mice. Int Immunol. 2014;26:159–72.

Ryan JJ, Bateman HR, Stover A, Gomez G, Norton SK, Zhao W, Schwartz LB, Lenk R, Kepley CL. Fullerene nanomaterials inhibit the allergic response. J Immunol. 2007;179:665–72.

Rydman EM, Ilves M, Koivisto AJ, Kinaret PA, Fortino V, Savinko TS, Lehto MT, Pulkkinen V, Vippola M, Hameri KJ, et al. Inhalation of rod-like carbon nanotubes causes unconventional allergic airway inflammation. Part Fibre Toxicol. 2014;11:48.

Sabareeswaran A, Ansar EB, Harikrishna Varma PR, Mohanan PV, Kumary TV. Effect of surface-modified superparamagnetic iron oxide nanoparticles (SPIONS) on mast cell infiltration: an acute in vivo study. Nanomedicine. 2016;12:1523–33.

Saito H, Ishizaka T, Ishizaka K. Mast cells and IgE: from history to today. Allergol Int. 2013;62:3–12.

Salamon P, Shefler I, Moshkovits I, Munitz A, Horwitz Klotzman D, Mekori YA, Hershko AY. IL-33 and IgE stimulate mast cell production of IL-2 and regulatory T cell expansion in allergic dermatitis. Clin Exp Allergy. 2017;47:1409–16.

Samoszuk M, Kanakubo E, Chan JK. Degranulating mast cells in fibrotic regions of human tumors and evidence that mast cell heparin interferes with the growth of tumor cells through a mechanism involving fibroblasts. BMC Cancer. 2005;5:121.

Secor VH, Secor WE, Gutekunst CA, Brown MA. Mast cells are essential for early onset and severe disease in a murine model of multiple sclerosis. J Exp Med. 2000;191:813–22.

Serra-Pages M, Torres R, Plaza J, Herrerias A, Costa-Farre C, Marco A, Jimenez M, Maurer M, Picado C, de Mora F. Activation of the Prostaglandin E2 receptor EP2 prevents house dust mite-induced airway hyperresponsiveness and inflammation by restraining mast cells' activity. Clin Exp Allergy. 2015;45:1590–600.

Shikotra A, Ohri CM, Green RH, Waller DA, Bradding P. Mast cell phenotype, TNF alpha expression and degranulation status in non-small cell lung cancer. Sci Rep. 2016;6:38352.

Shimokawa C, Kanaya T, Hachisuka M, Ishiwata K, Hisaeda H, Kurashima Y, Kiyono H, Yoshimoto T, Kaisho T, Ohno H. Mast cells are crucial for induction of group 2 innate lymphoid cells and clearance of helminth infections. Immunity. 2017;46:863–74 e864.

Slodka A, Wiktorska M, Brzezinska-Blaszczyk E. IgE by itself affects mature rat mast cell preformed and de novo-synthesized mediator release and amplifies mast cell migratory response. PLoS One. 2013;8:e79286.

Stasikowska-Kanicka O, Danilewicz M, Glowacka A, Wagrowska-Danilewicz M. Mast cells and eosinophils are involved in activation of ulcerative colitis. Adv Med Sci. 2012;57:230–6.

Suto H, Nakae S, Kakurai M, Sedgwick JD, Tsai M, Galli SJ. Mast cell-associated TNF promotes dendritic cell migration. J Immunol. 2006;176:4102–12.

Szebeni J, Alving CR, Rosivall L, Bunger R, Baranyi L, Bedocs P, Toth M, Barenholz Y. Animal models of complement-mediated hypersensitivity reactions to liposomes and other lipid-based nanoparticles. J Liposome Res. 2007;17:107–17.

Tahara K, Tadokoro S, Yamamoto H, Kawashima Y, Hirashima N. The suppression of IgE-mediated histamine release from mast cells following exocytic exclusion of biodegradable polymeric nanoparticles. Biomaterials. 2012;33:343–51.

Talahalli R, Zarini S, Sheibani N, Murphy RC, Gubitosi-Klug RA. Increased synthesis of leukotrienes in the mouse model of diabetic retinopathy. Invest Ophthalmol Vis Sci. 2010;51:1699–708.

Toms R, Weiner HL, Johnson D. Identification of IgE-positive cells and mast cells in frozen sections of multiple sclerosis brains. J Neuroimmunol. 1990;30:169–77.

Varricchi G, Galdiero MR, Loffredo S, Marone G, Iannone R, Marone G, Granata F. Are mast cells MASTers in cancer? Front Immunol. 2017;8:424.

Visciano C, Liotti F, Prevete N, Cali G, Franco R, Collina F, de Paulis A, Marone G, Santoro M, Melillo RM. Mast cells induce epithelial-to-mesenchymal transition and stem cell features in human thyroid cancer cells through an IL-8-Akt-Slug pathway. Oncogene. 2015;34:5175–86.

Wang HW, Tedla N, Lloyd AR, Wakefield D, McNeil PH. Mast cell activation and migration to lymph nodes during induction of an immune response in mice. J Clin Invest. 1998;102:1617–26.

Wingard CJ, Walters DM, Cathey BL, Hilderbrand SC, Katwa P, Lin S, Ke PC, Podila R, Rao A, Lust RM, et al. Mast cells contribute to altered vascular reactivity and ischemia-reperfusion injury following cerium oxide nanoparticle instillation. Nanotoxicology. 2011;5:531–45.

Xiao W, Nishimoto H, Hong H, Kitaura J, Nunomura S, Maeda-Yamamoto M, Kawakami Y, Lowell CA, Ra C, Kawakami T. Positive and negative regulation of mast cell activation by Lyn via the FcepsilonRI. J Immunol. 2005;175:6885–92.

Yamaki K, Yoshino S. Comparison of inhibitory activities of zinc oxide ultrafine and fine particulates on IgE-induced mast cell activation. Biometals. 2009;22:1031–40.

Yang PM, Chen HZ, Huang YT, Hsieh CW, Wung BS. Lycopene inhibits NF-kappaB activation and adhesion molecule expression through Nrf2-mediated heme oxygenase-1 in endothelial cells. Int J Mol Med. 2017;39:1533–450.

Zhang C, Baumgartner RA, Yamada K, Beaven MA. Mitogen-activated protein (MAP) kinase regulates production of tumor necrosis factor-alpha and release of arachidonic acid in mast cells. Indications of communication between p38 and p42 MAP kinases. J Biol Chem. 1997;272:13397–402.

Zhuravskii S, Yukina G, Kulikova O, Panevin A, Tomson V, Korolev D, Galagudza M. Mast cell accumulation precedes tissue fibrosis induced by intravenously administered amorphous silica nanoparticles. Toxicol Mech Methods. 2016;26:260–9.

Zylka MJ, Dong X, Southwell AL, Anderson DJ. Atypical expansion in mice of the sensory neuron-specific Mrg G protein-coupled receptor family. Proc Natl Acad Sci U S A. 2003;100:10043–8.

Chapter 5
Impact of Nanoparticles on Dendritic Cells

Rob J. Vandebriel and Henk van Loveren

Abstract Dendritic cells (DCs) are pivotal in the initiation of the adaptive immune response. DCs undergo maturation, a process central to their functioning, where immature DCs ingest protein antigens and mature DCs present peptides to naive T cells. Various types of nanoparticles (NPs) can influence the process of DC maturation and by that the immune response. Chemical composition, size, and surface modification have been shown to affect DC maturation; chemical composition and size effects on the direction (Th1/Th2/Th17) of the immune response have been observed. Further studies studying series of NP are required to further delineate the particle characteristics that determine these effects.

Keywords Dendritic cell · Maturation · Cell surface marker · Cytokine · DC-SIGN

5.1 Introduction

Dendritic cells (DCs) are sentinel cells that are pivotal in the initiation of adaptive immune responses (Banchereau and Steinman 1998). Moreover, they integrate various stimuli, such as from different pathogen-associated molecular patterns (PAMPs) and the cytokine milieu. PAMPs are detected by pattern recognition receptors (PRRs) that are highly expressed by DC. PRR can be located on the cell surface, in endocytic compartments, or in the cytoplasm. Important classes of PRR form the Toll-like receptors (TLR) and Nod-like receptors (NLR). Importantly, the nature of

R. J. Vandebriel (✉)
Centre for Health Protection, National Institute for Public Health and the Environment, Bilthoven, The Netherlands
e-mail: rob.vandebriel@rivm.nl

H. van Loveren
Department of Toxicogenomics, Maastricht University Medical Centre, Maastricht, The Netherlands

J. C. Bonner, J. M. Brown (eds.), *Interaction of Nanomaterials with the Immune System*, Molecular and Integrative Toxicology,
https://doi.org/10.1007/978-3-030-33962-3_5

the immune response following DC function is greatly influenced by the PRR (or combination of different PRR). In this way, DCs form an important link between the innate and adaptive immune response.

DCs appear as immature DCs that are very well capable of ingesting protein antigens and as mature DCs that are especially capable of presenting peptides to naive T cells. This process of DC maturation is central to the functioning of DC. DC maturation can be readily measured based on cell surface marker expression, using a FACS, and cytokine production, using ELISA-based assays. Various types of nanoparticles (NP) are able to influence the process of DC maturation and by that the immune function. The effects by NP on the degree and "direction" of DC maturation are central in the studies discussed in this chapter.

Several types of DC exist, the classical division being between myeloid DC (mDC) and plasmacytoid DC (pDC). mDCs are specialized antigen-processing and antigen-presenting cells, with high phagocytic activity as immature cells and high cytokine producing capacity as mature cells. They regulate T-cell responses both in the steady state and during infection. pDCs are specialized to respond to viral infection with the production of type I interferons. They can, however, also act as antigen-presenting cells and control T-cell responses.

Insights that are more recent have identified four main types of DC: classical DC, pDC, Langerhans cells, and monocyte-derived DC (reviewed by Satpathy et al. 2012). In this division, mDCs are not regarded as one of the main types of DC.

- Classical DCs (cDCs) encounter and capture antigens in peripheral tissues and migrate from these tissues via afferent lymphatics to the T-cell zone of secondary lymphoid organs to initiate adaptive immune responses. They can be separated in two subsets based on cell surface marker expression, in lymphoid tissue of mice $CD8\alpha^+$ and $CD4^+$. While $CD8\alpha^+$ DCs are highly efficient at cross-presenting exogenous antigens on MHC I to $CD8^+$ T cells, $CD4^+$ is efficient in priming $CD4^+$ T cells through MHC II-restricted presentation. In peripheral tissues, $CD8\alpha$ is not expressed on DC, but CD103 is expressed instead. Similarly, in peripheral tissues, CD4 is not expressed on DC, but CD11b is expressed instead. Human counterparts of these two subtypes of cDC have been identified in lymphoid and nonlymphoid organs.
- pDCs are not phagocytic and have a high rate of MHC II turnover, making them inefficient at presenting exogenous antigens to $CD4^+$ T cells.
- Langerhans cells have been viewed as DC population in the epidermis, although they resemble tissue-resident macrophages in several ways.
- Monocyte-derived DCs are similar to cDC in that they upregulate CD11c and MHC class II expression and efficiently stimulate naïve T cells.

The studies described below employ in vivo and in vitro studies. For in vitro studies, DCs either isolated from murine bone marrow (bone marrow DC or BMDC) or derived from human peripheral blood monocytes (monocyte-derived DC or MDDC, also named moDC) are used. moDCs are a popular cell type to study DC function in vitro, for various reasons: (1) since they can be cultured from monocytes isolated from human blood, they are readily available in large numbers; (2) no cross-species extrapolation is required; and (3) no laboratory animals are required. It should be noted, however, that the in vivo equivalent of moDC has been difficult to identify. The panel of cell surface markers used to measure DC maturation generally comprises HLA-DR (MHC class II) and the T-cell co-receptors CD40, CD80, and CD86, although additions to this panel are, for example, CD83, PD-L1, and DC-SIGN, sometimes included, while in other studies, the panel used is more limited. The cytokines measured to evaluate DC maturation are mostly IL-12p40 or IL-12p70, but also IL-10 and TNF-α.

5.2 Effect of NP Chemical Composition on DC Function

Instillation in the lungs of mice of 40- to 300-nm Ag NP resulted in increased expression of PDCA-1, a marker for pDC, but not CD11c, a marker for mDC (Bezemer et al. 2011). Expressions of the surface markers CD54 (ICAM-1), CD80, and CD86 on pDC were increased, while no effects on surface marker expression on mDC were seen.

Schanen et al. (2013) showed that 7- to 10-nm TiO_2 but not 3- to 5-nm CeO_2 induced maturation of moDC, as shown by CD80, CD83, CD86, and CCR7 surface expression. TiO_2 NP induced IL-12 production by DC, while CeO_2 NP induced IL-10 production. Differences in these responses could not be explained by differences in NP uptake. The authors also showed that TiO_2 NP induced ROS and NLRP3 inflammasome activation (see below) in DC, while CeO_2 NP did not. DC treated with TiO_2 NP induced a proliferative allogeneic $CD4^+$ T-cell response with Th1 cytokines, while treatment with CeO_2 induced a nonproliferative response with Th2 cytokines. In conclusion, TiO_2 and CeO_2 NP induce DC maturation (and ensuing immune responses) in clearly different directions.

The NF-κB pathway was implicated in DC maturation induced by 82-nm TiO_2 particles, as measured by CD80 and CD86 upregulation (Zhu et al. 2014).

Vandebriel et al. (2018) showed that DC maturation is more strongly induced by anatase TiO_2 NP than by rutile ones, based on CD83 and CD86 surface expression and IL-12p40 production (anatase and rutile are two crystal forms of TiO_2 that differ in their surface reactivity). In conclusion, for NP of similar size, chemical composition is highly important for their immune activity.

5.3 Effects of Particle Size on DC Function

A relatively early observation of considerable interest was that large (500 nm to 2 μm) polystyrene (PS) particles required cells to be transported from the injection site (footpads of mice) to the draining lymph nodes (LN), whereas small (20–200 nm)

PS particles were transported freely, through lymphatic drainage, to the draining LN where they were taken up by LN-resident cell such as $CD11c^{+}CD8\alpha^{-}$ and $CD11c^{+}CD8\alpha^{+}$ DC but also pDC and B cells (Manolova et al. 2008).

Vallhov et al. (2007) compared the effects of two types of mesoporous silica particles on moDC. The size difference between these particles is mainly related to pore size of 3.9 nm (AMS-6) and 2.8 nm (AMS-8). Both types reduced CD40 expression, while AMS-8 reduced CD80 and increased CD86 expression. Neither particle affected CD83 or HLA-DR expression.

Incubation with 30 and 50 nm, but not 70 and 120 nm, silica-coated iron oxide particles showed reduced viability of moDC (Kunzmann et al. 2011). In contrast, 20 and 50 nm commercially available superparamagnetic iron oxide NP did not affect viability. None of the particles induced TNF-α production by the DC.

Hardy et al. (2013) compared the uptake of 50-nm and 500-nm carboxylated PS particles in vivo after intratracheal instillation. The 50-nm particles were taken up in the lung by B cells, macrophages, and both $CD11^{+}$ and $CD103^{+}$ DCs. In the draining LN, they were only taken up by DC ($CD11b^{+}$, $CD103^{+}$, and $CD11b^{+}CD103^{+}$). The 50-nm particles induced the expression of CD40, CD80, and CD86. In an ovalbumin mouse allergy model, the 50-nm particles reduced responses associated with respiratory allergy (number of eosinophils in the BAL, number of leukocytes in the lung, OVA-specific serum IgE, IL-5 and IL-13 in the BAL). All these effects were seen to a much lesser extent by the 500-nm particles.

In the bronchoalveolar lavage fluid (BALF) of intranasally instilled mice, 20- and 50-nm PS particles were taken up by $CD11b^{high}$ DC, whereas 100-, 200-, and 1000-nm particles were not (Blank et al. 2013). In addition, $CD11b^{low}$ migratory DC showed higher uptake frequencies of 20- and 50-nm particles compared to larger ones.

In murine bone marrow-derived DC, a significantly higher co-localization of 20-nm PS particles with lysosomes was seen compared to 1-μm particles (Seydoux et al. 2014).

In a study by Fernández et al. (2015) that investigates the effects of both size and surface modification, the uptake by moDC of 3-nm Au nanoclusters (NC) and 12-nm Au NP and their effects on these cells were examined. NCs were taken up more efficiently than NP, and NCs with a zwitterionic[1] ligand (zw-NC) were taken up more efficiently than PEG-NC. Maturation was induced more strongly by zw-NC than by PEG-NC or PEG-NP, as measured by CD80, CD83, CD86, and HLA-DR expression. Next, zw-NC induced higher IL-12p70 and IFN-γ levels than PEG-NC, zw-NP, and PEG-NP, indicating a Th1-skewed response by zw-NC. Production of IL-10, which is suggestive of the induction of regulatory T cells, was also induced by zw-NC. Both zw-NP and PEG-NP induced production of IL-4, which together with a lack of induction of IL-12p70 and IFN-γ indicates a Th2-skewed response.

[1]A zwitterion is a molecule with two (or more) functional groups of which (at least) one has a positive charge and (at least) one a negative charge, while the net charge is zero.

Au NP of 10 nm and 50 nm impaired LPS-induced moDC maturation, seen as reduced expression of CD83, CD86, and HLA-DR and reduced allostimulation (Tomić et al. 2014), with the effect of the 10-nm particles being more prominent. The 10-nm particles inhibited IL-12p70, IL-17, and IL-23 production and induced IL-4 and IL-10 production (toward a Th2 response), whereas the 50-nm particle induced IL-17 and IL-23 production and inhibited IL-4 and IL-10 production (toward a Th17 response).

All these studies clearly reflect the strong effects on DC maturation based on size alone, a critical issue with regard to the activity of the NP.

5.4 Effect of Surface Modification on DC Function

Propyl cyanide surface modification of a mesoporous silica particle resulted in abrogation of the induction of CD86 expression by moDC and its production of IFN-γ, IL-4, and IL-13 (Vallhov et al. 2012).

Two similarly sized (240-nm) functionalized polystyrene particles induced IL-6, IL-12p40, and TNF-α production by moDC, with the particle having the strongest negative ζ-potential giving stronger induction (Frick et al. 2012). The induction of cell surface marker expression by moDC was similar for the two particles.

In a study by Fytianos et al. (2015) that investigates the effects of both surface modification and coating, COOH-modified PEG-coated Au NPs were taken up in moDC to a smaller extent than NH_2-modified PEG-coated, COOH-modified PVA-coated, and COOH-modified PEG+PVA-coated Au NP. The highest uptake was seen for NH_2-modified PVA-coated and NH_2-modified PEG+PVA-coated Au NP. The latter two induced higher IL-1β secretion, which is possibly related to inflammasome activation (see below). All NPs used in this study ranged between 20 and 80 nm in size. In a follow-up study by Seydoux et al. (2016) that compared COOH-modified PVA-coated Au NP to NH_2-modified PVA-coated Au NP after intranasal administration in vivo, NH_2-modified particles were shown to be preferentially taken up compared to their COOH-modified counterparts. In BALF, expression of CD40 but not CD86 was higher in CD11b$^+$ DC in animals treated with COOH-modified PVA-coated Au NP compared to NH_2-modified PVA-coated Au NP.

In a recent study by Gao et al. (2017), the effects of 130-nm cationic and anionic liposomes on murine bone marrow-derived DC were compared. Cationic but not anionic liposomes elevated the lysosomal pH and reduced antigen degradation, thereby increasing cross-presentation[2] and cross priming of CD8$^+$ cells. Thus, cationic liposomes are promising antigen delivery systems.

[2]Antigen cross-presentation is the process by which APCs internalize extracellular antigens and present them in the context of MHC class I to CD8$^+$ T cells.

5.5 Suppressive Effects on DC Maturation

Although no effects were seen on CD80, CD83, and CD86 expression by moDC, 30-nm PVA-coated SPIONs decreased the capacity of these cells to stimulate $CD4^+$ T cells (Blank et al. 2011). The production of IL-1β, IL-5, IL-6, IL-12p70, IFN-γ, and TNF-α by the moDC/CD4+ T cells coculture was strongly reduced, while the production of IL-10 was strongly increased. This may possibly relate to the induction of a more regulatory T-cell function or phenotype.

Mice were treated by pharyngeal aspiration with 1–3 μm × 1–4 nm MWCNT (Tkach et al. 2011). Preincubation of naive T cells with pulmonary DC from these animals resulted in reduced T-cell proliferative activity.

Zhang et al. (2011) showed that COOH-modified 18-nm quantum dots (QDs) were taken up by moDC, whereas PEG-modified QDs or NH_2-modified QDs were not; COOH-modified QD reduced CD80 and CD86 expression of LPS-stimulated moDC.

As mentioned above, Au NP of 10 nm and 50 nm impaired LPS-induced moDC maturation, seen as reduced expression of CD83, CD86, and HLA-DR and reduced allostimulation (Tomić et al. 2014).

5.6 DC and Lung Allergic Responses

Hardy et al. (2012) observed an inhibition of pulmonary inflammation by 50-nm glycine-coated PS NP; simultaneous administration of these NPs with OVA reduced the number of $OVA\text{-}FITC^+CD11c^+MHC\ II^{hi}$ and $OVA\text{-}FITC^+CD11c^+CD11b^+$ DC in the lungs.

As described above, Hardy et al. (2013) showed in an ovalbumin mouse allergy model that 50-nm carboxylated PS particles reduced responses associated with respiratory allergy.

5.7 Inflammasome Activation in DC

The capacity to activate the inflammasome has been reported for a wide range of NP (Sun et al. 2013). NLRP3 is the most widely studied inflammasome, among others because of its broad range of agents (pathogens, host molecules, and xenobiotics) that are able to activate this multiprotein complex. Activation of NLRP3 by pathogen-associated molecular patterns (PAMP) or danger-associated molecular patterns (DAMP) results in the activation of caspase 1 (cleaved from pro-caspase-1) that in turn cleaves (inactive) pro-IL-1β and pro-IL-18 to (active) IL-1β and IL-18. Winter et al. (2011) showed that 14-nm amorphous SiO_2 and 20- to 80-nm anatase/rutile TiO_2 particles induced maturation of murine bone marrow-derived DC, as measured by

CD80, CD86, and MHC class II surface expression. Both types of particles induced IL-1β secretion, which was shown to be dependent on NLRP3 and caspase-1.

Nakanishi et al. (2016) reported inflammasome activation in mouse bone marrow-derived DC following exposure to 30-, 70-, and 300-nm silica NP. The activation was shown to be mediated by purinergic signalling.[3] A clear size-dependent effect (30 nm > 70 nm > 300 nm) on the production of IL-1β, IL-18, ROS, and ATP was seen.

Winkler et al. (2017) showed inflammasome activation in mouse bone marrow-derived DC following exposure to 7- and 13-nm synthetic amorphous silica (SAS) but not 33- and 140-nm TiO_2 or 11- and 21-nm $FePO_4$. Inflammasome activation was shown to depend on macropinocytosis, endosomal TLR activation, and the adapter protein MyD88. Although Tlr2, Tlr4, Tlr7, and Tlr9 require MyD88 for downstream signaling, none of these receptors were required for inflammasome activation (Tlr3 that does not depend on MyD88 was not required either). Inflammasome activation was induced by SAS only; priming by an inflammatory trigger was not required. Both cDC and pDC showed maturation by SAS: for cDC, this was shown by increased CD40 and CD69 expression; for pDC, this was shown by increased CD40, CD69, and CD86 expression and reduced CD62L expression.

5.8 Targeting to DC-SIGN

DC-SIGN, also known as CD209, is a C-type lectin receptor present on the surface of DC. On myeloid and preplasmacytoid DC, DC-SIGN mediates DC rolling interactions with blood endothelium and activation of $CD4^+$ T cells as well as recognition of pathogen haptens. DC-SIGN is used to deliver antigens to DC.

Using particles consisting of a PEGylated PLGA core carrying a humanized antibody that targets DC-SIGN, Cruz et al. (2010) showed that 200-nm particles but not 2-μm particles effectively targeted DC. The particles were loaded with tetanus toxin (TT) peptide; 200-nm particles but not 2-μm particles induced proliferation of autologous TT-specific peripheral blood lymphocytes.

Later studies used α-fucosyl-β-alanyl amide functionalized 2-nm Au NP that targeted DC-SIGN and did not induce DC maturation (Arosio et al. 2014), while galactofuranose-functionalized 2-nm Au NP that also targeted DC-SIGN did induce DC maturation as shown by increased CD80, CD86, and HLA-DR expression (Chiodo et al. 2014).

In a study using NH_2- and COOH-functionalized PVA-coated 15-nm Au NP, DC-SIGN targeting resulted in increased MHC class II expression. CD1c, CD40, CD80, CD83, and CD86 expressions were not affected by this targeting (Fytianos et al. 2017). In summary, nanometer size and type of functionalization influence the response to DC-SIGN targeting particles.

[3] Purinergic signaling is a form of extracellular signaling mediated by purine nucleotides and nucleosides such as ATP.

5.9 Conclusions

- The literature reviewed in this chapter clearly shows the effects of NP exposure on DC maturation in vivo and in vitro.
- Studies that compare effects of different NP on DC maturation are detailed with regard to the number of parameters measured and the underlying mechanisms, but in most cases, the number of particles compared is rather limited. Nevertheless, roles of chemical composition, size, and surface modification on DC maturation (and sometimes ensuing immune responses) have been established. Several outcomes are highlighted here.
  - Regarding the role of chemical composition, the exposure of moDC to TiO_2 vs. CeO_2 NP resulted in different immune responses (proliferative, Th1 vs. nonproliferative, Th2). Anatase TiO_2 NP had a stronger effect on DC maturation than rutile TiO_2 NP.
  - Regarding the role of size, uptake of 20-nm and 50-nm PS particles by DC in vivo was higher than that of 100-, 200-, and 500-nm particles. Furthermore, 3-nm Au NCs were taken up more efficiently by moDC than 12-nm Au NP. While zwitterionic NC induced a Th1/Treg response, zwitterionic NP induced a Th2 response. Finally, 10-nm Au particles showed a stronger inhibiting effect on LPS-induced moDC maturation and allostimulation than 50-nm Au particles, with the 10-nm particles inducing a Th2 response and the 50-nm particles inducing a Th17 response.
  - Regarding the role of surface modification, NH_2-modified Au NPs coated with PEG, PVA, or both were, in general, taken up in moDC to a higher extent than COOH-modified ones. In vivo, COOH modification resulted in higher CD40 expression than NH_2 modification. Furthermore, uptake in moDC and their maturation were higher by zwitterionic NC compared to PEGylated NC. Finally, cationic liposomes, in contrast to anionic ones, elevated lysosomal pH and reduced antigen degradation in BMDC.
- Nanoparticle exposure may not only induce but may also suppress DC maturation.
- Inflammasome activation is induced also in DC. Both chemical composition and size were shown to play a role, while the role of surface modification was not investigated.[4]
- Targeting DC can be successful using NP coupled with antibodies against DC-SIGN.
- Not only the size of effect on DC maturation is determined by size and surface modification, but also the direction of the immune response is affected.

[4] The role of surface modification on inflammasome activation has, in fact, been shown, but not in DC.

Various applications of NP may require different effects on DC maturation, such as immune activation, immune suppression, immune deviation, and DC targeting. Alternatively, they should lack an effect on DC maturation. Further studies comparing the series of NP are required to further delineate which particle characteristics determine these effects.

References

Arosio D, Chiodo F, Reina JJ, Marelli M, Penadés S, van Kooyk Y, Garcia-Vallejo JJ, Bernardi A. Effective targeting of DC-SIGN by α-fucosylamide functionalized gold nanoparticles. Bioconjug Chem. 2014;25:2244–51.

Banchereau J, Steinman RM. Dendritic cells and the control of immunity. Nature. 1998;392:245–52.

Bezemer GF, Bauer SM, Oberdörster G, Breysse PN, Pieters RH, Georas SN, Williams MA. Activation of pulmonary dendritic cells and Th2-type inflammatory responses on instillation of engineered, environmental diesel emission source or ambient air pollutant particles in vivo. Innate Immun. 2011;3(2):150–66.

Blank F, Gerber P, Rothen-Rutishauser B, Sakulkhu U, Salaklang J, De Peyer K, Gehr P, Nicod LP, Hofmann H, Geiser T, Petri-Fink A, Von Garnier C. Biomedical nanoparticles modulate specific CD4+ T cell stimulation by inhibition of antigen processing in dendritic cells. Nanotoxicology. 2011;5:606–21.

Blank F, Stumbles PA, Seydoux E, Holt PG, Fink A, Rothen-Rutishauser B, Strickland DH, von Garnier C. Size-dependent uptake of particles by pulmonary antigen-presenting cell populations and trafficking to regional lymph nodes. Am J Respir Cell Mol Biol. 2013;49:67–77.

Chiodo F, Marradi M, Park J, Ram AF, Penadés S, van Die I, Tefsen B. Galactofuranose-coated gold nanoparticles elicit a pro-inflammatory response in human monocyte-derived dendritic cells and are recognized by DC-SIGN. ACS Chem Biol. 2014;9:383–9.

Cruz LJ, Tacken PJ, Fokkink R, Joosten B, Stuart MC, Albericio F, Torensma R, Figdor CG. Targeted PLGA nano- but not microparticles specifically deliver antigen to human dendritic cells via DC-SIGN in vitro. J Control Release. 2010;144:118–26.

Fernández TD, Pearson JR, Leal MP, Torres MJ, Blanca M, Mayorga C, Le Guével X. Intracellular accumulation and immunological properties of fluorescent gold nanoclusters in human dendritic cells. Biomaterials. 2015;43:1–12.

Frick SU, Bacher N, Baier G, Mailänder V, Landfester K, Steinbrink K. Functionalized polystyrene nanoparticles trigger human dendritic cell maturation resulting in enhanced CD4+ T cell activation. Macromol Biosci. 2012;12:1637–47.

Fytianos K, Rodriguez-Lorenzo L, Clift MJ, Blank F, Vanhecke D, von Garnier C, Petri-Fink A, Rothen-Rutishauser B. Uptake efficiency of surface modified gold nanoparticles does not correlate with functional changes and cytokine secretion in human dendritic cells in vitro. Nanomedicine. 2015;11:633–44.

Fytianos K, Chortarea S, Rodriguez-Lorenzo L, Blank F, von Garnier C, Petri-Fink A, Rothen-Rutishauser B. Aerosol delivery of functionalized gold nanoparticles target and activate dendritic cells in a 3D lung cellular model. ACS Nano. 2017;11:375–83.

Gao J, Ochyl LJ, Yang E, Moon JJ. Cationic liposomes promote antigen cross-presentation in dendritic cells by alkalizing the lysosomal pH and limiting the degradation of antigens. Int J Nanomedicine. 2017;12:1251–64.

Hardy CL, LeMasurier JS, Belz GT, Scalzo-Inguanti K, Yao J, Xiang SD, Kanellakis P, Bobik A, Strickland DH, Rolland JM, O'Hehir RE, Plebanski M. Inert 50-nm polystyrene nanoparticles that modify pulmonary dendritic cell function and inhibit allergic airway inflammation. J Immunol. 2012;188:1431–41.

Hardy CL, LeMasurier JS, Mohamud R, Yao J, Xiang SD, Rolland JM, O'Hehir RE, Plebanski M. Differential uptake of nanoparticles and microparticles by pulmonary APC subsets induces discrete immunological imprints. J Immunol. 2013;191:5278–90.

Kunzmann A, Andersson B, Vogt C, Feliu N, Ye F, Gabrielsson S, Toprak MS, Buerki-Thurnherr T, Laurent S, Vahter M, Krug H, Muhammed M, Scheynius A, Fadeel B. Efficient internalization of silica-coated iron oxide nanoparticles of different sizes by primary human macrophages and dendritic cells. Toxicol Appl Pharmacol. 2011;253:81–93.

Manolova V, Flace A, Bauer M, Schwarz K, Saudan P, Bachmann MF. Nanoparticles target distinct dendritic cell populations according to their size. Eur J Immunol. 2008;38:1404–13.

Nakanishi K, Tsukimoto M, Tanuma S, Takeda K, Kojima S. Silica nanoparticles activate purinergic signaling via P2X7 receptor in dendritic cells, leading to production of pro-inflammatory cytokines. Toxicol In Vitro. 2016;35:202–11.

Satpathy AT, Wu X, Albring JC, Murphy KM. Re(de)fining the dendritic cell lineage. Nat Immunol. 2012;13:1145–54.

Schanen BC, Das S, Reilly CM, Warren WL, Self WT, Seal S, Drake DR 3rd. Immunomodulation and T helper TH1/TH2 response polarization by CeO2 and TiO2 nanoparticles. PLoS One. 2013;8:e62816.

Seydoux E, Rothen-Rutishauser B, Nita IM, Balog S, Gazdhar A, Stumbles PA, Petri-Fink A, Blank F, von Garnier C. Size-dependent accumulation of particles in lysosomes modulates dendritic cell function through impaired antigen degradation. Int J Nanomedicine. 2014;9:3885–902.

Seydoux E, Rodriguez-Lorenzo L, Blom RA, Stumbles PA, Petri-Fink A, Rothen-Rutishauser BM, Blank F, von Garnier C. Pulmonary delivery of cationic gold nanoparticles boost antigen-specific CD4+ T Cell Proliferation. Nanomedicine. 2016;12:1815–26.

Sun B, Wang X, Ji Z, Li R, Xia T. NLRP3 inflammasome activation induced by engineered nanomaterials. Small. 2013;9:1595–607.

Tkach AV, Shurin GV, Shurin MR, Kisin ER, Murray AR, Young SH, Star A, Fadeel B, Kagan VE, Shvedova AA. Direct effects of carbon nanotubes on dendritic cells induce immune suppression upon pulmonary exposure. ACS Nano. 2011;5:5755–62.

Tomić S, Đokić J, Vasilijić S, Ogrinc N, Rudolf R, Pelicon P, Vučević D, Milosavljević P, Janković S, Anžel I, Rajković J, Rupnik MS, Friedrich B, Colić M. Size-dependent effects of gold nanoparticles uptake on maturation and antitumor functions of human dendritic cells in vitro. PLoS One. 2014;9(5):e96584.

Vallhov H, Gabrielsson S, Strømme M, Scheynius A, Garcia-Bennett AE. Mesoporous silica particles induce size dependent effects on human dendritic cells. Nano Lett. 2007;7:3576–82.

Vallhov H, Kupferschmidt N, Gabrielsson S, Paulie S, Strømme M, Garcia-Bennett AE, Scheynius A. Adjuvant properties of mesoporous silica particles tune the development of effector T cells. Small. 2012;8:2116–24.

Vandebriel RJ, Vermeulen JP, van Engelen LB, de Jong B, Verhagen LM, de la Fonteyne-Blankestijn LJ, Hoonakker ME, de Jong WH. The crystal structure of titanium dioxide nanoparticles influences immune activity in vitro and in vivo. Part Fibre Toxicol. 2018;15:9.

Winkler HC, Kornprobst J, Wick P, von Moos LM, Trantakis I, Schraner EM, Bathke B, Hochrein H, Suter M, Naegeli H. MyD88-dependent pro-interleukin-1β induction in dendritic cells exposed to food-grade synthetic amorphous silica. Part Fibre Toxicol. 2017;14:21.

Winter M, Beer HD, Hornung V, Krämer U, Schins RP, Förster I. Activation of the inflammasome by amorphous silica and TiO2 nanoparticles in murine dendritic cells. Nanotoxicology. 2011;5:326–40.

Zhang LW, Bäumer W, Monteiro-Riviere NA. Cellular uptake mechanisms and toxicity of quantum dots in dendritic cells. Nanomedicine (Lond). 2011;6:777–91.

Zhu R, Zhu Y, Zhang M, Xiao Y, Du X, Liu H, Wang S. The induction of maturation on dendritic cells by TiO2 and Fe3O4@TiO2 nanoparticles via NF-κB signaling pathway. Mater Sci Eng C Mater Biol Appl. 2014;39:305–14.

Chapter 6
Complement Activation by Nanomaterials

Dmitri Simberg and Seyed M. Moghimi

Abstract Complement represents one of the most important innate immune pathways of neutralization of invading pathogens. With years, there has been an increasing awareness of the role of complement in recognition and clearance of engineered nanomaterials including nanopharmaceuticals. Here, we review the main pathways of complement activation, the assays used to characterize complement activation by nanopharmaceuticals, briefly discuss biological/clinical implications of nanomaterial-mediated complement incitement processes, and strategies to avoid complement activation.

Keywords Complement · Classical pathway · Lectin pathway · Alternative pathway · Properdin · Antibody · Hypersensitivity · Clearance · Nanoparticle · Nanomedicine

---

D. Simberg (✉)
Translational Bio-Nanosciences Laboratory, Department of Pharmaceutical Sciences, Skaggs School of Pharmacy and Pharmaceutical Sciences, University of Colorado, Aurora, CO, USA

Colorado Center for Nanomedicine and Nanosafety, University of Colorado Anschutz Medical Campus, Aurora, CO, USA
e-mail: dmitri.simberg@cuanschutz.edu

S. M. Moghimi
Colorado Center for Nanomedicine and Nanosafety, University of Colorado Anschutz Medical Campus, Aurora, CO, USA

School of Pharmacy, King George VI Building, Newcastle University, Newcastle upon Tyne, UK

Division of Stratified Medicine, Biomarkers & Therapeutics, Institute of Cellular Medicine, Faculty of Medical Sciences, Newcastle University, Newcastle upon Tyne, UK

J. C. Bonner, J. M. Brown (eds.), *Interaction of Nanomaterials with the Immune System*, Molecular and Integrative Toxicology,
https://doi.org/10.1007/978-3-030-33962-3_6

6.1 Activation of the Complement System by Nanoparticles

The complement system accounts for about 5% of globulins in serum and is responsible for recognition, elimination, and destruction of pathogens (Ricklin et al. 2010). Activation of complement cascade leads to several events, the main ones being opsonization of a surface with various fragments of complement component 3 (C3), release of proinflammatory mediators C3a and C3a, and formation of membrane attack complex MAC (C5b-9). The initial steps of complement activation are the most critical, and the mechanistic details are relevant to both pathogens and nanoparticles. According to Fig. 6.1, activation of the complement on a foreign surface takes place via the classical pathway (CP), the lectin pathway (LP), or the alternative pathway (AP). CP and LP start with the recognition of a foreign surface by soluble mediators with broad specificities. For CP, the main mediators are antibodies of IgG

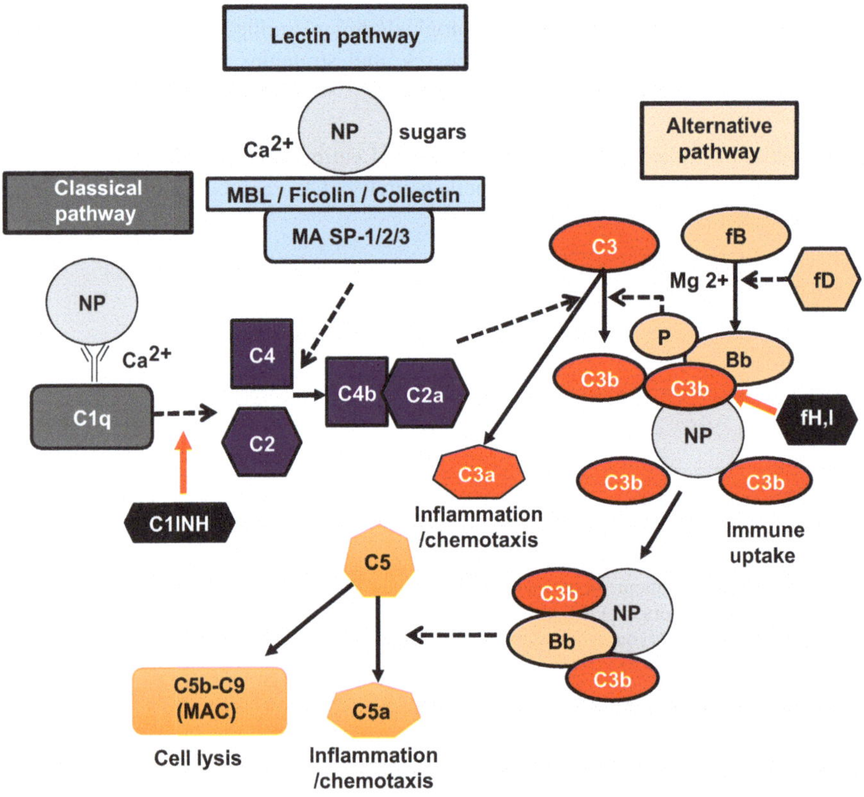

Fig. 6.1 Schematic representation of pathways of complement activation, C5 convertase, and MAC formation, as well as some of the regulatory proteins (red arrows)

or IgM type directed against surface determinants presented on pathogens and nanoparticles. The presentation of Fc domains leads to binding of C1 (C1qC1sC1r complex), with C1q being attached to the Fc portion. For successful CP activation, multivalency is a key (Nayak et al. 2012), therefore, IgM and/or multiple IgG molecules are needed. Binding of C1q leads to the cleavage of C2 and C4 by C1s/r and formation of C4bC2a, a classical C3 convertase, which is the serine protease enzyme responsible for complement activation. In some instances, C1q may bind to a surface without antibody participation and activate C1s/r. Activation of the LP also leads to the formation of the classical C4bC2a convertase, through binding of another type of mediator called lectins to sugars (mainly N-acetyl glucosamine and mannose) on the foreign surface. LP is somewhat different between mice and humans. In mice, the activation is primarily triggered via binding of mannose-binding lectins A/C or ficolin A to carbohydrates on the pathogen surface, leading to the activation of MBL-associated serum protease (MASP)-2 that cleaves C4 and C2. In humans, five different sugar recognition molecules have been identified that are able to initiate the LP: MBL, M-, L-, and H-ficolins, and collectin 11 (CL11 or CL-K1), but the downstream activation of the classical C3 convertase is believed to be similar in mice and in humans (Ali et al. 2012). Other MASPs in humans play activation role rather than direct cleavage of C4 and C2. Thus, MASP-2 is important for activation of MASP-1, while MASP-3 activates pro-Factor D, an important protease for the alternative pathway (discussed below). The main function of the complement convertase is the cleavage of the abundant serum C3 (~150 mg/dL) into C3a and C3b. The latter covalently attaches via a highly reactive thioester group to hydroxyl and amine groups on the foreign surface (Janssen et al. 2005).

It has long been known that CP and LP cannot account for the entire complement activation, and the majority of surface deposited C3b is generated through the alternative pathway (AP). The difference between the AP and the other pathways is that it requires the presence of C3 on the surface. C3b deposited through the CP or LP associates with serum factor B (Fig. 6.1), which is then cleaved by soluble factor D to form the AP convertase C3bBb, a magnesium-dependent serine protease that cleaves additional serum C3 into C3a and C3b. The AP convertase is intrinsically unstable, but is stabilized by properdin (P), which is present in blood at ~20 μg/ml (Kemper et al. 2010). Properdin forms a polymeric (dimer, trimer, and tetramer) complex (Alcorlo et al. 2013). Some studies demonstrated that binding of properdin can be the initial step in the activation of the AP (Nilsson and Nilsson Ekdahl 2012; Spitzer et al. 2007), but our studies with superparamagnetic iron oxide nanoparticles (SPIO) did not find a role of properdin in the AP initiation (Wang et al. 2016; Vu et al. 2019). More recent studies showed that many surfaces, including liposomes, trigger complement activation via surface deposition of meta-stable form of C3, C3(H_2O), with subsequent binding of FB and formation of C3(H_2O)Bb convertase (Klapper et al. 2014). This process is called "tick-over"(Bexborn et al. 2008) and is very pronounced in human sera. Therefore, in addition to the amplification of C3b deposited through other pathways, the AP is the self-activating pathway due to baseline hydrolysis of C3.

The complement system is tightly regulated via soluble and membrane-bound molecules such as C1-inhibitor, C4-binding protein, factors H and I, decay accelerating factor (CD55), and CD46 that modulate the activity of serine proteases, disassemble the convertase, or prevent MAC formation (reviewed in Ricklin et al. (2010) and Zipfel and Skerka (2009)). Noncomplement proteins such as apolipoproteins A-I and A-II also regulate complement activation apparently by minimizing the formation of the membrane attack complex (Hamad et al. 2013).

Over the past several years, the AP has been shown to be the essential pathway of complement activation in health and disease (Holers 2014). Due to its key role, the AP represents a unique therapeutic target that has been explored in many diseases (Holers and Thurman 2004; Ricklin 2012; Risitano et al. 2012). For many types of drug delivery/diagnostic nanoplatforms, including liposomes, iron oxides, and PLGA-lipid nanoparticles, the AP has been shown to be a critical pathway for complement activation (Wang et al. 2016; Vu et al. 2019; Salvador-Morales et al. 2009; Cunningham et al. 1979; Moghimi et al. 2006; Banda et al. 2014). Our comprehensive study of superparamagnetic iron oxide nanoworms, ultrasmall iron oxide nanoparticle Feraheme, liposomal doxorubicin (Doxil), and liposomal irinotecan (Onivyde) using multiple sera and plasma samples from healthy donors and breast cancer patients showed no involvement of the CP and predominant role of the AP and the LP in the level of deposition of C3 on nanoparticles (Vu et al. 2019).

6.2 Common Assays for Measuring Complement Activation

Classical complementology traditionally used immunoassays for detection of complement split products, as well as functional assays designed to test intactness of certain components and pathways of the system (reviewed in Kirschfink and Mollnes (2003)). All these assays are applicable for measuring complement activation by nanopharmaceuticals, with the caveat that after incubation of nanoparticles in biological media, often times there is a need to remove nanoparticles, for example, by ultracentrifugation in order to avoid interference with the assay. Enzyme-linked immunosorbent assay (ELISA) and Western blotting are most useful for detecting cleavage products of the complement factors using antibodies. These assays are excellent for measuring total degree of complement activation (C5a, sC5b-9) and also to dissect the involvement of specific pathways. Thus, measurement of fluid phase Bb can detect activation of the AP, whereas measurement of C4d (degradation product of C4b) and C4c can determine the generation of the classical C3 convertase and hence the involvement of the CP and LP. Studies at the Nanotechnology Characterization Lab at the National Cancer Institute showed that measurement of C3 split products (C3c, C3a), which show up as small peptides in Western blots, can be a reliable indicator of complement activation in plasma (Neun and Dobrovolskaia 2011). On the other hand, these assays are designed to analyze the upstream rather than the downstream pathways; therefore, effector molecules such as C5a and sC5b-9 need to be measured (Wolf-Grosse et al. 2017).

A special consideration in these assays is given to the type of biological media. Thus, careful preparation of serum is critical in order to retain the functionality of the complement system (Lachmann 2010). Recent reports show that whole blood assay using recombinant hirudin (lepirudin) anticoagulated blood is preferable to mimic conditions of in vivo environment (Kirschfink and Mollnes 2003; Wolf-Grosse et al. 2017). Indeed, lepirudin is the only anticoagulant that minimally affects complement activity, as opposed to heparin and EDTA/citrate that severely inhibit all pathways.

Inhibition assays have been extensively used to characterize the pathways of activation by nanoparticles. Thus, the involvement of the AP can be tested using a neutralizing antiproperdin antibody (Pauly et al. 2014), and we demonstrated robust complement inhibition by SPIO nanoworms through this approach (Banda et al. 2014). Chelation of Ca^{2+} by EGTA/Mg^{2+} can detect the involvement of CP and LP, whereas the use of C1 inhibitor, which is a natural protease inhibitor of C1, can block the CP. The main functional assay testing hemolytic effect of the terminal pathway is the sheep erythrocyte lysis assay (CH50), where anti-sheep RBC antibodies trigger the CP and subsequent lysis. This assay has been used for measurement of complement consumption after exposure of nanoparticles to serum (Meerasa et al. 2011). Despite the sensitivity of the CH50 for measurement of the residual complement activity, this assay should be used cautiously since it is skewed toward the CP, and somewhat the AP, and may not reflect the function of the LP.

Lastly, the role of different pathways can be dissected using mouse or human sera deficient for complement factors. Using complement-deficient mouse sera, we determined the critical role of lectin pathway as the initiating pathway and the AP as the amplification pathway for SPIO nanoworms (Banda et al. 2014). Regarding mouse sera, there are disparities in nanoparticle-mediated complement activation and C3 opsonization processes between rodents and humans. For instance, dextran-coated SPIO nanoworms trigger complement activation differently in mice and humans (Banda et al. 2014). Another example is complement activation by poly(2-methyl-2-oxazoline)-coated nanoparticles in human sera, whereas these nanoparticles do not activate mouse complement and, moreover, resist rapid uptake by murine macrophages (Tavano et al. 2018). For studies of human complement pathways, the word of caution is that complement-depleted and complement-deficient sera reveal the pathway of activation in one subject, whereas there is a significant variability among individuals in terms of the level of activation and the variability in activation pathway (Vu et al. 2019; Benasutti et al. 2017).

6.3 Measurement of Deposition of Complement Factors on Nanoparticles

As opposed to the above-mentioned traditional assays, we developed an immuno-dot blot assay (illustrated in Fig. 6.2) that measures and quantifies deposition of complement factors (e.g., C3, properdin, C1q) on the nanoparticle surface (Wang et al. 2016,

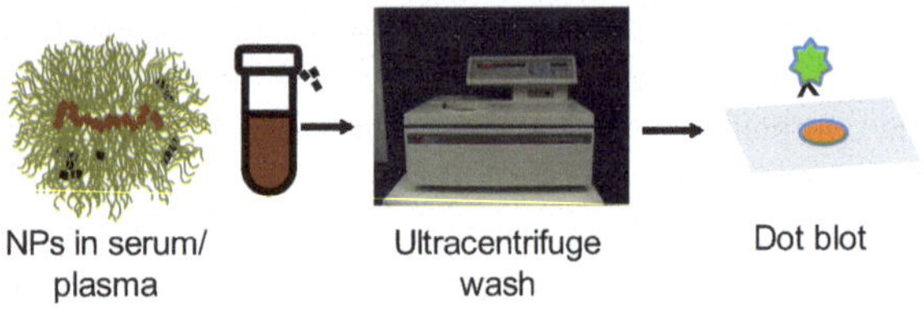

Fig. 6.2 Immuno-dot blot assay to detect deposition of complement proteins on nanoparticle surface

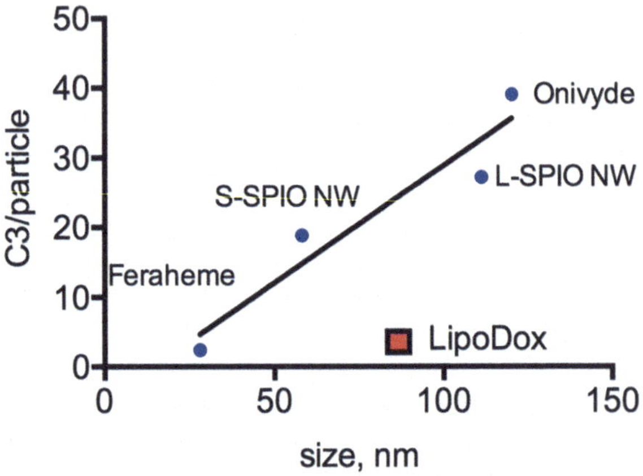

Fig. 6.3 Deposition of complement C3 on nanoparticles as a function of size and surface properties. S-SPIO is small superparamagnetic iron oxides, L-SPIO is large superparamagnetic iron oxides. (Obtained with permission from Benasutti et al. (2017))

2017; Benasutti et al. 2017; Chen et al. 2017). The assay produces a better linear response than Western blot. In a typical assay, a solution of nanoparticles at 1 mg/mL (iron or drug) of SPIO NWs, Feraheme, LipoDox, or Onivyde is incubated at a 1:3 volume ratio with serum or plasma for 30 min in a 37 °C water bath. Following incubation, samples are washed five times with PBS at 4 °C using ultracentrifuge. To determine the amount of specific proteins bound to nanoparticles, the nanoparticle pellet is resuspended in PBS, and 2 μL of each sample is pipetted in triplicates onto a nitrocellulose membrane (0.45-μm pore, Bio-Rad). Standard twofold dilutions of purified protein standards are applied to the same membrane (2 μL dots in triplicate). The membranes are blocked at room temperature, probed with the primary antibody for 1 h at room temperature, and followed by washes (3×) with PBS-T. Finally, the membranes are probed with corresponding secondary antibody labeled with IRDye 680 or IRDye 800 and scanned using the Odyssey infrared imager (Li-COR Biosciences, Lincoln, NE). For quantification, the background of 16-bit grayscale images is subtracted and integrated densities of the dots are measured with ImageJ software. The number of protein molecules per dot is determined from standard dilutions of a protein of interest on the same membrane. Number of protein molecules per mg nanoparticle can be calculated by dividing the number of molecules/dot by the amount of nanoparticles (mg Fe or mg drug) applied per dot. The molar concentration of nanoparticles and liposomes per milligram of material can be determined experimentally or derived theoretically (Benasutti et al. 2017).

Using this assay, an interesting conclusion was obtained when we compared C3 levels as a function of nanoparticle size and surface properties (Fig. 6.3).

Smaller nanoparticles (e.g., 30-nm Feraheme) showed much fewer C3 molecules per nanoparticle than larger iron oxide nanoparticles. The reason for this is probably related to the size constraints: each C3 molecule is ~15 nm in the longest dimension, and C3bBb Properdin complex is over 30 nm in diameter (Chen et al. 2017). At the same time, 90-nm PEGylated liposomal doxorubicin showed much fewer C3 per nanoparticle than 110 nm minimally PEGylated liposomal irinotecan Onivyde, likely due to the PEGylation of the former (Benasutti et al. 2017).

6.4 Emerging Role of Natural Antibodies in Complement Activation

In addition to the CP activation described in Fig. 6.1, surface-deposited antibodies can promote complement activation through other pathways. While the CP involves C1q binding to the Fc portion of closely adjacent antibodies (Duncan and Winter 1988), the LP may be triggered via binding of mannose-binding lectin, collectins and ficolins to a galactosyl glycoform of IgG (corresponding to 30% of total serum IgG), oligomannose structures on IgM, and the polymeric form of serum IgA (Banda et al. 2008; Malhotra et al. 1995). In addition, IgG can trigger C3 deposition by acting as a scaffold for initial C3b binding (Schenkein and Ruddy 1981; Russell and Mansa 1989). Indeed, absorbed IgG was shown to bring C3b molecules to a surface (Andersson et al. 2005). Moreover, IgG-bound C3b is protected against factor I, which together with factor H cleaves C3b into iC3b and C3dg (Fries et al. 1984). It has also been shown that C3b can covalently bind via the thioester to circulating antibodies, and IgG-bound C3b can efficiently mediate the AP convertase assembly and trigger complement activation on erythrocyte surface (Lutz et al. 1993a, b). We recently demonstrated that natural serum IgG triggers complement activation by four different nanoparticle types, including SPIO and liposomes (Vu et al. 2019). Moreover, we observed that IgG binds more efficiently to the protein corona rather than to the pristine nanoparticle surface. The epitopes that are recognized by IgG are poorly understood. Since only a few IgG molecules are bound per nanoparticle (Vu et al. 2019), and considering that a 110-nm diameter nanoparticle can bind up to 400 IgG molecules (based on the surface area occupancy), it appears that there are only so many epitopes per nanoparticle available for antibody binding. While there is a substantial evidence of circulating anti-PEG (Wang et al. 2007; Hamada et al. 2008; Chen et al. 2016) and antidextran antibodies (Pedersen et al. 2010; Chacko and Appukuttan 2003), the correlation of these antibodies with complement activation by nanoparticles is still unclear (Banda et al. 2014; Neun et al. 2018). Recent report points to ~44% incidence of anti-PEG antibodies in general population (Chen et al. 2016), and our recent study showed that about 20% of healthy donors have antidextran antibodies (Banda et al. 2014). Natural antibodies may recognize epitopes of self-proteins (presumably unfolded or denatured) (Holodick et al. 2017) and proteins are known to undergo denaturation on nanoparticle surface (Deng et al. 2011; Mortimer et al. 2014). Binding of antiphospholipid antibodies to different phospholipid headgroups in liposomes has been suggested to trigger complement activation through the classical pathway (Moghimi et al. 2012). However, this requires at least two IgG molecules within 10–40 nm of each other to form a stable platform for C1q binding and C1 activation. On the other hand, β-2-glycoprotein-1 deposition is a prerequisite for antiphospholipid antibody binding to liposomes and complement activation (Jones et al. 1992). It is highly likely that a few surface bound antiphospholipid antibodies could directly aid assembly of alternative pathway convertases and accelerate C3 opsonization.

6.5 Role of Protein Corona in Complement Activation

Numerous studies have been focused on proteins that adsorb to nanoparticles in biological milieu (*aka* protein corona) (Monopoli et al. 2011; Cedervall et al. 2007; Karmali and Simberg 2011). Studies involving model protein systems demonstrated the effects of the corona on nanoparticle targeting (Salvati et al. 2013), immune recognition (Mortimer et al. 2014), and toxicity (Deng et al. 2011). While proteomic studies identified hundreds of proteins, including complement proteins adsorbed on nanoparticles in serum/plasma (Karmali and Simberg 2011), the immunological significance of the binding cannot be easily inferred from these studies. The notion of protein corona is somewhat confusing due to its biological and physicochemical complexity and dynamic nature (Chen et al. 2017; Saptarshi et al. 2013). Shotgun mass spectrometry studies can miss important aspects of functionality of the bound proteins. For example, the proteomic data reveal no information on the denaturation status of proteins, which could be key to understanding why natural IgG binds to nanosurfaces. Using core-shell SPIO nanoworms, we determined that protein corona plays a critical role in complement activation on the surface of nanoparticles. Our experiments demonstrated: (1) C3b covalently binds to the proteins, (2) C3 dissociates from the particles together with the soft corona; (3) if exposed to fresh serum, the particles are efficiently reopsonized with more complement C3 (Fig. 6.4); (4) complement is triggered via immunoglobulin, which binds to the corona. The shedding of C3 and reopsonization means that complement inactivation (e.g., cleavage of C3b into iC3b by factor I) may not stop the opsonization.

6.6 Proinflammatory Response Elicited Due to Complement Activation on Nanoparticles

A significant amount of work has been dedicated to the biological effects of nanopharmaceuticals upon exposure to the extracellular milieu. The main implications of complement activation by engineered nanomaterials can be divided into several categories: (1) increased immune uptake by macrophages and white blood cells (Fig. 6.5, Moghimi et al. (2006), Song et al. (2007), and Hamad et al. (2010)); (2) proinflammatory response (Escamilla-Rivera et al. 2019; Gabizon et al. 2016); and

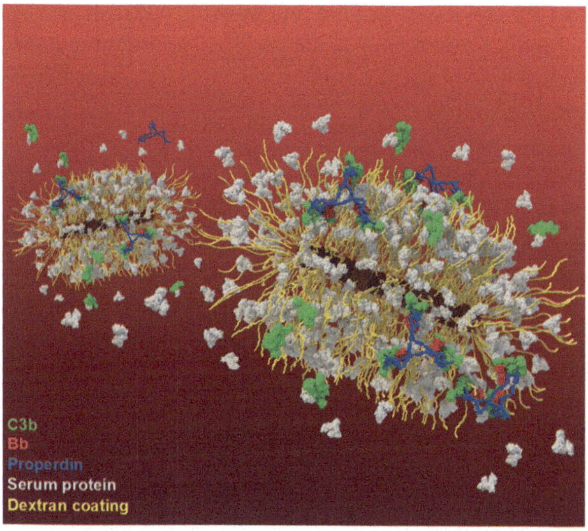

Fig. 6.4 Illustration of the exchangeable corona on dextran-coated SPIO leading to unstable binding and deposition of complement factors

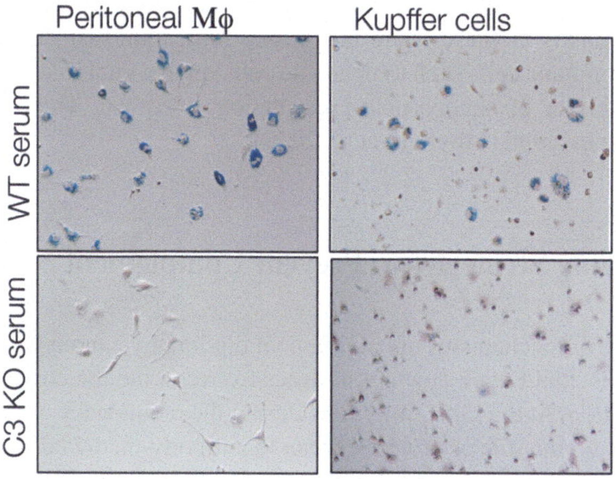

Fig. 6.5 Role of complement C3 in the uptake of SPIO nanoworms by mouse peritoneal macrophages and liver Kupffer cells. SPIO NWs were added to freshly isolated macrophages (0.1 mg/mL Fe) in media supplemented with 0.1% serum from wild-type or C3 knockout mice (C57/BL6 background). Cells were washed 4 h later and stained with Prussian blue

(3) improved adaptive response (antibody generation) toward vaccines (Thomas et al. 2011; Reddy et al. 2007). These biological effects are a consequence of efficient opsonization by C3b/iC3b, as well as release of highly proinflammatory C3a and C5a anaphylatoxins. C5a is known to stimulate release of TNF-α, IL-1β, and IL-6 from blood monocytes, as well as to potentiate the response of neutrophils and

macrophages to LPS (reviewed in Guo and Ward (2005)). It was established that nano-sized amphiphile-coated iron oxides trigger proinflammatory response and cytokine release in whole blood. The effect was blocked with inhibitors of the complement cascade (Wolf-Grosse et al. 2017). Recent data suggest that nanopharmaceuticals may trigger tumor growth via complement activation and C5a release in vivo (Moghimi 2014), and it was observed in mice with empty PEGylated liposomes (Gabizon et al. 2016); these events may ultimately explain the lack of therapeutic efficiency of Doxil in some patients (Petersen et al. 2016). On the other hand, it is not clear whether formation of MAC and subsequent cell damage play any role in the deleterious effects triggered by complement activation. Another controversial topic is the role of complement in infusion reactions observed in a significant proportion of patients (over 10% for Doxil® and over 30% for Onpattro®). These reactions are idiosyncratic and non-IgE-dependent, where typical symptoms (e.g., facial swelling, skin rash, cough, chest pain, and cardiovascular distress) vary from mild to severe depending on the individual, nanoparticle dose and the rate of infusion (Moghimi and Simberg 2018). Infusion reactions to nanoparticles cannot be predicted using standard allergy tests. Moreover, the suggested animal model of infusion reaction (pig) is overly sensitive and has been criticized (Moghimi 2014). While activation of complement in patients administered with Doxil was observed a while ago, and was suggested to be a causative factor of infusion reaction (Chanan-Khan et al. 2003), the ultimate evidence that complement is involved in clinical infusion reactions is lacking. There is a strong indication that macrophages (and perhaps other immune cells such as dendritic cells) play a vital role in infusion reaction, where the rate of nanoparticle presentation is critical. This topic has been extensively discussed in (Moghimi et al. 2019).

6.7 Emerging Strategies to Prevent Complement Activation

The knowledge of mechanisms of complement binding by nanomaterials and nanopharmaceuticals could open new possibilities to overcome the challenges of complement activation. At the same time, considering the redundancy of the system, and recent findings on the role of protein corona in antibody-mediated triggering of the AP (Vu et al. 2019) and the exchangeable dynamics of C3 opsonization (Chen et al. 2017), it is a daunting task to design a nanoparticle surface with virtually no complement activation. However, some attempts have tried to address these shortfalls. For instance, since PEG can minimize protein adsorption, keeping PEG chains 10–15 Å apart may generate ideal spacing for optimal protein exclusion effect. This approach has shown some degree of success in minimizing complement activation by single wall carbon nanotubes coated with poly(maleic anhydride-alt-1-octadecene) bearing two molecules of mPEG5000 (Andersen et al. 2013), but there are reports that PEGylated particles still activate complement (Hamada et al. 2008; Moghimi 2014). Backfilling the gaps within the PEG layer could improve the antifouling efficiency and improve resistance to complement activation (Dai et al. 2014). Another strategy could be a co-administration of the systemic complement inhibitor compstatin (Ricklin et al. 2010) with nanoparticles or the use of surface conjugated complement inhibitors (e.g., factor H Wu et al. (2011)). We recently and others described the promising use of natural complement inhibitors derived from membrane-bound convertase inhibitor CD55 (decay accelerating factor) and compstatin with preclinical and clinical nanoparticles (Wolf-Grosse et al. 2017; Gifford et al. 2019). Furthermore, our data demonstrate that the AP activation critically depends on properdin binding. Properdin normally binds to the surface following initial deposition of C3b, but could also recruit convertase components by first binding to the surface (Hourcade 2006). Elucidating mechanisms of properdin binding could provide additional strategies to block complement activation on nanosurfaces.

In terms of nanoparticle surface design considerations, some successful attempts include a hexosome formulation from citrem and glycerylmonooleate (Banda et al. 2014). Here, the lack of complement activation may have been due to recruitment of factor H (Wibroe et al. 2015), where the amino acid residues in the complement control protein domain 20 have been suggested to bind to the glycerol side chain and carboxyl moiety of sialic acid (Blaum et al. 2015). This process is analogous to the mechanism where some virulent bacterial pathogens bypass complement activation (Langford-Smith et al. 2015). Subsequent studies have also shown that other citrem-containing nonlamellar liquid crystalline nanoassemblies also behave as poor complement activating entities (Azmi et al. 2016). Earlier we showed that sulfonation and carboxymethylation of cross-linked dextran iron oxide nanoparticles (CLIO) could not overcome complement activation (Wang et al. 2016), thus implying that this approach may not be universally applicable to every nanoparticle. Other successful biomimetic approaches in overcoming complement activation have included surface enrichment with a phage peptide that recruits factor H (Wu et al. 2011) and surface functionalization with peptides that attenuate IgM deposition (Guan et al. 2018). Lastly, the rapidly developing field of complement therapeutics (Morgan and Harris 2015) may offer interesting solutions for inhibition of complement by nanopharmaceuticals in clinical practice.

6.8 Conclusions

Considering the important role of the complement system in orchestrating and bridging various responses during immune and inflammatory reactions, the interactions between complement proteins and nanopharmaceuticals of different sizes, morphology, structural complexities, and surface characteristics require careful and systematic evaluation. Despite recent developments, we are still in need of better understanding of the universal role of protein corona in complement activation as well as the role of complement system in regulation and synchronization of nanoparticle opsonization, macrophage clearance and responses, and disease progression.

Acknowledgments This work was supported by the grants from the National Institutes of Health EB022040, CA194058, and CA174560 to DS. S.M.M. acknowledges financial support by International Science and Technology Cooperation of Guangdong Province (reference 2015A050502002) and Guangzhou City (reference 201620160403005O) with RiboBio Co, Ltd., China.

Financial Disclosure The authors declare no conflict of interest.

References

Alcorlo M, Tortajada A, Rodriguez de Cordoba S, Llorca O. Structural basis for the stabilization of the complement alternative pathway C3 convertase by properdin. Proc Natl Acad Sci U S A. 2013;110:13504–9.

Ali YM, Lynch NJ, Haleem KS, Fujita T, Endo Y, Hansen S, Holmskov U, Takahashi K, Stahl GL, Dudler T, Girija UV, Wallis R, Kadioglu A, Stover CM, Andrew PW, Schwaeble WJ. The lectin pathway of complement activation is a critical component of the innate immune response to pneumococcal infection. PLoS Pathog. 2012;8:e1002793.

Andersen AJ, Robinson JT, Dai H, Hunter AC, Andresen TL, Moghimi SM. Single-walled carbon nanotube surface control of complement recognition and activation. ACS Nano. 2013;7:1108–19.

Andersson J, Ekdahl KN, Lambris JD, Nilsson B. Binding of C3 fragments on top of adsorbed plasma proteins during complement activation on a model biomaterial surface. Biomaterials. 2005;26:1477–85.

Azmi ID, Wibroe PP, Wu LP, Kazem AI, Amenitsch H, Moghimi SM, Yaghmur A. A structurally diverse library of safe-by-design citrem-phospholipid lamellar and non-lamellar liquid crystalline nano-assemblies. J Control Release. 2016;239(1–9):1.

Banda NK, Wood AK, Takahashi K, Levitt B, Rudd PM, Royle L, Abrahams JL, Stahl GL, Holers VM, Arend WP. Initiation of the alternative pathway of murine complement by immune complexes is dependent on N-glycans in igg antibodies. Arthritis Rheum. 2008;58:3081–9.

Banda NK, Mehta G, Chao Y, Wang G, Inturi S, Fossati-Jimack L, Botto M, Wu L, Moghimi SM, Simberg D. Mechanisms of complement activation by dextran-coated superparamagnetic iron oxide (SPIO) nanoworms in mouse versus human serum. Part Fibre Toxicol. 2014;11:64.

Benasutti H, Wang G, Vu VP, Scheinman R, Groman E, Saba L, Simberg D. Variability of complement response toward preclinical and clinical nanocarriers in the general population. Bioconjug Chem. 2017;28:2747–55.

Bexborn F, Andersson PO, Chen H, Nilsson B, Ekdahl KN. The tick-over theory revisited: formation and regulation of the soluble alternative complement C3 convertase (C3(H2o)Bb). Mol Immunol. 2008;45:2370–9.

Blaum BS, Hannan JP, Herbert AP, Kavanagh D, Uhrin D, Stehle T. Structural basis for sialic acid-mediated self-recognition by complement factor H. Nat Chem Biol. 2015;11:77–82.

Cedervall T, Lynch I, Lindman S, Berggard T, Thulin E, Nilsson H, Dawson KA, Linse S. Understanding the nanoparticle-protein corona using methods to quantify exchange rates and affinities of proteins for nanoparticles. Proc Natl Acad Sci U S A. 2007;104:2050–5.

Chacko BK, Appukuttan PS. Dextran-binding human plasma antibody recognizes bacterial and yeast antigens and is inhibited by glucose concentrations reached in diabetic sera. Mol Immunol. 2003;39:933–9.

Chanan-Khan A, Szebeni J, Savay S, Liebes L, Rafique NM, Alving CR, Muggia FM. Complement activation following first exposure to pegylated liposomal doxorubicin (Doxil): possible role in hypersensitivity reactions. Ann Oncol. 2003;14:1430–7.

Chen BM, Su YC, Chang CJ, Burnouf PA, Chuang KH, Chen CH, Cheng TL, Chen YT, Wu JY, Roffler SR. Measurement of pre-existing IgG and IgM antibodies against polyethylene glycol in healthy individuals. Anal Chem. 2016;88:10661–6.

Chen F, Wang G, Griffin JI, Brenneman B, Banda NK, Holers VM, Backos DS, Wu L, Moghimi SM, Simberg D. Complement proteins bind to nanoparticle protein corona and undergo dynamic exchange in vivo. Nat Nanotechnol. 2017;12:387–93.

Cunningham CM, Kingzette M, Richards RL, Alving CR, Lint TF, Gewurz H. Activation of human complement by liposomes: a model for membrane activation of the alternative pathway. J Immunol. 1979;122:1237–42.

Dai Q, Walkey C, Chan WC. Polyethylene glycol backfilling mitigates the negative impact of the protein corona on nanoparticle cell targeting. Angew Chem Int Ed Engl. 2014;53:5093–6.

Deng ZJ, Liang M, Monteiro M, Toth I, Minchin RF. Nanoparticle-induced unfolding of fibrinogen promotes Mac-1 receptor activation and inflammation. Nat Nanotechnol. 2011;6:39–44.

Duncan AR, Winter G. The binding site for C1q on igg. Nature. 1988;332:738–40.

Escamilla-Rivera V, Solorio-Rodriguez A, Uribe-Ramirez M, Lozano O, Lucas S, Chagolla-Lopez A, Winkler R, De Vizcaya-Ruiz A. Plasma protein adsorption on Fe3O4-PEG nanoparticles activates the complement system and induces an inflammatory response. Int J Nanomedicine. 2019;14:2055–67.

Fries LF, Gaither TA, Hammer CH, Frank MM. C3b covalently bound to igg demonstrates a reduced rate of inactivation by factors H and I. J Exp Med. 1984;160:1640–55.

Gabizon AA, Patil Y, La-Beck NM. New insights and evolving role of pegylated liposomal doxorubicin in cancer therapy. Drug Resist Updat. 2016;29:90–106.

Gifford G, Vu VP, Banda NK, Holers VM, Wang G, Groman EV, Backos D, Scheinman R, Moghimi SM, Simberg D. Complement therapeutics meets nanomedicine: overcoming human complement activation and leukocyte uptake of nanomedicines with soluble domains of CD55. J Control Release. 2019;302:181–9.

Guan J, Shen Q, Zhang Z, Jiang Z, Yang Y, Lou M, Qian J, Lu W, Zhan C. Enhanced immunocompatibility of ligand-targeted liposomes by attenuating natural IgM absorption. Nat Commun. 2018;9:2982.

Guo RF, Ward PA. Role of C5a in inflammatory responses. Annu Rev Immunol. 2005;23:821–52.

Hamad I, Al-Hanbali O, Hunter AC, Rutt KJ, Andresen TL, Moghimi SM. Distinct polymer architecture mediates switching of complement activation pathways at the nanosphere-serum interface: implications for stealth nanoparticle engineering. ACS Nano. 2010;4:6629–38.

Hamad I, Hunter AC, Moghimi SM. Complement monitoring of pluronic 127 gel and micelles: suppression of copolymer-mediated complement activation by elevated serum levels of Hdl, Ldl, and apolipoproteins Ai and B-100. J Control Release. 2013;170:167–74.

Hamada I, Hunter AC, Szebeni J, Moghimi SM. Poly(ethylene glycol)s generate complement activation products in human serum through increased alternative pathway turnover and a MASP-2-dependent process. Mol Immunol. 2008;46:225–32.

Holers VM. Complement and its receptors: new insights into human disease. Annu Rev Immunol. 2014;32:433–59.

Holers VM, Thurman JM. The alternative pathway of complement in disease: opportunities for therapeutic targeting. Mol Immunol. 2004;41:147–52.

Holodick NE, Rodriguez-Zhurbenko N, Hernandez AM. Defining natural antibodies. Front Immunol. 2017;8:872.

Hourcade DE. The role of Properdin in the assembly of the alternative pathway C3 convertases of complement. J Biol Chem. 2006;281:2128–32.

Janssen BJC, Huizinga EG, Raaijmakers HCA, Roos A, Daha MR, Nilsson-Ekdahl K, Nilsson B, Gros P. Structures of complement component C3 provide insights into the function and evolution of immunity. Nature. 2005;437:505–11.

Jones JV, James H, Tan MH, Mansour M. Antiphospholipid antibodies require beta 2-glycoprotein I (apolipoprotein H) as cofactor. J Rheumatol. 1992;19:1397–402.

Karmali PP, Simberg D. Interactions of nanoparticles with plasma proteins: implication on clearance and toxicity of drug delivery systems. Expert Opin Drug Deliv. 2011;8:343–57.

Kemper C, Atkinson JP, Hourcade DE. Properdin: emerging roles of a pattern-recognition molecule. Annu Rev Immunol. 2010;28:131–55.

Kirschfink M, Mollnes TE. Modern complement analysis. Clin Diagn Lab Immunol. 2003;10:982–9.

Klapper Y, Hamad OA, Teramura Y, Leneweit G, Nienhaus GU, Ricklin D, Lambris JD, Ekdahl KN, Nilsson B. Mediation of a non-proteolytic activation of complement component C3 by phospholipid vesicles. Biomaterials. 2014;35:3688–96.

Lachmann PJ. Preparing serum for functional complement assays. J Immunol Methods. 2010;352:195–7.

Langford-Smith A, Day AJ, Bishop PN, Clark SJ. Complementing the sugar code: role of gags and sialic acid in complement regulation. Front Immunol. 2015;6:25.

Lutz HU, Nater M, Stammler P. Naturally occurring anti-band 3 antibodies have a unique affinity for C3. Immunology. 1993a;80:191–6.

Lutz HU, Stammler P, Fasler S. Preferential formation of C3b-igg complexes in vitro and in vivo from nascent C3b and naturally occurring anti-band 3 antibodies. J Biol Chem. 1993b;268:17418–26.

Malhotra R, Wormald MR, Rudd PM, Fischer PB, Dwek RA, Sim RB. Glycosylation changes of igg associated with rheumatoid arthritis can activate complement via the mannose-binding protein. Nat Med. 1995;1:237–43.

Meerasa A, Huang JG, Gu FX. CH50: a revisited hemolytic complement consumption assay for evaluation of nanoparticles and blood plasma protein interaction. Curr Drug Deliv. 2011;8:290–8.

Moghimi SM. Cancer nanomedicine and the complement system activation paradigm: anaphylaxis and tumour growth. J Control Release. 2014;190:556–62.

Moghimi SM, Simberg D. Translational gaps in animal models of human infusion reactions to nanomedicines. Nanomedicine (Lond). 2018;13:973–5.

Moghimi SM, Hamad I, Andresen TL, Jorgensen K, Szebeni J. Methylation of the phosphate oxygen moiety of phospholipid-methoxy(polyethylene glycol) conjugate prevents pegylated liposome-mediated complement activation and anaphylatoxin production. FASEB J. 2006;20:2591–3.

Moghimi SM, Hunter AC, Andresen TL. Factors controlling nanoparticle pharmacokinetics: an integrated analysis and perspective. Annu Rev Pharmacol Toxicol. 2012;52:481–503.

Moghimi SM, Simberg D, Skotland T, Yaghmur A, Hunter C. The interplay between blood proteins, complement, and macrophages on nanomedicine performance and responses. J Pharmacol Exp Ther. 2019;370:581.

Monopoli MP, Bombelli FB, Dawson KA. Nanobiotechnology: nanoparticle coronas take shape. Nat Nanotechnol. 2011;6:11–2.

Morgan BP, Harris CL. Complement, a target for therapy in inflammatory and degenerative diseases. Nat Rev Drug Discov. 2015;14:857–77.

Mortimer GM, Butcher NJ, Musumeci AW, Deng ZJ, Martin DJ, Minchin RF. Cryptic epitopes of albumin determine mononuclear phagocyte system clearance of nanomaterials. ACS Nano. 2014;8:3357–66.

Nayak A, Pednekar L, Reid KB, Kishore U. Complement and non-complement activating functions of C1q: a prototypical innate immune molecule. Innate Immun. 2012;18:350–63.

Neun BW, Dobrovolskaia MA. Qualitative analysis of total complement activation by nanoparticles. Methods Mol Biol. 2011;697:237–45.

Neun BW, Barenholz Y, Szebeni J, Dobrovolskaia MA. Understanding the role of anti-PEG antibodies in the complement activation by Doxil in vitro. Molecules. 2018;23:1700.

Nilsson B, Nilsson Ekdahl K. The tick-over theory revisited: is C3 a contact-activated protein? Immunobiology. 2012;217:1106–10.

Pauly D, Nagel BM, Reinders J, Killian T, Wulf M, Ackermann S, Ehrenstein B, Zipfel PF, Skerka C, Weber BH. A novel antibody against human properdin inhibits the alternative complement system and specifically detects properdin from blood samples. PLoS One. 2014;9:e96371.

Pedersen MB, Zhou X, Larsen EK, Sorensen US, Kjems J, Nygaard JV, Nyengaard JR, Meyer RL, Boesen T, Vorup-Jensen T. Curvature of synthetic and natural surfaces is an important target feature in classical pathway complement activation. J Immunol. 2010;184:1931–45.

Petersen GH, Alzghari SK, Chee W, Sankari SS, La-Beck NM. Meta-analysis of clinical and preclinical studies comparing the anticancer efficacy of liposomal versus conventional non-liposomal doxorubicin. J Control Release. 2016;232:255–64.

Reddy ST, van der Vlies AJ, Simeoni E, Angeli V, Randolph GJ, O'Neil CP, Lee LK, Swartz MA, Hubbell JA. Exploiting lymphatic transport and complement activation in nanoparticle vaccines. Nat Biotechnol. 2007;25:1159–64.

Ricklin D. Manipulating the mediator: modulation of the alternative complement pathway C3 convertase in health, disease and therapy. Immunobiology. 2012;217:1057–66.

Ricklin D, Hajishengallis G, Yang K, Lambris JD. Complement: a key system for immune surveillance and homeostasis. Nat Immunol. 2010;11:785–97.

Risitano AM, Notaro R, Pascariello C, Sica M, del Vecchio L, Horvath CJ, Fridkis-Hareli M, Selleri C, Lindorfer MA, Taylor RP, Luzzatto L, Holers VM. The complement receptor 2/factor H fusion protein Tt30 protects paroxysmal nocturnal hemoglobinuria erythrocytes from complement-mediated hemolysis and C3 fragment. Blood. 2012;119:6307–16.

Russell MW, Mansa B. Complement-fixing properties of human Iga antibodies. Alternative pathway complement activation by plastic-bound, but not specific antigen-bound, Iga. Scand J Immunol. 1989;30:175–83.

Salvador-Morales C, Zhang L, Langer R, Farokhzad OC. Immunocompatibility properties of lipid-polymer hybrid nanoparticles with heterogeneous surface functional groups. Biomaterials. 2009;30:2231–40.

Salvati A, Pitek AS, Monopoli MP, Prapainop K, Bombelli FB, Hristov DR, Kelly PM, Aberg C, Mahon E, Dawson KA. Transferrin-functionalized nanoparticles lose their targeting capabilities when a biomolecule corona adsorbs on the surface. Nat Nanotechnol. 2013;8:137–43.

Saptarshi SR, Duschl A, Lopata AL. Interaction of nanoparticles with proteins: relation to bioreactivity of the nanoparticle. J Nanobiotechnol. 2013;11:26.

Schenkein HA, Ruddy S. The role of immunoglobulins in alternative complement pathway activation by zymosan. I. Human igg with specificity for zymosan enhances alternative pathway activation by zymosan. J Immunol. 1981;126:7–10.

Song H, Qiao F, Atkinson C, Holers VM, Tomlinson S. A complement C3 inhibitor specifically targeted to sites of complement activation effectively ameliorates collagen-induced arthritis in DBA/1J mice. J Immunol. 2007;179:7860–7.

Spitzer D, Mitchell LM, Atkinson JP, Hourcade DE. Properdin can initiate complement activation by binding specific target surfaces and providing a platform for de novo convertase assembly. J Immunol. 2007;179:2600–8.

Tavano R, Gabrielli L, Lubian E, Fedeli C, Visentin S, Polverino De Laureto P, Arrigoni G, Geffner-Smith A, Chen F, Simberg D, Morgese G, Benetti EM, Wu L, Moghimi SM, Mancin F, Papini E. C1q-mediated complement activation and C3 opsonization trigger recognition of stealth poly(2-methyl-2-oxazoline)-coated silica nanoparticles by human phagocytes. ACS Nano. 2018;12:5834.

Thomas SN, van der Vlies AJ, O'Neil CP, Reddy ST, Yu SS, Giorgio TD, Swartz MA, Hubbell JA. Engineering complement activation on polypropylene sulfide vaccine nanoparticles. Biomaterials. 2011;32:2194–203.

Vu VP, Gifford GB, Chen F, Benasutti H, Wang G, Groman EV, Scheinman R, Saba L, Moghimi SM, Simberg D. Immunoglobulin deposition on biomolecule corona determines complement opsonization efficiency of preclinical and clinical nanoparticles. Nat Nanotechnol. 2019;14:260–8.

Wang X, Ishida T, Kiwada H. Anti-PEG IgM elicited by injection of liposomes is involved in the enhanced blood clearance of a subsequent dose of pegylated liposomes. J Control Release. 2007;119:236–44.

Wang G, Chen F, Banda NK, Holers VM, Wu L, Moghimi SM, Simberg D. Activation of human complement system by dextran-coated iron oxide nanoparticles is not affected by dextran/Fe ratio, hydroxyl modifications, and crosslinking. Front Immunol. 2016;7:418.

Wang G, Griffin JI, Inturi S, Brenneman B, Banda NK, Holers VM, Moghimi SM, Simberg D. In vitro and in vivo differences in murine third complement component (C3) opsonization and macrophage/leukocyte responses to antibody-functionalized iron oxide nanoworms. Front Immunol. 2017;8:151.

Wibroe PP, Mat Azmi ID, Nilsson C, Yaghmur A, Moghimi SM. Citrem modulates internal nanostructure of glyceryl monooleate dispersions and bypasses complement activation: towards development of safe tunable intravenous lipid nanocarriers. Nanomed Nanotechnol Biol Med. 2015;11:1909–14.

Wolf-Grosse S, Rokstad AM, Ali S, Lambris JD, Mollnes TE, Nilsen AM, Stenvik J. Iron oxide nanoparticles induce cytokine secretion in a complement-dependent manner in a human whole blood model. Int J Nanomedicine. 2017;12:3927–40.

Wu YQ, Qu H, Sfyroera G, Tzekou A, Kay BK, Nilsson B, Nilsson Ekdahl K, Ricklin D, Lambris JD. Protection of nonself surfaces from complement attack by factor H-binding peptides: implications for therapeutic medicine. J Immunol. 2011;186:4269–77.

Zipfel PF, Skerka C. Complement regulators and inhibitory proteins. Nat Rev Immunol. 2009;9:729–40.

Chapter 7
Translocation, Biodistribution, and Fate of Nanomaterials in the Body

Melisa Bunderson-Schelvan, Andrij Holian, Kevin L. Trout, and Raymond F. Hamilton

Abstract The potential health effects associated with the widespread use of engineered nanomaterials are broadly concerning to developers, researchers, and policymakers alike. While extensive studies have been undertaken to understand the immediate consequences of exposure to these materials, the long-term effects associated with potential translocation to secondary and even tertiary organs are less well understood. Further, mechanisms underlying the toxic properties of nanomaterials after translocating to distal organs, if any, are largely unknown at this time. Here, we describe the current state of knowledge regarding translocation of commonly studied and primarily nonmedical engineered nanomaterials following exposure and discuss potential mechanisms and consequences to health outcomes. Current evidence suggests that nanomaterials are redistributed, at least in part, to secondary organs throughout the body. While the properties associated with any given nanomaterial may influence its redistribution, the most likely explanation for adverse effects following translocation involves increased inflammation leading to pathology. Whether more persistent nanomaterials will eventually be cleared from the body after exposure remains an important question. Further, information regarding the effects of systemic trafficking on the basic characteristics of nanomaterials would help predict whether nanomaterials will accumulate in specific organs and/or cause pathologies in secondary or even tertiary locations.

Keywords Nanomaterial biodistribution · Inhalation · NLRP3 inflammasome · Particle trafficking · Secondary organ exposures

M. Bunderson-Schelvan (✉) · A. Holian (✉) · K. L. Trout · R. F. Hamilton
Department of Biomedical and Pharmaceutical Sciences, Center for Environmental Health Sciences, University of Montana, Missoula, MT, USA
e-mail: Melisa.Schelvan@umontana.edu; andrij.holian@umontana.edu; kevin.trout@umontana.edu; ray.hamilton@umontana.edu

J. C. Bonner, J. M. Brown (eds.), *Interaction of Nanomaterials with the Immune System*, Molecular and Integrative Toxicology,
https://doi.org/10.1007/978-3-030-33962-3_7

7.1 Introduction

Since engineered nanomaterials are increasingly used in a wide range of industries, numerous studies have been conducted to improve our understanding of the occupational exposure risk associated with their manufacturing, handling, or end product use. Current nanotechnologies primarily target the nanoelectronics, biology, and medical industries (Niska et al. 2018), resulting in the production of numerous metal-based and other nanoparticles, including those containing Ag, Al_2O, Fe_2O_3, Fe_3O_4, SiO_2, TiO_2, and ZnO, with the global production of these products reaching several tons per year (Schmid and Riediker 2008). In addition, metal oxide nanomaterials are increasingly being used in consumer goods such as sunscreens and ointments. As such, nanomaterials within these products have become a significant source of potential human exposures in the environment (Subramaniam et al. 2018). While development of these materials is rapid and distribution is becoming more widespread, there is a common concern among researchers and policy-makers that our current understanding of the potential health effects associated with exposure to these materials is insufficient. Progress has been made in elucidating the factors that contribute to nanomaterial pathologies; however, gaps remain in our knowledge of how the nanoparticles are distributed in biological systems, as well as their ultimate fate in the body and potential pathological outcomes. Further, certain nanomaterials are known to translocate to secondary organs such as the liver, central nervous system, heart, kidneys, and other tissues (Kermanizadeh et al. 2015).

The use of engineered nanomaterials in the medical industry in particular has been the focus of a great deal of research, as the possibilities for novel drug-delivery systems or even new therapeutics increase. As such, the safety of engineered nanomaterials in the medical industry has been studied more extensively than the safety of these materials after accidental or occupational exposures. Medically directed exposures to engineered nanomaterials likely occur via dermal (injection) or ingestion (drugs) routes, whereas occupational and environmental exposures occur primarily via inhalation; further, occupational and environmental exposures have the added risk of co-exposures and/or long-term exposures that are undetected or unmonitored. While there are many similarities between exposure to engineered nanomaterials while under the care of a physician and occupational/environmental exposures, in addition to other types of unintentional or unrecognized exposures, there are clear distinctions that must be recognized and considered in the context of risk assessment. A primary distinction is likely the route of exposure, as the likelihood of secondary exposures following translocation to other system organs is higher for inhalation exposures than for dermal or ingestion, for example. Furthermore, limited and expensive analysis methods have made it difficult to quantify environmental exposures to the public. Therefore, modeling is primarily used to quantify environmental concentrations of nanomaterials, with a high degree of uncertainty (Kuhlbusch et al. 2018). Here, we discuss the consequences of environmental and occupational, but not medical, exposures to engineered nanomaterials, as well as evidence for biodistribution and the ultimate fate of these particles following initial exposures. The purpose of this chapter is to provide an update from recent literature (Kermanizadeh et al. 2015) and identify future research needs.

7.2 Pulmonary Exposures

Environmental and occupational exposure to nanomaterials is most likely to occur via the pulmonary system, with differing potentials for translocation to secondary/tertiary organs being dependent on nanomaterial attributes.

7.2.1 Types of Pulmonary Exposures

In the literature, pulmonary exposures are generally classified as inhalation, intratracheal, intranasal, or pharyngeal aspiration. Translocation is dependent, in part, on solubility and the method of exposure, potentially increasing with increasing solubility of nanoparticles (Shinohara et al. 2017). Inhalation exposure has been reported to result in less translocation than intratracheal, intranasal, and pharyngeal aspiration exposures (Kermanizadeh et al. 2015). To date, most nanomaterials are poorly soluble, suggesting that they do not rapidly translocate to secondary tissues in a manner similar to that of soluble drugs. Rather, translocation of insoluble nanomaterials is likely slower, with mechanisms less clear. Nonetheless, the potential for nanomaterial translocation increases as solubility increases (Kermanizadeh et al. 2015).

7.2.1.1 Inhalation Exposures

Translocation to Secondary/Tertiary Organs

Studies examining the movement of nanomaterials to secondary organs are relatively rare and have focused primarily on a small number of nanomaterial types, including iridium-based nanomaterials; single- and multi-walled carbon nanotubes (MWCNT); gold-based nanomaterials; titanium dioxide and zinc oxide nanomaterials; as well as uranium-, silver-, cerium-based nanomaterials and a few others (see Table 7.1).

Iridium-based nanomaterials and MWCNT have been shown to have potential for translocating to secondary organs. Iridium nanomaterials translocated to the skin, heart, liver, kidneys, and brain after a 24-h exposure period (Geiser et al. 2014), with smaller particles translocating to a greater extent than large particles. The mechanisms underlying the translocation of iridium are somewhat size-dependent, as fewer 80-nm iridium nanomaterials accumulated in the liver, spleen, kidneys, heart, brain, and blood than 20-nm materials (Kreyling et al. 2009).

Table 7.1 Summary of experimental evidence for secondary organ exposure with nanomaterials

Nanomaterial	Route	Model	Secondary organ involvement observed	Exposure time	Reference
Ir	Inh	Ms	Skin, hrt, liv, kid, brain	24 h to 7 d	Geiser et al. (2014)
MWCNT	Inh	Ms	Diaphragm, chest wall, liv, kid, hrt, brain	1 d to 1 y	Mercer et al. (2013)
Au	Inh	Hu	Blood, urine	1 d to 3 m	Miller et al. (2017)
Au	Inh	Rt	Esophagus, tongue, kid, hrt, aorta, brain	15 d	Yu et al. (2007)
Au	Inh	Rt	Kid	13 w	Sung et al. (2011)
Au	Inh	Rt	Esophagus, kid, aorta, brain	15 d	Balasubramanian et al. (2013)
Au	Inh	Rt	Spleen, liv, brain, testis, blood	1, 3, & 24 d	Han et al. (2015)
TiO_2	Inh	Rt	Mediastinal lymph nodes (LN)	5 d	van Ravenzwaay et al. (2009)
TiO_2	Inh	Rt	Blood, LN, liv, kid, spleen	14 d	Pujalte et al. (2017)
TiO_2	Inh	Rt	Liv, & spleen only	28 d	Gate et al. (2017)
ZnO	Inh	Ms	None detected	2 w	Adamcakova-Dodd et al. (2014)
Ur	Inh	Rt	Blood, kid, bones, olfactory bulb (OB)	10 m to 24 h	Petitot et al. (2013)
CeO_2	Inh	Rt	Skin, brain, spleen, liv, kid, testis	28 d	Geraets et al. (2012)
CeO_2	Inh	Ms	Kid, liv, hrt, brain	7, 14, & 28 d	Aalapati et al. (2014)
Ag	Inh	Rt	Liv	13 w	Sung et al. (2009)
Ag	Inh	Rt	None detected	4 d	Braakhuis et al. (2014)
Ag	Inh	Rt	None detected	28 d	Ji et al. (2007)
Ag	Pa	Ms	Hrt, thymus, spleen	1 d	Hussain et al. (2013)
C_{14} MWCNT	PA	Ms	Spleen, bone marrow (BM)	1 y	Czarny et al. (2014)
Gd_2O_3	PA	Ms	Liv, kid, hrt, spleen	24 h	Abid et al. (2013)
Au	IT	Ms	Liv	1 h to 3 w	Sadauskas et al. (2009)
Au	IT	Rt	Blood, liv, kid, skin	24 h	Semmler-Behnke et al. (2008)
Au	IT	Rt	Liv, spleen, kid, hrt, brain, uterus, blood, urine	24 h	Kreyling et al. (2014)
CeO_2	IT	Rt	Liv, spleen, kid, hrt, brain, testis, stomach	28 d	He et al. (2010)
CeO_2	IT	Rt	Liv only	28 d	Nalabotu et al. (2011)
Polystyrene	IT	Rt	Blood, liv	24 h	Chen et al. (2006)
Diamond	IT	Ms	Liv, hrt, bone, spleen	1 h to 48 h	Zhang et al. (2010)
TiO2	IT	Ms	Liv, hrt	24 h	Husain et al. (2015)

(continued)

Table 7.1 (continued)

Nanomaterial	Route	Model	Secondary organ involvement observed	Exposure time	Reference
Printex 90	IT	Ms	Offspring affected	7–18 d	Jackson et al. (2012)
TiO_2	IN	Ms	Brain	30 d	Wang et al. (2008)
Polystyrene	IN	Ms	LN	24 h	Blank et al. (2013)
Ag	IN	Ms	Spleen, nasal cavity, renal medulla	1 and 7 d	Genter et al. (2012)
Cu	IN	Ms	Brain	21 d	Bai et al. (2014)
Cu	IN	Ms	Liv, kid, OB	1 w	Liu et al. (2009)
Fe_2O_3	IN	Ms	OB	4 h to 30 d	Wang et al. (2016)
MWCNT	IN	Review	LN, spleen, liv, BM	Up to 1 y	Jacobsen et al. (2017)
TiO_2	Oral	Rt	Liv, spleen, lung	5 d	Geraets et al. (2014)
TiO_2	Oral	Rt	None detected	96 h	MacNicoll et al. (2015)
TiO_2	Oral	Rt	Thyroid, ovary, spleen	5 d	Tassinari et al. (2014)
TiO_2	Oral	Rt	None detected	24 h	Cho et al. (2013)
ZnO	Oral	Rt	Liv, kid, spleen, brain	24 h	Cho et al. (2013)
CeO_2	Oral	Rt	Liv, kid, brain, spleen, hrt, blood	28 d	Kumari et al. (2014a)
CeO_2	Oral	Rt	Liv, kid, brain, spleen, hrt, blood	28 d	Kumari et al. (2014b)
CeO_2	Oral	Rt	GI tract only	24 h	Park et al. (2018)
ZnO	Oral	Ms	Liv, serum, kid	7 w	Wang et al. (2017)
TiO_2	Dermal	Hu	None detected	24 h	Crosera et al. (2015)
Ag	IV	Rt	Testis, liv, lung, spleen, brain, hrt, kid	5 d	Lankveld et al. (2010)
Ag	IV	Rt	Liv, lung, spleen, kid, brain, urine	1, 7, 14 & 28 d	Dziendzikowska et al. (2012)
Au	IV	Rt	Lung, liv, spleen, kid, testis	1 d, 1w, 1 m, & 2 m	Balasubramanian et al. (2010)
CeO_2	IV	Rt	All tissues tested, liv, spleen mainly	24 h	Park et al. (2018)
TiO2	IP	Ms	Liv, spleen, kid	10 d	Silva et al. (2017)

Abbreviations: *Route* (*Inh* inhalation, *PA* pharyngeal aspiration, *IT* intratracheal, *IN* intranasal, *IV* intravenous, and *IP* intraperitoneal). *Model* (*Hu* human, *Ms* mouse, and *Rt* rat)

However, size-independent retention of iridium nanoparticles has been measured and varies widely among animals; iridium was homogenously distributed in the liver and kidney (Buckley et al. 2017).

MWCNT have been detected in the diaphragm, chest wall, liver, kidneys, heart, and brain as early as 1 day after exposure, and by 336 days after exposure, the content of these materials in extra-pulmonary organs significantly increased (Mercer et al. 2013). Further, approximately 7% of carbon nanotubes deposited within the lung translocate to near pulmonary regions, while approximately 1% make their way to distal organs (Jacobsen et al. 2017).

In addition to iridium-based nanomaterials and MWCNT, gold nanomaterials have been shown to translocate to secondary organs. In humans, gold was detected in the blood and urine within 1 day after exposure and remained detectable after 3 months (Miller et al. 2017). A 15-day exposure to gold nanomaterials resulted in accumulation in the esophagus, tongue, kidney, aorta, spleen, heart, and blood (Yu et al. 2007). Further, the concentration of gold in the kidneys of rats exposed to gold nanomaterials increased in a dose-dependent manner when exposed for 13 weeks; however, no dose-dependent increase was found in the blood, liver, or olfactory bulb (Sung et al. 2011). In another study, translocation of gold nanomaterials to numerous secondary organs in rats was measured in small quantities, including brain, aorta, kidneys, and esophagus (Balasubramanian et al. 2013). Rats exposed to small gold nanomaterials had gold in the spleen on days, 1, 3, and 28 after exposure, while gold was measured in the liver, brain, testis, and blood on day 1 but not on days 3 and 28 after exposure; exposure to larger nanomaterials did not result in translocation to secondary organs (Han et al. 2015).

The apparent translocation of titanium dioxide and zinc oxide nanomaterials to secondary organs is less dramatic than other types of nanomaterials. A 5-day exposure to high concentrations of titanium dioxide in rats resulted in detectable levels in the lung and mediastinal lymph nodes, but not in the liver, kidney, spleen, or basal brain with olfactory bulb (van Ravenzwaay et al. 2009). An inhalation study of titanium dioxide nanomaterials in rats suggested translocation to the systemic circulation, with particles detected in the blood, lymph nodes, liver, kidney, and spleen (Pujalte et al. 2017). However, another study did not detect titanium dioxide particles in the blood, kidneys, or brain after day 28 post-exposure despite showing detectable levels in the spleen and liver, which were higher in elderly rats than in healthy young adult rats (Gate et al. 2017). ZnO nanomaterials, which are highly soluble, did not result in any significant increases in Zn in the blood, spleen, and kidneys, heart, or brain after a 2-week exposure (Adamcakova-Dodd et al. 2014).

Exposure to uranium nanomaterials for 10 min, 4 h, and 24 h resulted in significant concentrations in the blood within 10 min post-exposure, with the highest levels observed in the kidneys, bones, and olfactory bulb (Petitot et al. 2013). In contrast, cerium dioxide nanomaterials translocated to secondary organs with an order of dose ranking as follows: epididymis > brain > spleen > liver > kidney > testis (Geraets et al. 2012). Further, Ce accumulated in a time-dependent manner in the kidney, liver, heart, and brain of mice exposed to Ce-nanomaterials for 28 days; these concentrations were significantly reduced by 28 days post-exposure (Aalapati et al. 2014).

Silver nanomaterials have been reported to dissolve in the acidic environment of the phagolysosome (Zhu et al. 2017) and have been shown to primarily translocate to the lungs and liver in rats, with little toxicity associated with translocation (Sung et al. 2009). A 4-day inhalation exposure study using two differently sized silver nanomaterials demonstrated minimal risk of translocation to secondary organs (spleen, kidneys, testis, and the brain) (Braakhuis et al. 2014). Similarly, no significant adverse health effects were observed for silver nanomaterials in rats despite silver-induced effects in distal organs (Ji et al. 2007).

Silver nanoparticles (20 nm) were rapidly transported to the olfactory bulb in rats exposed via inhalation; however, in rats exposed to larger (110 nm) particles, a gradual increase in the translocation of silver to the olfactory bulb was noted (Patchin et al. 2016).

7.2.1.2 Pharyngeal Aspiration

Several studies have been conducted to examine the tissue distribution of nanomaterials following a bolus administration of sample via pharyngeal aspiration (PA). While these studies are valuable from a research perspective, one must be careful not to misinterpret the results, as bolus-type exposures do not represent actual exposure scenarios. Therefore, the observed physiological outcomes, including patterns of redistribution, may not be comparable to real-life inhalation exposures if the exposure is of a smaller amount or occurs over a protracted period of time.

Translocation to Secondary/Tertiary Organs

Less research has been conducted on the translocation of nanomaterials following exposure via pharyngeal aspiration than for other types of pulmonary exposure. Studies examining gold nanomaterials, C^{14}-radiolabeled multi-walled carbon nanotubes, and lanthanide-labelled gadolinium (iii) oxide nanomaterials have been conducted.

Mice administered gold nanomaterials showed high gold levels in the lung following pharyngeal exposure, which was lower when the mice were pretreated with LPS. Further, gold levels were higher in the heart and thymus of non-LPS-primed animals than the spleens of LPS-primed animals where the gold was higher. The authors suggest that the particle distribution may be dependent on the health of the animals (Hussain et al. 2013). After 1 year of administering C_{14}-radiolabeled multi-walled carbon nanotubes, it was determined that they accumulate in the spleen and bone marrow (Czarny et al. 2014). Most lanthanide-labelled gadolinium (III) oxide nanomaterials were retained in the lungs, with some translocating to the liver, kidney, heart, and spleen (Abid et al. 2013).

7.2.1.3 Intratracheal Instillation

Translocation to Secondary/Tertiary Organs

In mice, 1.4% of 2 nm gold nanomaterials intratracheally instilled translocated to the liver; this did not occur for 40 and 100 nm nanomaterials (Sadauskas et al. 2009). In another study, as much as 25% of gold nanomaterials (1.4 and 18 nm in size) was removed via mucociliary clearance and excreted in the feces, with significant quantities of 1.4 nm particles found in the blood, kidneys, liver, skin, and carcass (bones, soft tissue and fat with organs removed) (Semmler-Behnke et al. 2008). Smaller gold nanomaterials have been shown to cross the air–blood barrier to a greater extent than larger sized particles, thereby translocating to the liver, spleen, kidneys, heart, brain, uterus, carcass, blood, and urine by 24 h post-exposure (Kreyling et al. 2014). A recent study ranked the order of persistent retention of gold nanoparticles in secondary organs and tissues as liver > spleen > kidneys > skeleton > blood > uterus > heart > brain (Kreyling et al. 2018). Further, translocation of negatively charged nanomaterials was greater than that of positively charged nanomaterials (Kreyling et al. 2014).

Exposure to ceria nanoparticles resulted in translocation to the systemic circulation, thereby accumulating in extra-pulmonary organs despite most (63.9%) of the deposited material remaining in the lung 28 days after administration (He et al. 2010). Rats administered cerium nanomaterials and then euthanized 28 days later showed significant levels of cerium in the liver upon evaluation (Nalabotu et al. 2011). In contrast, radio-labeled, polystyrene nanomaterials were eliminated from the lungs within 24 h of exposure; further, in the presence of LPS, translocation was increased, with an increase in the recovered radioactivity found in secondary organs (blood, liver) (Chen et al. 2006). Irradiated diamond nanomaterials are reportedly transported to the spleen, bone, liver, and heart but not to the brain in mice (Zhang et al. 2010).

Titanium dioxide nanomaterials were shown to translocate to the heart and liver following intratracheal instillation in mice 24 h post-exposure (Husain et al. 2015). Translocation through the placenta following intratracheal instillation was examined after exposure to Printex 90 in mice, with limited direct translocation observed; however, secondary effects in the offspring were noted (Jackson et al. 2012).

7.2.1.4 Intranasal Installation

Translocation to Secondary/Tertiary Organs

Titanium dioxide nanomaterials administered to mice accumulated in most parts of the brain after 30 days of treatment (Wang et al. 2008). Mice exposed to polystyrene nanomaterials of different sizes had high quantities of particles in the lymph nodes after 24 h, with the majority of the materials engulfed by macrophages, independent of size (Blank et al. 2013). Silver nanomaterials administered to mice showed aggregated accumulation of silver in the spleen, nasal cavity, and renal medulla (Genter et al. 2012). Mice exposed to high- (40 mg/kg/bw) and medium-sized doses (10 mg/kg/bw) of copper nanomaterials presented with significantly higher levels of copper in the brain than animals exposed to low (1 mg/kg/bw) doses of the nanomaterials (Bai et al. 2014). Administration of nano- and submicron-sized Fe_2O_3 nanoparticles resulted in translocation of the nanomaterials to the olfactory bulb in a size- and time-dependent manner (Wang et al. 2016). It has been shown that pristine MWCNT remain in the lung for extended periods of time (months or years) after being deposited in the lungs. These particles may be cleared via the mucociliary escalator, by which they are transferred to the GI tract, with minimal uptake. However, "a significant fraction of MWCNT translocate from the alveolar space to the near pulmonary region including lymph nodes, subpleura and pleura (<7% of the pulmonary deposited dose) and to distal organs including liver, spleen and bone marrow (~1%)." The primary site of deposition for MWCNT is likely the liver, where they may persist for at least a year following exposure (Jacobsen et al. 2017).

7.2.2 *Mechanisms of Particle Translocation After Pulmonary Exposure*

Overall, mechanisms underlying the translocation of engineered nanomaterials are not completely understood. However, the primary interaction in the lung between the distal deposition of nanomaterials and lung cells occurs with the alveolar macrophage (AM) internalizing particles in the alveolar space (Bunderson-Schelvan et al. 2016). If the particles do not make it to the distal regions of the lung, they become trapped in the mucosa regions of the upper airways and trafficked up through the mucociliary escalator to relative clearance from the body within 24 h (Sturm 2014). However, when carbon nanotubes deposit in the lung, they can alternatively be cleared by AM and travel to the gastrointestinal tract via the mucociliary escalator and excreted (Jacobsen et al. 2017). Interestingly, intravenous injections of carbon nanotubes are cleared relatively quickly, while the major target organs remain the same; an exception is particular single-walled carbon nanotubes (SWCNT) that are functionalized and accumulate in the lung tissue (Jacobsen et al. 2017). The mucociliary clearance can also remove the particles out of the body completely by sneezing or coughing (Beule 2010). In addition, mucociliary clearance of particles can result in particle deposition in the paranasal sinus areas with an unknown fate.

In addition to the mucociliary escalator, one potential pathway out of the lung is for AM to internalize nanomaterials and move through the draining lymph nodes (brachial, mediastinal, and mesenteric lymph nodes) (Kirby et al. 2009), eventually ending up in the spleen white pulp area (see Figs. 7.1 and 7.2). It is unknown if the particles in the spleen are trapped or if there is a way for the foreign material to be eliminated from the organ. On the other hand, large aggregated nanomaterials (e.g., SWCNT) in the lung, which are difficult for AM to internalize, are most likely remaining in the lung for a long period of time, creating a greater potential for disease progression (Sturm 2015).

Exceptions to the AM interaction with nanomaterials in the lung would include soluble particles such as ZnO, CuO, Ag, and/or extremely small, disaggregated particles <10 nm such as Au. In these instances, it is possible that the nanomaterials get into the circulatory system where they can be dispersed throughout the body (Kermanizadeh et al. 2015).

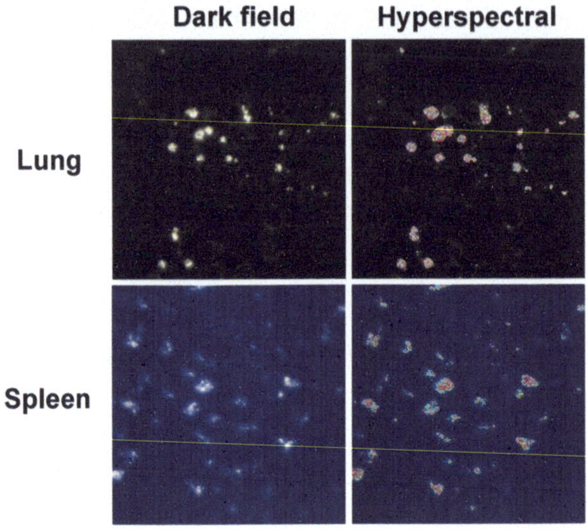

Fig. 7.1 Translocation of $La_2Ti_2O_7$ nanomaterial from lung to spleen following a 7-day exposure in Balb/c mice following oropharyngeal instillation. Dark field and hyperspectral (CytoViva) imaging confirm particle retention and shifts in spectral content (color change) indicate changes to the nanomaterial during translocation

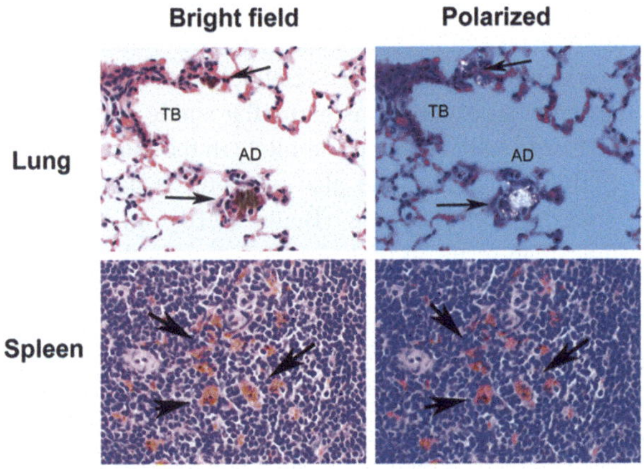

Fig. 7.2 Translocation of $La_2Ti_2O_7$ nanomaterial from lung to spleen following a 7-day exposure in Balb/c mice by PA. Bright field and polarized light imaging confirm particle retention (arrows upper panels) and translocation by macrophage cells (arrows lower panels pointing to red cells). TB terminal bronchi, AD alveolar duct

7.3 Oral Exposures

Various nanomaterials are used to prolong the shelf life of commercial products within the food industry. Other dietary exposures may occur via food packaging, coloring, flavoring, and nutrient enhancement (Smolkova et al. 2015). Consequently, food safety issues include exposure during food preparation, as well as after and during digestion (Szakal et al. 2014). Passage through the gastrointestinal tract may alter the physicochemical properties of certain nanoparticles (Yada et al. 2014). In addition, there is growing evidence suggesting that both ingested and inhaled engineered nanomaterials impact the gastrointestinal microbiota (Pietroiusti et al. 2017).

7.3.1 Translocation to Secondary/Tertiary Organs

Translocation of nanomaterials following ingestion is similar to that observed for inhalation exposures, with soluble nanomaterials translocating to a greater extent than less soluble particles (Kermanizadeh et al. 2015). Repeated oral administration of titanium dioxide nanomaterials showed very low to undetectable amounts of the materials in secondary target organs after 5 days of exposure (Geraets et al. 2014). Rats exposed to titanium dioxide nanomaterials did not have significantly more accumulation in the liver, kidneys, spleen, heart, or brain than water-treated rats (MacNicoll et al. 2015). Further, in rats exposed to titanium dioxide nanomaterials for 5 days, the highest levels of titanium were measured in the thyroid, ovaries, and spleen (Tassinari et al. 2014).

Translocation of highly soluble zinc oxide nanomaterials is greater than that of titanium dioxide nanomaterials when orally ingested; titanium dioxide nanomaterials did not appear to translocate to secondary organs, whereas zinc was measured in the liver, kidney, spleen, and brain (Cho et al. 2013).

Significant amounts of cerium were detected in the liver, kidneys, brain, spleen, heart, and whole blood in rats repeatedly exposed to cerium oxide nanomaterials (Kumari et al. 2014a). Cerium dioxide nanomaterials were maximally distributed to the liver, kidneys, spleen, and heart of rats in a size- and dose-dependent manner (Kumari et al. 2014b). Oral exposure to cerium oxide nanoparticles in rats showed low levels of cerium in the blood and secondary tissues, with most cerium detected in the feces; these data suggest that cerium oxide nanoparticles are minimally absorbed in the gastrointestinal tract (Park et al. 2018).

Translocation of zinc oxide and gold nanomaterials has also been evaluated following oral exposures. Gastrointestinal administration of nano-ZnO particles for 7 weeks resulted in increased concentrations of zinc in the liver, serum, and kidneys (Wang et al. 2017). The rate of absorption for gold nanomaterials in intestinal epithelial cells increased as the mean particle diameter decreased; however, the mean accumulation did not change (Yao et al. 2015).

7.4 Dermal Exposures

Metal and metal oxide nanoparticles are commonly used in cosmetic products. The dermis is deceptively complex, effectively modulating the absorption of the various nanoparticle types, although particles less than 600 Da passively cross the dermal barrier (Niska et al. 2018). There is some disagreement in the literature regarding the extent of penetration through the stratum corneum of the epidermis (Niska et al. 2018).

7.4.1 Translocation to Secondary/Tertiary Organs

In general, uptake of nanomaterials from dermal exposures is thought to be low (Kermanizadeh et al. 2015). For example, titanium was found in the epidermal layer after 24 h of exposure to titanium dioxide nanoparticles, but was undetected in the dermal layer (Crosera et al. 2015).

7.4.2 Mechanisms of Particle Translocation

Surface charge of the nanomaterials can affect the efficiency by which they are taken up and subsequently transported to secondary organs following a dermal exposure. For example, gold nanomaterials with a positive surface charge penetrate the skin more efficiently than negative or neutrally charged particles, though all three types penetrate the epidermis. As such, penetration of gold nanomaterials alters the stratum corneum. Nanomaterials primarily penetrate the epidermis via endocytosis and exocytosis processes (Hao et al. 2017).

7.5 Other Exposure Route Studies

Several translocation studies have exposed animals to various nanomaterials via intravenous (IV) or intraperitoneal (IP) injections. Intravenous administration studies of silver, gold, and cerium oxide nanoparticles have been conducted. The distribution of silver in rats administered silver nanoparticles intravenously for 5 consecutive days was observed in all organs evaluated, including the liver, lungs, spleen, brain, heart, kidneys, and testes (Lankveld et al. 2010). Further, silver nanoparticles administered intravenously translocated from the blood to main organs in a dose-dependent manner, with the highest concentration measured in the liver after 21 days (Dziendzikowska et al. 2012). A single intravenous injection of gold nanoparticles in rats resulted in accumulation of gold in the liver and spleen from day 1 and in the kidneys and testis from 1 month post-injection. Gold accumulated in the lungs 1 day after the injection, but rapidly decreased; no gold accumulation was measured in the brain.

Changes in gene expression involving detoxification, lipid metabolism, cell cycle, defense response, and circadian rhythm were noted (Balasubramanian et al. 2010). Cerium oxide nanoparticles administered to rats via a single intravenous injection resulted in high concentrations of cerium in all tissues, with the liver and spleen being the primary target organs (Park et al. 2018).

Titanium dioxide is present in numerous products that can be used daily. Mice administered titanium dioxide via intraperitoneal injections showed an increase mononuclear inflammatory infiltrates in the visceral abdominal fat. Titanium dioxide nanoparticles significantly accumulated in the liver, spleen, and kidneys following intraperitoneal administration (Silva et al. 2017). Overall, exposure to titanium dioxide nanoparticles appears to be relatively innocuous.

7.6 Toxicity

7.6.1 Pulmonary Exposures

There have been extensive studies on the outcomes and mechanisms in response to pulmonary exposures to a wide variety of nanomaterials that have been published and are covered in other chapters. A few more recent studies with metal and metal oxide nanoparticles are summarized below including impacts on secondary tissues.

Toxic responses following pulmonary exposures to copper nanoparticles have been reported in rodents. High doses (40 mg/kg body weight instilled 3× per week) of nasally inhaled copper nanoparticles resulted in pathological changes in mice; however, a low dose (1 mg/kg body weight) resulted in no obvious pathological damage. The copper particles primarily accumulated in the liver, kidneys, and olfactory bulbs (Liu et al. 2009). However, following a 5-day exposure to various concentrations and exposure times of copper oxide nanoparticles, dose-dependent lung inflammation was observed starting at a concentration of 2.4 mg/m^3; limited inflammation was still observed 22 days later (Gosens et al. 2016). Further, evidence in mice suggests that inhalation of copper nanoparticles may contribute to the development of lung fibrosis (Lai et al. 2018).

Aluminum oxide nanoparticles are widely used nanomaterials, attributable to their unique semi-conductor properties. Repeated inhalation exposures to aluminum oxide nanoparticles in rats resulted in a dose-dependent increase in aluminum in the lung that was associated with alveolar macrophage accumulation during both the exposure period and recovery; an increase in total cell and neutrophil counts, lactate dehydrogenase, tumor necrosis factor alpha, and interleukin 6 was also observed (Kim et al. 2018). Further, in a study examining sex-dependent differences in response to aluminum oxide nanoparticle exposure, only female mice exhibited depressive-like behavior, with differential changes in gene expression relating to neurotransmitter and voltage-gated ion channels (Zhang et al. 2015). Exposure to aluminum oxide nanoparticles for 7 days was shown to cause emphysema and small airway remodeling in mice; this was accompanied by enhanced inflammation and apoptosis (Li et al. 2017).

Zinc oxide nanoparticles reportedly interfere with the autophagy pathway. A zinc oxide nanoparticle-induced model of acute lung injury suggested that exposure to these nanomaterials causes accumulation of autophagic vacuoles and elevated microtubule-associated proteins 1A/1B light chain 3B-11 and p62, leading to inflammation and oxidative stress; these processes were inhibited by 2-methyladenine, which is a classical autophagy inhibitor (Jiang et al. 2018). As such, nanomaterials comprised of zinc oxide may result in toxic effects associated with inflammation, oxidative stress, and autophagy.

A reduction in antioxidant potential has been noted following exposure to magnesium oxide nanoparticles. Specifically, these nanoparticles caused intracellular depletion of glutathione, without a concentration-dependent increase in reactive oxygen species in human lung epithelial cells (Akhtar et al. 2018). Therefore, exposure to magnesium oxide nanomaterials may contribute to oxidative stress-related pathologies.

Inflammation is likely a large contributor to the toxic response following exposure to silicon dioxide nanoparticles, an effect that is compounded in patients with existing conditions. In particular, silicon dioxide nanoparticles intranasally instilled in mice resulted in an increase in inflammatory cell counts and proinflammatory cytokines in the bronchoalveolar lavage fluid; this effect was compounded in the presence of lipopolysaccharide, suggesting that exposure to silicon dioxide nanoparticles may exacerbate existing pulmonary disease (Ko et al. 2018). As such, silicon dioxide-induced inflammation is a likely contributor to pathologies observed following exposure to these particles (Han et al. 2016).

Despite evidence of translocation to secondary organs of cerium oxide nanoparticles, intratracheal administration of cerium oxide to rats did not appear to affect the spleen, kidney, or heart, despite upregulation of various factors and enzymes, including fibroblast growth factor-basic, haptoglobin, IgA, IL-11, matrix metalloproteinase 9, myoglobin, serum amyloid protein, serum glutamic oxaloacetic acid, and transaminase thrombopoietin (Nalabotu et al. 2011). More recently, cerium oxide nanoparticles were shown to increase the formation of 8-hydroxydeoxyguanosine (8-OHdG) in rat lungs after both intratracheal instillation and inhalation exposures (Li et al. 2018). Further, tubular degeneration, characterized by cytoplasmic vacuoles and tubular cell hypertrophy and tubular cell necrosis, was observed in the kidneys of male mice exposed to cerium oxide nanomaterials via inhalation; these changes correlated to increased blood urea nitrogen and creatinine, suggesting reduced renal function (Aalapati et al. 2014). Additional research should be conducted to more fully understand the toxicity associated with exposure to cerium oxide nanoparticles.

7.6.2 Oral Exposure

Oral exposure to nanoparticles may disrupt nutrient absorption via changes in the microvilli and/or transporter genes; further, it is likely that foodstuffs directly and indirectly affect the absorption of nanoparticles. As such, the toxicity of orally ingested nanoparticles likely differs significantly from other exposure types (Cao et al. 2016).

End-organ damage and alterations in homeostasis, but not genotoxic effects, have been observed following oral administration of certain nanomaterials. Evidence of slight liver damage was observed in rats orally administered silver nanoparticles for 13 weeks; silver was dose-dependently detected in all tissues examined (Kim et al. 2010). Highly soluble zinc oxide nanoparticles administered orally are more toxic than less soluble forms and may interfere with homeostasis of zinc ions in the body (Wang et al. 2017). Oral administration of copper oxide nanoparticles to rats significantly altered antioxidant enzyme activity, with a decrease in glutathione, catalase, and superoxide dismutase activity and a concomitant increase in products of lipid peroxidation, suggesting that acute toxicity in the liver might be attributable to oxidative stress (Anreddy 2018). Accumulation of gold nanomaterials in intestinal epithelial cells decreases mitochondrial membrane potential (Yao et al. 2015). Finally, genotoxic effects at the DNA, gene, or chromosome level were not observed in rats orally administered low doses of CeO_2 or a high dose of $BaSO_4$ for 3–6 months (Cordelli et al. 2017).

7.6.3 Dermal Exposure

Limited evidence has been reported demonstrating serious pathologies following dermal exposure of various nanomaterials. In particular, Ag, TiO_2, and CeO_2 nanoparticles did not cause irritation in a human skin equivalent model derived from keratinocytes (Miyani and Hughes 2017). Repetitive dermal exposure to ZnO-NPs enhanced cyto- but not genotoxicity, with intracellular nanoparticles accumulating up to 6 weeks (Ickrath et al. 2017).

7.6.4 Other Exposures

Injection of silver nanoparticles (0.25 mg/kg) in Balb/C mice resulted in pathological changes to the liver, spleen, brain, heart, lungs, kidneys, and testicles, as well as changes in sperm quality and blood–brain barrier permeability (Moradi-Sardareh et al. 2018). Further, exposure to silver nanomaterials via inhalation resulted in microbial activation (Patchin et al. 2016). In patients at risk of stroke, gold particles were detected in carotid artery specimens taken during surgery, suggesting a mechanism(s) between environmental nanoparticle exposure and cardiovascular disease (Miller et al. 2017).

Hepatic and genotoxic effects have been observed following exposure to certain nanomaterials. Specifically, MWCNT caused microfoci of necrosis, enlarged single hepatocytes, increased Kupffer cells, and binucleate hepatocytes in mice 6 weeks after exposure; these microscopic changes persisted 4 months after exposure (Cao et al. 2014). Changes in gene expression involving detoxification, lipid metabolism, cell cycle, defense response, and circadian rhythm were noted after a single intravenous injection of gold nanoparticles (Balasubramanian et al. 2010).

7.7 Potential Pathology of Secondary and Tertiary Nanomaterial Exposures

Due to the lack of research in the area, very little is known about the pathological potential of secondary nanomaterial exposures. We can presume, however, that the known effects of these particle exposures in primary organs could apply to secondary/tertiary exposures. For example, the primary effect of nanomaterial exposure is inflammation, regardless of the primary organ exposed (Armstead et al. 2015; Baisch et al. 2014; Hamilton Jr et al. 2013, 2018; Shvedova et al. 2008). The primary driver of this inflammatory response appears to be activation of the NLRP3 inflammasome and autophagy in the AM (Stern et al. 2012), resulting from lysosomal dysfunction and release of catalytic enzymes into the cytoplasm of AM (Bunderson-Schelvan et al. 2016).

Following the development of sustained inflammation, fibrosis is the next most frequent pathology to be seen by nanomaterial researchers looking at in vivo exposure models (Mercer et al. 2011; Sayan and Mossman 2016; Rahman et al. 2017). One of the few studies of secondary nanomaterial pathology showed that inhaled gold nanoparticles can collect at sites of sustained inflammation outside of the lung at sites of vascular disease (Miller et al. 2017). Still another pathology, autoimmune disease is sometimes associated with nanomaterial exposures (Kononenko et al. 2015). Unfortunately, this is the most under-researched aspect of nanomaterial effects on biological models.

The link from nanomaterial exposure to cancer has always been tenuous, but there is some evidence that certain sustained nanomaterial exposures could be involved in cancer promotion (Nakayama 2018). Coincidently, some modified SWCNT, and MWCNT have been tested as a vehicle for anticancer measures (Xiao et al. 2009; Wu et al. 2009). The nanomaterial cancer relationship is obviously complex. The nanomaterial most linked to cancer have been the carbon moieties due to a fiber structure resembling asbestos (Donaldson et al. 2006). However, it stands to reason that known cancer causers such as SiO_2 and As in nanoparticle format will be just as dangerous to the biologic system. Some basic elements of cancer promotion are affiliated with nanomaterial exposures, such as carcinogenesis (Rahman et al. 2017), inflammasome activation (Sayan and Mossman 2016), lysosomal dysfunction (Kirkegaard and Jaattela 2009), inflammation (Valavanidis et al. 2013), autophagy (Appelqvist et al. 2013), and cell death (Kirkegaard and Jaattela 2009). Therefore, more research is need to sort out these complex relationships and if secondary nanomaterial exposures pose any cancer risks.

7.7.1 Changes in Nanomaterial Physical Characteristics as a Result of Translocation in a Biological System

Another under-researched area concerning the translocation of nanomaterials is how the interaction of particle and biological system changes the nanomaterial over time. The most basic understanding of this interaction is in regard to the soluble nanoparticle, where the nanomaterial changes phases from solid to a dissolved form (e.g., ionized) upon contact with some biological medium. Examples of this are Ag, ZnO, CuO, and other metal oxides. This could be the most complex nanomaterial transformation because the metal ions are free to move about the entire body reacting and interacting with various tissues.

Another possibility is much less understood. This involves the change of nanomaterials due to exposure to catalytic enzymes and acidic pH, which occur in the phagocytic cells internalizing the particles. The nanomaterial properties will vary from type to type, with some nanomaterials being more or less susceptible to degradation or change. For example, gold nanomaterials are practically indestructible in a biological system (Almeida et al. 2011). However, a particle like $LaTi_2O_7$ can be drastically altered over a period of 7 days in a PA mouse model (Figs. 7.1 and 7.2). This particle was changed when it was transported out of the lung to the spleen over time. Figure 7.1 shows the dark field hyperspectral imaging of these tissues with a CytoViva analysis done to identify unique spectra occurring with the particle. Both the dark field image and the CytoViva image show particle deposition in the lung (primary organ) and spleen (secondary organ). The color differences between the lung and spleen particles (CytoViva) indicate that a surface molecular change occurred as the particle relocated to the spleen. This observation is reinforced by evidence in Fig. 7.2, where the polarized light is no longer effective at identifying the particle within the macrophage cells (unpublished preliminary data). Fortunately, the tools for identifying particles in tissues such as CytoViva hyperspectral imaging and Raman imaging are now available to further the understanding of how nanomaterials change and redistribute following exposure.

Lastly, bioaccumulation of nanomaterials over time can change the particle characteristics. A 90-day exposure study of montmorillonite (MMT) with observation of the lung at 7, 28, and 90 days post-exposure revealed the MMT became more 'tissue-like' over time, based on a CytoViva analysis (unpublished proprietary data). In this case, the particle was not eliminated, but more like assimilated over a period of time. The process involved in this particle transformation is not understood at this time, but this could apply to many internalized nanomaterials. Again, more research is needed in this area.

7.8 Discussion

Studies have suggested that workplace exposures to certain nanomaterials may exceed the highest suggested occupational exposure limit (OEL) (Mihalache et al. 2017). Further, nanomaterial risk assessment has been somewhat confused by the plethora of possible exposures, including intentional or directed exposures, such as those that occur during the use of medical diagnostic procedures, occupational exposures, or nondirected exposures (e.g., cosmetics) (Warheit 2018). Studies assessing exposure via dermal/ingestion, as well as inhalation exposure studies for downstream and end-of-life uses are limited. However, in occupational settings, all three routes (dermal/ingestion/inhalation) can be relevant when determining exposure. In the case of suspensions/liquids, dermal exposure is likely even if inhalation exposure is unlikely. Increased exposure may result from "manual activities such as cleaning and maintenance, collection/harvesting, spraying, and finishing, as well as those involving feeding into a process and handling of powders outside enclosures" (Basinas et al. 2018). Here, we reviewed the potential for translocation of various nanomaterials to secondary organs following inhalation, oral, and dermal exposures.

Environmental exposure to nanomaterials differs from medically directed exposures in numerous ways. Research has shown that nanoparticle aging can have a dramatic impact on uptake, and, therefore, toxicity (Valsami-Jones and Lynch 2015). Further, there is significant variation regarding environmental persistence of the different nanomaterials, with soluble nanomaterials such as silver, copper, and certain other metal oxides (e.g., zinc) being the least persistent. The current understanding of environmental exposures is complicated by a lack of information regarding the nanomaterial form. For example, exposure to original or pristine nanomaterials likely differs from exposure to nanomaterials that have been incorporated into products and then released into the environment where they become aged and/or transformed (Mitrano et al. 2015). Environmental matrices such as water systems, agglomeration of particles in ambient air, or soil type make it difficult to identify the nanomaterials involved in an environmental exposure, and there is an important gap in knowledge regarding the environmental fate of nanomaterials, including environmental concentrations and dosimetry. Nanomaterial release studies do not generally do an adequate job of describing the nanomaterial, and few data exist regarding the use of nanomaterials in the various products available to consumers (Valsami-Jones and Lynch 2015). These environmentally related variables make it difficult to accurately assess the overall impact of environmental exposures to the various nanomaterials.

Discrepancies in the effects of nanomaterials are widely reported, likely attributable to the plethora of forms and synthetic methods used to generate these products. While the intrinsic properties of nanoparticles have been relatively well characterized for many types, different mediums make the characterization of extrinsic properties more difficult and less exact (Gao and Lowry 2018). Nano-specific effects such as bioavailability, toxicity, and overall fate have been observed; however, these effects are not always consistent for the nanomaterial type, form, and manufacturing process. As such, it is important to understand the life cycle of nanomaterials in order to fully characterize risk following environmental exposures.

Elimination of nanomaterials within a biological system is also likely dependent on individual nanomaterial properties, as well as type. Evidence suggests that nanoparticles (soluble and/or extremely small nanomaterials) taken up by the respiratory tract may translocate to the bloodstream, resulting in dispersion to secondary organs (Wu and Tang 2018). In particular, accumulation in the kidneys is extremely likely for very small nanoparticles and may result in adverse effects. This is an area that needs more research. Further, as many nanoparticles are contaminated with metals, exposure may interfere with the homeostasis of trace elements, as well as neurotransmitters and enzymes that are sensitive to metals in the brain (Wu and Tang 2018). Current evidence suggests that exposure to synthesized nanomaterials results in less toxicity than exposure to the equivalent dissolved materials (e.g., metals), but more than the equivalent bulk materials (Lead et al. 2018; Mihalache et al. 2017).

Several studies have reported the distribution of various nanomaterials to secondary organs, with corresponding adverse effects; however, a complete picture of the overall consequences to human health remains lacking. Issues regarding the impact of environmental factors on nanomaterial form, dose, and uptake into biological systems following environmental exposures remain to be fully elucidated.

7.9 Conclusion

A brief overview of the studies presented here supports the notion that nanomaterial exposures, regardless of route of exposure, type, size and solubility, result in a redistribution of the particles to some degree in secondary organs throughout the body. Some generalizations can be made, however. The inhalation of nanomaterials is the most likely way to distribute particles throughout the body (IV exposures are not considered as a significant source of environmental exposure), when contrasted with oral and dermal exposures. The dermal exposure is the most limiting of nanomaterial redistribution following exposures. Of the nanomaterials tested, the more soluble and smaller the particle is, the more likely it is to distribute throughout the body regardless of exposure route. These particles are the most likely to enter the blood circulation, and that will result in the distribution of nanomaterial throughout the entire body.

For inhaled insoluble nanomaterials, the alveolar macrophage (AM) is the most likely agent of transport (see Fig. 7.3). When the particles are phagocytosed by AM, they can go to one of several locations. The mucociliary escalator can transport the AM with particle cargo out of the body entirely (sneeze or cough), into the GI tract, or possibly even olfactory bulb via sinus cavity and eventually the brain. Another possibility is that AM can traffic with the particles to interstitial regions of the lung through epithelial gaps, entrapping the particulates in the lung for long periods of time. Yet another possibility is transporting the nanomaterial to the lymph nodes and on to the spleen through the lymphatic system. Accumulation of these particles in the spleen is an open question that needs to be researched.

Lastly, it is known that most of these nanomaterials cause inflammation in various biological systems. It can be postulated that chronic inflammation eventually leads to pathology. The big questions that still need to be addressed by research are (1) does the nanomaterial of interest eventually clear out of the body after exposure; (2) does the nanomaterial of interest cause pathologies in the primary, secondary, and/or tertiary exposure locations; (3) do the nanomaterials accumulate in specific organs; and (4) does trafficking throughout the body modify the nanomaterials into something fundamentally different, something more tissue-like, or something that

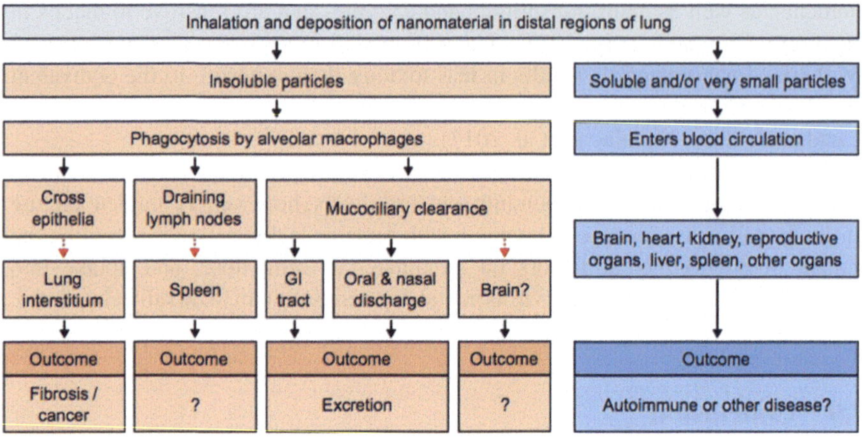

Fig. 7.3 Proposed biodistribution of nanomaterials in the distal regions of the lung. Inhalation is a nanomaterial exposure route of major concern. Insoluble particles deposited in the distal lung are phagocytosed by alveolar macrophages, which cross epithelia into the lung interstitium, translocate to the spleen via draining lymph nodes, or are excreted by mucociliary clearance. Mucociliary mechanisms may also be a route to the brain, potentially via paranasal sinuses and the olfactory bulb. Nanomaterials that are soluble or small enough to enter the blood circulation (e.g., diameter <10 nm approximately) have high biodistribution to multiple organ systems. Pathological outcomes vary with particle translocation, but a common underlying factor is inflammation. Red dotted arrows indicate hypothesized particle translocations that require further research

needs to be walled-up like SiO_2 or asbestos? Clearly, more research needs to be done, because exposure to various nanomaterials is going to increase over time as they are developed and utilized in a number of common applications.

Acknowledgments This work was partially supported by NIH Research Grant: R01 ES023209.

References

Aalapati S, Ganapathy S, Manapuram S, Anumolu G, Prakya BM. Toxicity and bio-accumulation of inhaled cerium oxide nanoparticles in CD1 mice. Nanotoxicology. 2014;8(7):786–98. https://doi.org/10.3109/17435390.2013.829877.

Abid AD, Anderson DS, Das GK, Van Winkle LS, Kennedy IM. Novel lanthanide-labeled metal oxide nanoparticles improve the measurement of in vivo clearance and translocation. Part Fibre Toxicol. 2013;10:1. https://doi.org/10.1186/1743-8977-10-1.

Adamcakova-Dodd A, Stebounova LV, Kim JS, Vorrink SU, Ault AP, O'Shaughnessy PT, Grassian VH, Thorne PS. Toxicity assessment of zinc oxide nanoparticles using sub-acute and sub-chronic murine inhalation models. Part Fibre Toxicol. 2014;1:15. https://doi.org/10.1186/1743-8977-11-15.

Akhtar MJ, Ahamed M, Alhadlaq HA, Alrokayan SA. MgO nanoparticles cytotoxicity caused primarily by GSH depletion in human lung epithelial cells. J Trace Elem Med Biol. 2018;50:283–90. https://doi.org/10.1016/j.jtemb.2018.07.016.

Almeida JP, Chen AL, Foster A, Drezek R. In vivo biodistribution of nanoparticles. Nanomedicine (Lond). 2011;6(5):815–35. https://doi.org/10.2217/nnm.11.79.

Anreddy RNR. Copper oxide nanoparticles induces oxidative stress and liver toxicity in rats following oral exposure. Toxicol Rep. 2018;5:903–4. https://doi.org/10.1016/j.toxrep.2018.08.022.
Appelqvist H, Waster P, Kagedal K, Ollinger K. The lysosome: from waste bag to potential therapeutic target. J Mol Cell Biol. 2013;5(4):214–26. https://doi.org/10.1093/jmcb/mjt022.
Armstead AL, Minarchick VC, Porter DW, Nurkiewicz TR, Li B. Acute inflammatory responses of nanoparticles in an intra-tracheal instillation rat model. PLoS One. 2015;10(3):e0118778. https://doi.org/10.1371/journal.pone.0118778.
Bai R, Zhang L, Liu Y, Li B, Wang L, Wang P, Autrup H, Beer C, Chen C. Integrated analytical techniques with high sensitivity for studying brain translocation and potential impairment induced by intranasally instilled copper nanoparticles. Toxicol Lett. 2014;226(1):70–80. https://doi.org/10.1016/j.toxlet.2014.01.041.
Baisch BL, Corson NM, Wade-Mercer P, Gelein R, Kennell AJ, Oberdorster G, Elder A. Equivalent titanium dioxide nanoparticle deposition by intratracheal instillation and whole body inhalation: the effect of dose rate on acute respiratory tract inflammation. Part Fibre Toxicol. 2014;11:5. https://doi.org/10.1186/1743-8977-11-5.
Balasubramanian SK, Jittiwat J, Manikandan J, Ong CN, Yu LE, Ong WY. Biodistribution of gold nanoparticles and gene expression changes in the liver and spleen after intravenous administration in rats. Biomaterials. 2010;31(8):2034–42. https://doi.org/10.1016/j.biomaterials.2009.11.079.
Balasubramanian SK, Poh KW, Ong CN, Kreyling WG, Ong WY, Yu LE. The effect of primary particle size on biodistribution of inhaled gold nano-agglomerates. Biomaterials. 2013;34(22):5439–52. https://doi.org/10.1016/j.biomaterials.2013.03.080.
Basinas I, Jimenez AS, Galea KS, Tongeren MV, Hurley F. A systematic review of the routes and forms of exposure to engineered nanomaterials. Ann Work Expo Health. 2018;62(6):639–62. https://doi.org/10.1093/annweh/wxy048.
Beule AG. Physiology and pathophysiology of respiratory mucosa of the nose and the paranasal sinuses. GMS Curr Top Otorhinolaryngol Head Neck Surg. 2010;9:Doc07. https://doi.org/10.3205/cto000071.
Blank F, Stumbles PA, Seydoux E, Holt PG, Fink A, Rothen-Rutishauser B, Strickland DH, von Garnier C. Size-dependent uptake of particles by pulmonary antigen-presenting cell populations and trafficking to regional lymph nodes. Am J Respir Cell Mol Biol. 2013;49(1):67–77. https://doi.org/10.1165/rcmb.2012-0387OC.
Braakhuis HM, Gosens I, Krystek P, Boere JA, Cassee FR, Fokkens PH, Post JA, van Loveren H, Park MV. Particle size dependent deposition and pulmonary inflammation after short-term inhalation of silver nanoparticles. Part Fibre Toxicol. 2014;11:49. https://doi.org/10.1186/s12989-014-0049-1.
Buckley A, Warren J, Hodgson A, Marczylo T, Ignatyev K, Guo C, Smith R. Slow lung clearance and limited translocation of four sizes of inhaled iridium nanoparticles. Part Fibre Toxicol. 2017;14(1):5. https://doi.org/10.1186/s12989-017-0185-5.
Bunderson-Schelvan M, Hamilton R, Trout K, Jessop F, Gulumian M, Holian A. Approaching a unified theory for particle-induced inflammation. In: Otsuki T, Yoshioka Y, Holian A, editors. Biological effects of fibrous and particulate substances. Current topics in environmental health and preventive medicine. Tokyo: Springer Japan; 2016. p. 51–76. https://doi.org/10.1007/978-4-431-55732-6_3.
Cao Y, Jacobsen NR, Danielsen PH, Lenz AG, Stoeger T, Loft S, Wallin H, Roursgaard M, Mikkelsen L, Moller P. Vascular effects of multiwalled carbon nanotubes in dyslipidemic ApoE−/− mice and cultured endothelial cells. Toxicol Sci. 2014;138(1):104–16. https://doi.org/10.1093/toxsci/kft328.
Cao Y, Li J, Liu F, Li X, Jiang Q, Cheng S, Gu Y. Consideration of interaction between nanoparticles and food components for the safety assessment of nanoparticles following oral exposure: a review. Environ Toxicol Pharmacol. 2016;46:206–10. https://doi.org/10.1016/j.etap.2016.07.023.
Chen J, Tan M, Nemmar A, Song W, Dong M, Zhang G, Li Y. Quantification of extrapulmonary translocation of intratracheal-instilled particles in vivo in rats: effect of lipopolysaccharide. Toxicology. 2006;222(3):195–201. https://doi.org/10.1016/j.tox.2006.02.016.

Cho WS, Duffin R, Bradley M, Megson IL, MacNee W, Lee JK, Jeong J, Donaldson K. Predictive value of in vitro assays depends on the mechanism of toxicity of metal oxide nanoparticles. Part Fibre Toxicol. 2013;10(1):55. https://doi.org/10.1186/1743-8977-10-55.

Cordelli E, Keller J, Eleuteri P, Villani P, Ma-Hock L, Schulz M, Landsiedel R, Pacchierotti F. No genotoxicity in rat blood cells upon 3-or 6-month inhalation exposure to CeO2 or BaSO4 nanomaterials. Mutagenesis. 2017;32(1):13–22. https://doi.org/10.1093/mutage/gew005.

Crosera M, Prodi A, Mauro M, Pelin M, Florio C, Bellomo F, Adami G, Apostoli P, De Palma G, Bovenzi M, Campanini M, Filon FL. Titanium dioxide nanoparticle penetration into the skin and effects on HaCaT cells. Int J Environ Res Public Health. 2015;12(8):9282–97. https://doi.org/10.3390/ijerph120809282.

Czarny B, Georgin D, Berthon F, Plastow G, Pinault M, Patriarche G, Thuleau A, L'Hermite MM, Taran F, Dive V. Carbon nanotube translocation to distant organs after pulmonary exposure: insights from in situ (14)C-radiolabeling and tissue radioimaging. ACS Nano. 2014;8(6):5715–24. https://doi.org/10.1021/nn500475u.

Donaldson K, Aitken R, Tran L, Stone V, Duffin R, Forrest G, Alexander A. Carbon nanotubes: a review of their properties in relation to pulmonary toxicology and workplace safety. Toxicol Sci. 2006;92(1):5–22. https://doi.org/10.1093/toxsci/kfj130.

Dziendzikowska K, Gromadzka-Ostrowska J, Lankoff A, Oczkowski M, Krawczynska A, Chwastowska J, Sadowska-Bratek M, Chajduk E, Wojewodzka M, Dusinska M, Kruszewski M. Time-dependent biodistribution and excretion of silver nanoparticles in male Wistar rats. J Appl Toxicol. 2012;32(11):920–8. https://doi.org/10.1002/jat.2758.

Gao XY, Lowry GV. Progress towards standardized and validated characterizations for measuring physicochemical properties of manufactured nanomaterials relevant to nano health and safety risks. Nano. 2018;9:14–30. https://doi.org/10.1016/j.impact.2017.09.002.

Gate L, Disdier C, Cosnier F, Gagnaire F, Devoy J, Saba W, Brun E, Chalansonnet M, Mabondzo A. Biopersistence and translocation to extrapulmonary organs of titanium dioxide nanoparticles after subacute inhalation exposure to aerosol in adult and elderly rats. Toxicol Lett. 2017;265:61–9. https://doi.org/10.1016/j.toxlet.2016.11.009.

Geiser M, Stoeger T, Casaulta M, Chen S, Semmler-Behnke M, Bolle I, Takenaka S, Kreyling WG, Schulz H. Biokinetics of nanoparticles and susceptibility to particulate exposure in a murine model of cystic fibrosis. Part Fibre Toxicol. 2014;11:19. https://doi.org/10.1186/1743-8977-11-19.

Genter MB, Newman NC, Shertzer HG, Ali SF, Bolon B. Distribution and systemic effects of intranasally administered 25 nm silver nanoparticles in adult mice. Toxicol Pathol. 2012;40(7):1004–13. https://doi.org/10.1177/0192623312444470.

Geraets L, Oomen AG, Schroeter JD, Coleman VA, Cassee FR. Tissue distribution of inhaled micro- and nano-sized cerium oxide particles in rats: results from a 28-day exposure study. Toxicol Sci. 2012;127(2):463–73. https://doi.org/10.1093/toxsci/kfs113.

Geraets L, Oomen AG, Krystek P, Jacobsen NR, Wallin H, Laurentie M, Verharen HW, Brandon EF, de Jong WH. Tissue distribution and elimination after oral and intravenous administration of different titanium dioxide nanoparticles in rats. Part Fibre Toxicol. 2014;11:30. https://doi.org/10.1186/1743-8977-11-30.

Gosens I, Cassee FR, Zanella M, Manodori L, Brunelli A, Costa AL, Bokkers BG, de Jong WH, Brown D, Hristozov D, Stone V. Organ burden and pulmonary toxicity of nano-sized copper (II) oxide particles after short-term inhalation exposure. Nanotoxicology. 2016;10(8):1084–95. https://doi.org/10.3109/17435390.2016.1172678.

Hamilton RF Jr, Wu Z, Mitra S, Shaw PK, Holian A. Effect of MWCNT size, carboxylation, and purification on in vitro and in vivo toxicity, inflammation and lung pathology. Part Fibre Toxicol. 2013;10(1):57. https://doi.org/10.1186/1743-8977-10-57.

Hamilton RF Jr, Tsuruoka S, Wu N, Wolfarth M, Porter DW, Bunderson-Schelvan M, Holian A. Length, but not reactive edges, of cup-stack MWCNT is responsible for toxicity and acute lung inflammation. Toxicol Pathol. 2018;46:62–74. https://doi.org/10.1177/0192623317732303.

Han SG, Lee JS, Ahn K, Kim YS, Kim JK, Lee JH, Shin JH, Jeon KS, Cho WS, Song NW, Gulumian M, Shin BS, Yu IJ. Size-dependent clearance of gold nanoparticles from lungs of

Sprague-Dawley rats after short-term inhalation exposure. Arch Toxicol. 2015;89(7):1083–94. https://doi.org/10.1007/s00204-014-1292-9.

Han H, Park YH, Park HJ, Lee K, Um K, Park JW, Lee JH. Toxic and adjuvant effects of silica nanoparticles on ovalbumin-induced allergic airway inflammation in mice. Respir Res. 2016;17(1):60. https://doi.org/10.1186/s12931-016-0376-x.

Hao F, Jin X, Liu QS, Zhou Q, Jiang G. Epidermal penetration of gold nanoparticles and its underlying mechanism based on human reconstructed 3D Episkin model. ACS Appl Mater Interfaces. 2017;9(49):42577–88. https://doi.org/10.1021/acsami.7b13700.

He X, Zhang H, Ma Y, Bai W, Zhang Z, Lu K, Ding Y, Zhao Y, Chai Z. Lung deposition and extrapulmonary translocation of nano-ceria after intratracheal instillation. Nanotechnology. 2010;21(28):285103. https://doi.org/10.1088/0957-4484/21/28/285103.

Husain M, Wu D, Saber AT, Decan N, Jacobsen NR, Williams A, Yauk CL, Wallin H, Vogel U, Halappanavar S. Intratracheally instilled titanium dioxide nanoparticles translocate to heart and liver and activate complement cascade in the heart of C57BL/6 mice. Nanotoxicology. 2015;9(8):1013–22. https://doi.org/10.3109/17435390.2014.996192.

Hussain S, Vanoirbeek JA, Haenen S, Haufroid V, Boland S, Marano F, Nemery B, Hoet PH. Prior lung inflammation impacts on body distribution of gold nanoparticles. Biomed Res Int. 2013;2013:923475. https://doi.org/10.1155/2013/923475.

Ickrath P, Wagner M, Scherzad A, Gehrke T, Burghartz M, Hagen R, Radeloff K, Kleinsasser N, Hackenberg S. Time-dependent toxic and genotoxic effects of zinc oxide nanoparticles after long-term and repetitive exposure to human mesenchymal stem cells. Int J Environ Res Public Health. 2017;14(12):1590. https://doi.org/10.3390/ijerph14121590.

Jackson P, Hougaard KS, Vogel U, Wu D, Casavant L, Williams A, Wade M, Yauk CL, Wallin H, Halappanavar S. Exposure of pregnant mice to carbon black by intratracheal instillation: toxicogenomic effects in dams and offspring. Mutat Res. 2012;745(1–2):73–83. https://doi.org/10.1016/j.mrgentox.2011.09.018.

Jacobsen NR, Moller P, Clausen PA, Saber AT, Micheletti C, Jensen KA, Wallin H, Vogel U. Biodistribution of carbon nanotubes in animal models. Basic Clin Pharmacol Toxicol. 2017;121(Suppl 3):30–43. https://doi.org/10.1111/bcpt.12705.

Ji JH, Jung JH, Kim SS, Yoon JU, Park JD, Choi BS, Chung YH, Kwon IH, Jeong J, Han BS, Shin JH, Sung JH, Song KS, Yu IJ. Twenty-eight-day inhalation toxicity study of silver nanoparticles in Sprague-Dawley rats. Inhal Toxicol. 2007;19(10):857–71. https://doi.org/10.1080/08958370701432108.

Jiang X, Tang Q, Zhang J, Wang H, Bai L, Meng P, Qin X, Xu G, Bose DD, Wang B, Chen C, Zou Z. Autophagy-dependent release of zinc ions is critical for acute lung injury triggered by zinc oxide nanoparticles. Nanotoxicology. 2018:1–24. https://doi.org/10.1080/17435390.2018.1513094.

Kermanizadeh A, Balharry D, Wallin H, Loft S, Moller P. Nanomaterial translocation--the biokinetics, tissue accumulation, toxicity and fate of materials in secondary organs--a review. Crit Rev Toxicol. 2015;45(10):837–72. https://doi.org/10.3109/10408444.2015.1058747.

Kim YS, Song MY, Park JD, Song KS, Ryu HR, Chung YH, Chang HK, Lee JH, Oh KH, Kelman BJ, Hwang IK, Yu IJ. Subchronic oral toxicity of silver nanoparticles. Part Fibre Toxicol. 2010;7:20. https://doi.org/10.1186/1743-8977-7-20.

Kim YS, Chung YH, Seo DS, Choi HS, Lim CH. Twenty-eight-day repeated inhalation toxicity study of aluminum oxide nanoparticles in male Sprague-Dawley rats. Toxicol Res. 2018;34(4):343–54. https://doi.org/10.5487/TR.2018.34.3.343.

Kirby AC, Coles MC, Kaye PM. Alveolar macrophages transport pathogens to lung draining lymph nodes. J Immunol. 2009;183(3):1983–9. https://doi.org/10.4049/jimmunol.0901089.

Kirkegaard T, Jaattela M. Lysosomal involvement in cell death and cancer. Biochim Biophys Acta. 2009;1793(4):746–54. https://doi.org/10.1016/j.bbamcr.2008.09.008.

Ko JW, Lee HJ, Shin NR, Seo YS, Kim SH, Shin IS, Kim JS. Silicon dioxide nanoparticles enhance endotoxin-induced lung injury in mice. Molecules. 2018;23(9):2247. https://doi.org/10.3390/molecules23092247.

Kononenko V, Narat M, Drobne D. Nanoparticle interaction with the immune system. Arh Hig Rada Toksikol. 2015;66(2):97–108. https://doi.org/10.1515/aiht-2015-66-2582.

Kreyling WG, Semmler-Behnke M, Seitz J, Scymczak W, Wenk A, Mayer P, Takenaka S, Oberdorster G. Size dependence of the translocation of inhaled iridium and carbon nanoparticle aggregates from the lung of rats to the blood and secondary target organs. Inhal Toxicol. 2009;21(Suppl 1):55–60. https://doi.org/10.1080/08958370902942517.

Kreyling WG, Hirn S, Moller W, Schleh C, Wenk A, Celik G, Lipka J, Schaffler M, Haberl N, Johnston BD, Sperling R, Schmid G, Simon U, Parak WJ, Semmler-Behnke M. Air-blood barrier translocation of tracheally instilled gold nanoparticles inversely depends on particle size. ACS Nano. 2014;8(1):222–33. https://doi.org/10.1021/nn403256v.

Kreyling WG, Moller W, Holzwarth U, Hirn S, Wenk A, Schleh C, Schaffler M, Haberl N, Gibson N, Schittny JC. Age-dependent rat lung deposition patterns of inhaled 20 nanometer gold nanoparticles and their quantitative biokinetics in adult rats. ACS Nano. 2018;12(8):7771–90. https://doi.org/10.1021/acsnano.8b01826.

Kuhlbusch TAJ, Wijnhoven SWP, Haase A. Nanomaterial exposures for worker, consumer and the general public. Nano. 2018;10:11–25. https://doi.org/10.1016/j.impact.2017.11.003.

Kumari M, Kumari SI, Grover P. Genotoxicity analysis of cerium oxide micro and nanoparticles in Wistar rats after 28 days of repeated oral administration. Mutagenesis. 2014a;29(6):467–79. https://doi.org/10.1093/mutage/geu038.

Kumari M, Kumari SI, Kamal SS, Grover P. Genotoxicity assessment of cerium oxide nanoparticles in female Wistar rats after acute oral exposure. Mutat Res Genet Toxicol Environ Mutagen. 2014b;775-776:7–19. https://doi.org/10.1016/j.mrgentox.2014.09.009.

Lai X, Zhao H, Zhang Y, Guo K, Xu Y, Chen S, Zhang J. Intranasal delivery of copper oxide nanoparticles induces pulmonary toxicity and fibrosis in C57BL/6 mice. Sci Rep. 2018;8(1):4499. https://doi.org/10.1038/s41598-018-22556-7.

Lankveld DP, Oomen AG, Krystek P, Neigh A, Troost-de Jong A, Noorlander CW, Van Eijkeren JC, Geertsma RE, De Jong WH. The kinetics of the tissue distribution of silver nanoparticles of different sizes. Biomaterials. 2010;31(32):8350–61. https://doi.org/10.1016/j.biomaterials.2010.07.045.

Lead JR, Batley GE, Alvarez PJJ, Croteau MN, Handy RD, McLaughlin MJ, Judy JD, Schirmer K. Nanomaterials in the environment: behavior, fate, bioavailability, and effects-an updated review. Environ Toxicol Chem. 2018;37(8):2029–63. https://doi.org/10.1002/etc.4147.

Li X, Yang H, Wu S, Meng Q, Sun H, Lu R, Cui J, Zheng Y, Chen W, Zhang R, Aschner M, Chen R. Suppression of PTPN6 exacerbates aluminum oxide nanoparticle-induced COPD-like lesions in mice through activation of STAT pathway. Part Fibre Toxicol. 2017;14(1):53. https://doi.org/10.1186/s12989-017-0234-0.

Li YS, Ootsuyama Y, Kawasaki Y, Morimoto Y, Higashi T, Kawai K. Oxidative DNA damage in the rat lung induced by intratracheal instillation and inhalation of nanoparticles. J Clin Biochem Nutr. 2018;62(3):238–41. https://doi.org/10.3164/jcbn.17-70.

Liu Y, Gao Y, Zhang L, Wang T, Wang J, Jiao F, Li W, Liu Y, Li Y, Li B, Chai Z, Wu G, Chen C. Potential health impact on mice after nasal instillation of nano-sized copper particles and their translocation in mice. J Nanosci Nanotechnol. 2009;9(11):6335–43.

MacNicoll A, Kelly M, Aksoy H, Kramer E, Bouwmeester H, Chaudhry Q. A study of the uptake and biodistribution of nano-titanium dioxide using in vitro and in vivo models of oral intake. J Nanopart Res. 2015;17(2):66. https://doi.org/10.1007/s11051-015-2862-3.

Mercer RR, Hubbs AF, Scabilloni JF, Wang L, Battelli LA, Friend S, Castranova V, Porter DW. Pulmonary fibrotic response to aspiration of multi-walled carbon nanotubes. Part Fibre Toxicol. 2011;8:21. https://doi.org/10.1186/1743-8977-8-21.

Mercer RR, Scabilloni JF, Hubbs AF, Wang L, Battelli LA, McKinney W, Castranova V, Porter DW. Extrapulmonary transport of MWCNT following inhalation exposure. Part Fibre Toxicol. 2013;10:38. https://doi.org/10.1186/1743-8977-10-38.

Mihalache R, Verbeek J, Graczyk H, Murashov V, van Broekhuizen P. Occupational exposure limits for manufactured nanomaterials, a systematic review. Nanotoxicology. 2017;11(1):7–19. https://doi.org/10.1080/17435390.2016.1262920.

Miller MR, Raftis JB, Langrish JP, McLean SG, Samutrtai P, Connell SP, Wilson S, Vesey AT, Fokkens PHB, Boere AJF, Krystek P, Campbell CJ, Hadoke PWF, Donaldson K, Cassee FR,

Newby DE, Duffin R, Mills NL. Inhaled nanoparticles accumulate at sites of vascular disease. ACS Nano. 2017;11(5):4542–52. https://doi.org/10.1021/acsnano.6b08551.
Mitrano DM, Motellier S, Clavaguera S, Nowack B. Review of nanomaterial aging and transformations through the life cycle of nano-enhanced products. Environ Int. 2015;77:132–47. https://doi.org/10.1016/j.envint.2015.01.013.
Miyani VA, Hughes MF. Assessment of the in vitro dermal irritation potential of cerium, silver, and titanium nanoparticles in a human skin equivalent model. Cutan Ocul Toxicol. 2017;36(2):145–51. https://doi.org/10.1080/15569527.2016.1211671.
Moradi-Sardareh H, Basir HRG, Hassan ZM, Davoudi M, Amidi F, Paknejad M. Toxicity of silver nanoparticles on different tissues of Balb/C mice. Life Sci. 2018;211:81–90. https://doi.org/10.1016/j.lfs.2018.09.001.
Nakayama M. Macrophage recognition of crystals and nanoparticles. Front Immunol. 2018;9:103. https://doi.org/10.3389/fimmu.2018.00103.
Nalabotu SK, Kolli MB, Triest WE, Ma JY, Manne ND, Katta A, Addagarla HS, Rice KM, Blough ER. Intratracheal instillation of cerium oxide nanoparticles induces hepatic toxicity in male Sprague-Dawley rats. Int J Nanomedicine. 2011;6:2327–35. https://doi.org/10.2147/IJN.S25119.
Niska K, Zielinska E, Radomski MW, Inkielewicz-Stepniak I. Metal nanoparticles in dermatology and cosmetology: interactions with human skin cells. Chem Biol Interact. 2018;295:38–51. https://doi.org/10.1016/j.cbi.2017.06.018.
Park K, Park J, Lee H, Choi J, Yu WJ, Lee J. Toxicity and tissue distribution of cerium oxide nanoparticles in rats by two different routes: single intravenous injection and single oral administration. Arch Pharm Res. 2018;41(11):1108–16. https://doi.org/10.1007/s12272-018-1074-7.
Patchin ES, Anderson DS, Silva RM, Uyeminami DL, Scott GM, Guo T, Van Winkle LS, Pinkerton KE. Size-dependent deposition, translocation, and microglial activation of inhaled silver nanoparticles in the rodent nose and brain. Environ Health Perspect. 2016;124(12):1870–5. https://doi.org/10.1289/EHP234.
Petitot F, Lestaevel P, Tourlonias E, Mazzucco C, Jacquinot S, Dhieux B, Delissen O, Tournier BB, Gensdarmes F, Beaunier P, Dublineau I. Inhalation of uranium nanoparticles: respiratory tract deposition and translocation to secondary target organs in rats. Toxicol Lett. 2013;217(3):217–25. https://doi.org/10.1016/j.toxlet.2012.12.022.
Pietroiusti A, Bergamaschi E, Campagna M, Campagnolo L, De Palma G, Iavicoli S, Leso V, Magrini A, Miragoli M, Pedata P, Palombi L, Iavicoli I. The unrecognized occupational relevance of the interaction between engineered nanomaterials and the gastro-intestinal tract: a consensus paper from a multidisciplinary working group. Part Fibre Toxicol. 2017;14(1):47. https://doi.org/10.1186/s12989-017-0226-0.
Pujalte I, Dieme D, Haddad S, Serventi AM, Bouchard M. Toxicokinetics of titanium dioxide (TiO2) nanoparticles after inhalation in rats. Toxicol Lett. 2017;265:77–85. https://doi.org/10.1016/j.toxlet.2016.11.014.
Rahman L, Jacobsen NR, Aziz SA, Wu D, Williams A, Yauk CL, White P, Wallin H, Vogel U, Halappanavar S. Multi-walled carbon nanotube-induced genotoxic, inflammatory and pro-fibrotic responses in mice: investigating the mechanisms of pulmonary carcinogenesis. Mutat Res. 2017;823:28–44. https://doi.org/10.1016/j.mrgentox.2017.08.005.
Sadauskas E, Jacobsen NR, Danscher G, Stoltenberg M, Vogel U, Larsen A, Kreyling W, Wallin H. Biodistribution of gold nanoparticles in mouse lung following intratracheal instillation. Chem Cent J. 2009;3:16. https://doi.org/10.1186/1752-153X-3-16.
Sayan M, Mossman BT. The NLRP3 inflammasome in pathogenic particle and fibre-associated lung inflammation and diseases. Part Fibre Toxicol. 2016;13(1):51. https://doi.org/10.1186/s12989-016-0162-4.
Schmid K, Riediker M. Use of nanoparticles in Swiss Industry: a targeted survey. Environ Sci Technol. 2008;42(7):2253–60.
Semmler-Behnke M, Kreyling WG, Lipka J, Fertsch S, Wenk A, Takenaka S, Schmid G, Brandau W. Biodistribution of 1.4- and 18-nm gold particles in rats. Small. 2008;4(12):2108–11. https://doi.org/10.1002/smll.200800922.

Shinohara N, Zhang G, Oshima Y, Kobayashi T, Imatanaka N, Nakai M, Sasaki T, Kawaguchi K, Gamo M. Kinetics and dissolution of intratracheally administered nickel oxide nanomaterials in rats. Part Fibre Toxicol. 2017;14(1):48. https://doi.org/10.1186/s12989-017-0229-x.

Shvedova AA, Kisin E, Murray AR, Johnson VJ, Gorelik O, Arepalli S, Hubbs AF, Mercer RR, Keohavong P, Sussman N, Jin J, Yin J, Stone S, Chen BT, Deye G, Maynard A, Castranova V, Baron PA, Kagan VE. Inhalation vs. aspiration of single-walled carbon nanotubes in C57BL/6 mice: inflammation, fibrosis, oxidative stress, and mutagenesis. Am J Physiol Lung Cell Mol Physiol. 2008;295(4):L552–65. https://doi.org/10.1152/ajplung.90287.2008.

Silva AH, Locatelli C, Filho UP, Gomes BF, de Carvalho Junior RM, de Gois JS, Borges DL, Creczynski-Pasa TB. Visceral fat increase and signals of inflammation in adipose tissue after administration of titanium dioxide nanoparticles in mice. Toxicol Ind Health. 2017;33(2):147–58. https://doi.org/10.1177/0748233715613224.

Smolkova B, El Yamani N, Collins AR, Gutleb AC, Dusinska M. Nanoparticles in food. Epigenetic changes induced by nanomaterials and possible impact on health. Food Chem Toxicol. 2015;77:64–73. https://doi.org/10.1016/j.fct.2014.12.015.

Stern ST, Adiseshaiah PP, Crist RM. Autophagy and lysosomal dysfunction as emerging mechanisms of nanomaterial toxicity. Part Fibre Toxicol. 2012;9:20. https://doi.org/10.1186/1743-8977-9-20.

Sturm R. Clearance of carbon nanotubes in the human respiratory tract-a theoretical approach. Ann Transl Med. 2014;2(5):46. https://doi.org/10.3978/j.issn.2305-5839.2014.04.12.

Sturm R. Nanotubes in the human respiratory tract – deposition modeling. Z Med Phys. 2015;25(2):135–45. https://doi.org/10.1016/j.zemedi.2014.08.002.

Subramaniam VD, Prasad SV, Banerjee A, Gopinath M, Murugesan R, Marotta F, Sun XF, Pathak S. Health hazards of nanoparticles: understanding the toxicity mechanism of nanosized ZnO in cosmetic products. Drug Chem Toxicol. 2018:1–10. https://doi.org/10.1080/01480545.2018.1491987.

Sung JH, Ji JH, Park JD, Yoon JU, Kim DS, Jeon KS, Song MY, Jeong J, Han BS, Han JH, Chung YH, Chang HK, Lee JH, Cho MH, Kelman BJ, Yu IJ. Subchronic inhalation toxicity of silver nanoparticles. Toxicol Sci. 2009;108(2):452–61. https://doi.org/10.1093/toxsci/kfn246.

Sung JH, Ji JH, Park JD, Song MY, Song KS, Ryu HR, Yoon JU, Jeon KS, Jeong J, Han BS, Chung YH, Chang HK, Lee JH, Kim DW, Kelman BJ, Yu IJ. Subchronic inhalation toxicity of gold nanoparticles. Part Fibre Toxicol. 2011;8:16. https://doi.org/10.1186/1743-8977-8-16.

Szakal C, Roberts SM, Westerhoff P, Bartholomaeus A, Buck N, Illuminato I, Canady R, Rogers M. Measurement of nanomaterials in foods: integrative consideration of challenges and future prospects. ACS Nano. 2014;8(4):3128–35. https://doi.org/10.1021/nn501108g.

Tassinari R, Cubadda F, Moracci G, Aureli F, D'Amato M, Valeri M, De Berardis B, Raggi A, Mantovani A, Passeri D, Rossi M, Maranghi F. Oral, short-term exposure to titanium dioxide nanoparticles in Sprague-Dawley rat: focus on reproductive and endocrine systems and spleen. Nanotoxicology. 2014;8(6):654–62. https://doi.org/10.3109/17435390.2013.822114.

Valavanidis A, Vlachogianni T, Fiotakis K, Loridas S. Pulmonary oxidative stress, inflammation and cancer: respirable particulate matter, fibrous dusts and ozone as major causes of lung carcinogenesis through reactive oxygen species mechanisms. Int J Environ Res Public Health. 2013;10(9):3886–907. https://doi.org/10.3390/ijerph10093886.

Valsami-Jones E, Lynch I. How safe are nanomaterials? Science. 2015;350(6259):388–9. https://doi.org/10.1126/science.aad0768.

van Ravenzwaay B, Landsiedel R, Fabian E, Burkhardt S, Strauss V, Ma-Hock L. Comparing fate and effects of three particles of different surface properties: nano-TiO(2), pigmentary TiO(2) and quartz. Toxicol Lett. 2009;186(3):152–9. https://doi.org/10.1016/j.toxlet.2008.11.020.

Wang J, Liu Y, Jiao F, Lao F, Li W, Gu Y, Li Y, Ge C, Zhou G, Li B, Zhao Y, Chai Z, Chen C. Time-dependent translocation and potential impairment on central nervous system by intranasally instilled TiO(2) nanoparticles. Toxicology. 2008;254(1–2):82–90. https://doi.org/10.1016/j.tox.2008.09.014.

Wang B, Wang Q, Chen H, Zhou X, Wang H, Wang H, Zhang J, Feng W. Size-dependent translocation pattern, chemical and biological transformation of nano- and submicron-sized ferric oxide particles in the central nervous system. J Nanosci Nanotechnol. 2016;16(6):5553–61.

Wang C, Cheng K, Zhou L, He J, Zheng X, Zhang L, Zhong X, Wang T. Evaluation of long-term toxicity of oral zinc oxide nanoparticles and zinc Sulfate in mice. Biol Trace Elem Res. 2017;178(2):276–82. https://doi.org/10.1007/s12011-017-0934-1.

Warheit DB. Hazard and risk assessment strategies for nanoparticle exposures: how far have we come in the past 10 years? F1000Res. 2018;7:376. https://doi.org/10.12688/f1000research.12691.1.

Wu T, Tang M. Review of the effects of manufactured nanoparticles on mammalian target organs. J Appl Toxicol. 2018;38(1):25–40. https://doi.org/10.1002/jat.3499.

Wu W, Li R, Bian X, Zhu Z, Ding D, Li X, Jia Z, Jiang X, Hu Y. Covalently combining carbon nanotubes with anticancer agent: preparation and antitumor activity. ACS Nano. 2009;3(9):2740–50. https://doi.org/10.1021/nn9005686.

Xiao Y, Gao X, Taratula O, Treado S, Urbas A, Holbrook RD, Cavicchi RE, Avedisian CT, Mitra S, Savla R, Wagner PD, Srivastava S, He H. Anti-HER2 IgY antibody-functionalized single-walled carbon nanotubes for detection and selective destruction of breast cancer cells. BMC Cancer. 2009;9:351. https://doi.org/10.1186/1471-2407-9-351.

Yada RY, Buck N, Canady R, DeMerlis C, Duncan T, Janer G, Juneja L, Lin MS, McClements DJ, Noonan G, Oxley J, Sabliov C, Tsytsikova L, Vazquez-Campos S, Yourick J, Zhong QX, Thurmond S. Engineered nanoscale food ingredients: evaluation of current knowledge on material characteristics relevant to uptake from the gastrointestinal tract. Compr Rev Food Sci Food Saftey. 2014;13(4):730–44. https://doi.org/10.1111/1541-4337.12076.

Yao M, He L, McClements DJ, Xiao H. Uptake of gold nanoparticles by intestinal epithelial cells: impact of particle size on their absorption, accumulation, and toxicity. J Agric Food Chem. 2015;63(36):8044–9. https://doi.org/10.1021/acs.jafc.5b03242.

Yu LE, Yung LYL, Ong CN, Tan YL, Balasubramaniam KS, Hartono D, Shui GH, Wenk MR, Ong WY. Translocation and effects of gold nanoparticles after inhalation exposure in rats. Nanotoxicology. 2007;1(3):235–42. https://doi.org/10.1080/17435390701763108.

Zhang X, Yin J, Kang C, Li J, Zhu Y, Li W, Huang Q, Zhu Z. Biodistribution and toxicity of nanodiamonds in mice after intratracheal instillation. Toxicol Lett. 2010;198(2):237–43. https://doi.org/10.1016/j.toxlet.2010.07.001.

Zhang X, Xu Y, Zhou L, Zhang C, Meng Q, Wu S, Wang S, Ding Z, Chen X, Li X, Chen R. Sex-dependent depression-like behavior induced by respiratory administration of aluminum oxide nanoparticles. Int J Environ Res Public Health. 2015;12(12):15692–705. https://doi.org/10.3390/ijerph121215011.

Zhu L, Guo D, Sun L, Huang Z, Zhang X, Ma W, Wu J, Xiao L, Zhao Y, Gu N. Activation of autophagy by elevated reactive oxygen species rather than released silver ions promotes cytotoxicity of polyvinylpyrrolidone-coated silver nanoparticles in hematopoietic cells. Nanoscale. 2017;9(17):5489–98. https://doi.org/10.1039/c6nr08188f.

Chapter 8
Oxidative Stress and Redox Modifications in Nanomaterial–Cellular Interactions

Tong Zhang, Matthew J. Gaffrey, Wei-Jun Qian, and Brian D. Thrall

Abstract The rapidly increasing development of engineered nanomaterials (ENMs) for multiple industries and societal applications has raised concern over the increased potential for human exposures. While knowledge gaps still exist regarding the fundamental mechanisms by which biological systems interact with and respond to various ENMs, it is now well established that many ENMs stimulate reactive oxygen species (ROS), either through direct redox chemistry or indirectly by modulating endogenous cellular metabolism. Indeed, many of the reported adverse biological responses to ENMs, such as the activation of antioxidant defense systems, inflammation, endoplasmic reticulum (ER) stress responses, and cell death, involve oxidative stress as a key event. Here, we summarize recent findings that support the central role of oxidative stress in ENM–cellular interactions, with an emphasis on phagocytic cells as an important target immune cell type. Emerging evidence enabled by new proteomics-based technologies indicates that redox-mediated cellular response pathways triggered by many ENMs may be initiated through the selective formation of protein redox modifications. The postulated role of protein redox modifications as molecular initiating events represents a novel conceptual framework for understanding specificity in ENM-induced oxidative stress responses and for discerning differences between cellular adaptive signaling and early toxicity.

Keywords Engineered nanomaterials · Nanotoxicology · Oxidative stress · Redox modifications · Proteomics · ER stress · Inflammation

---

T. Zhang · M. J. Gaffrey · W.-J. Qian · B. D. Thrall (✉)
Biological Sciences Division, Pacific Northwest National Laboratory, Richland, WA, USA
e-mail: brian.thrall@pnnl.gov

J. C. Bonner, J. M. Brown (eds.), *Interaction of Nanomaterials with the Immune System*, Molecular and Integrative Toxicology,
https://doi.org/10.1007/978-3-030-33962-3_8

8.1 Introduction

The rapid growth of nanotechnology and use of engineered nanomaterials (ENMs) in numerous applications from consumer products to medical devices have driven concern over the potential for human exposure and safety. For instance, studies in experimental systems have shown that ENMs interact with the immune system, modulating normal innate and adaptive immune responses (Halamoda-Kenzaoui and Bremer-Hoffmann 2018). At the frontline of the immune system, macrophages and other phagocytic cells mediate the clearance, distribution, and toxicity of ENMs (Boraschi et al. 2017), and therefore are widely used as cellular models to rapidly assess potential hazard and immune response to ENMs (Liu et al. 2015; Duan et al. 2016; Han et al. 2017; Chen et al. 2018; Nakayama 2018).

The ability of ENMs to promote production of reactive oxygen species (ROS) and induce cellular oxidative stress has emerged as an important predictive principle in nanotoxicology (Burello and Worth 2011; Xia et al. 2006; Kodali and Thrall 2015). The 'oxidative stress hypothesis' postulates that the potential toxicity of ENMs can be predicted from first principles of the physical and chemical properties of the nanomaterial, along with hierarchical analysis of the ability of an ENM to generate oxidative stress in cells. Indeed, it is well established that ENMs composed of transition metals, such as iron, copper, and vanadium oxides, are capable of directly catalyzing ROS formation through Haber-Weiss and Fenton reactions (Kodali and Thrall 2015; Sharma et al. 2014; Angelé-Martínez et al. 2017). In addition, a broader variety of ENMs indirectly induce cellular oxidative stress by modulating endogenous metabolic pathways linked to intracellular ROS generation machinery, such as the mitochondrial electron transfer chain, protein folding sites in the endoplasmic reticulum (ER), and phagosomal nicotinamide adenine dinucleotide phosphate (NADPH) oxidases (NOX) complexes (Fu et al. 2014; Khanna et al. 2015). The overproduction of ROS disrupts redox hemostasis, triggering adaptive mechanisms that influence a broad spectrum of cellular responses. These responses, ranging from induction of the antioxidant gene responses, pro-inflammatory cytokine production, and ultimately apoptosis, are positively correlated with ROS production according to a hierarchical oxidative stress model (Meng et al. 2009; Zhang et al. 2012). Unfortunately, assays that are commonly used to broadly measure cellular redox status (e.g., redox-sensitive dyes and total glutathione assays) generally lack the sensitivity and specificity needed to provide insights into the specific pathways or sites of oxidative modifications induced by ENMs. It can also be challenging to distinguish when a particular cell response to an ENM exposure represents an adaptive response versus an early indicator of an adverse outcome. As previously noted (Donaldson et al. 2009), exposure to particles that cause similar overall levels of oxidative stress can result in widely divergent biological responses. Thus, for nanotoxicology to progress toward a more predictive and mechanistic framework, there is a need to better understand how the chemical nature and early molecular targets of ENM-induced oxidative stress within the cell are dependent on the physicochemical attributes of the ENM.

Over the last decade, mechanisms by which ROS function as second messengers to modulate specific cell signaling processes have emerged. Redox posttranslational modifications (PTMs) on proteins have received extensive attention due to the fact that these modifications play a significant role in the regulation of the structure and function of the target proteins (Yang et al. 2016; Moldogazieva et al. 2018). In particular, because of the unique redox capacity of the nucleophilic thiol group, protein cysteines are selective targets of multiple forms of oxidative modifications, including S-glutathionylation (SSG), S-nitrosylation (SNO), sulfenylation (SOH), sulfinylation (SO_2H), and sulfonylation (SO_3H) (Klomsiri et al. 2011; Duan et al. 2017). Increasing evidence indicates these redox modifications occur nonstochastically throughout the proteome (Duan et al. 2016; Yang et al. 2016; Kramer et al. 2015), allowing site-specific modifications on specific proteins to facilitate signaling specificity. A growing body of evidence has demonstrated the physiological importance of protein redox modifications such as SSG in macrophages and innate immune signaling (Ullevig et al. 2013, 2016; Weigert et al. 2018). More recently, protein redox modifications have also been implicated in ENM-induced biological responses (Duan et al. 2016; Bouallegui et al. 2018; Sundarraj et al. 2017; Srikanth et al. 2016). Herein, we briefly review the general role of ROS in the context of ENM–cellular interactions and highlight emerging examples of where ROS-mediated protein modifications may play key roles in regulating specific cellular response pathways. Recent analytical advances such as mass spectrometry-based redox proteomics in profiling pathways and processes in response to ENMs are also highlighted. As these measurement capabilities continue to advance (Costa and Fadeel 2016; Fadeel et al. 2018; Zhang et al. 2018a), the resulting knowledge of the early molecular targets of redox stress provides an opportunity to build mechanistic and predictive models for ENM–cellular interactions. Studies utilizing macrophages are particularly emphasized, given the essential role of this cell type in innate immunity and their extensive use in nanosafety assessment studies.

8.2 Direct and Indirect Mechanisms of Cellular Redox Stress

Elevated cellular oxidative stress can result from either an excessive production of ROS, or through disruption or depletion of antioxidant protective systems (Grimsrud et al. 2008). A broad range of biological sources of ROS exist, including hydrogen peroxide (H_2O_2, the most common and stable form), hypochlorous acid (HOCl), and free radicals such as hydroxyl ($^{\bullet}OH$) and superoxide ($O_2^{\bullet-}$). Many ENMs hold the potential for directly catalyzing ROS production in cellular systems (Fig. 8.1; for review, see (Kodali and Thrall 2015)) through Haber-Weiss and Fenton reactions (redox cycling), whereby $^{\bullet}OH$ is generated from H_2O_2 and $O_2^{\bullet-}$. For instance, ENMs comprised of transition metals have been shown to produce highly reactive $^{\bullet}OH$ radical through these direct catalytic mechanisms (Sharma et al. 2014; Angelé-Martínez et al. 2017; He et al. 2012). ROS production can also be catalyzed through ENM-mediated photocatalysis, as recently demonstrated with emerging classes of

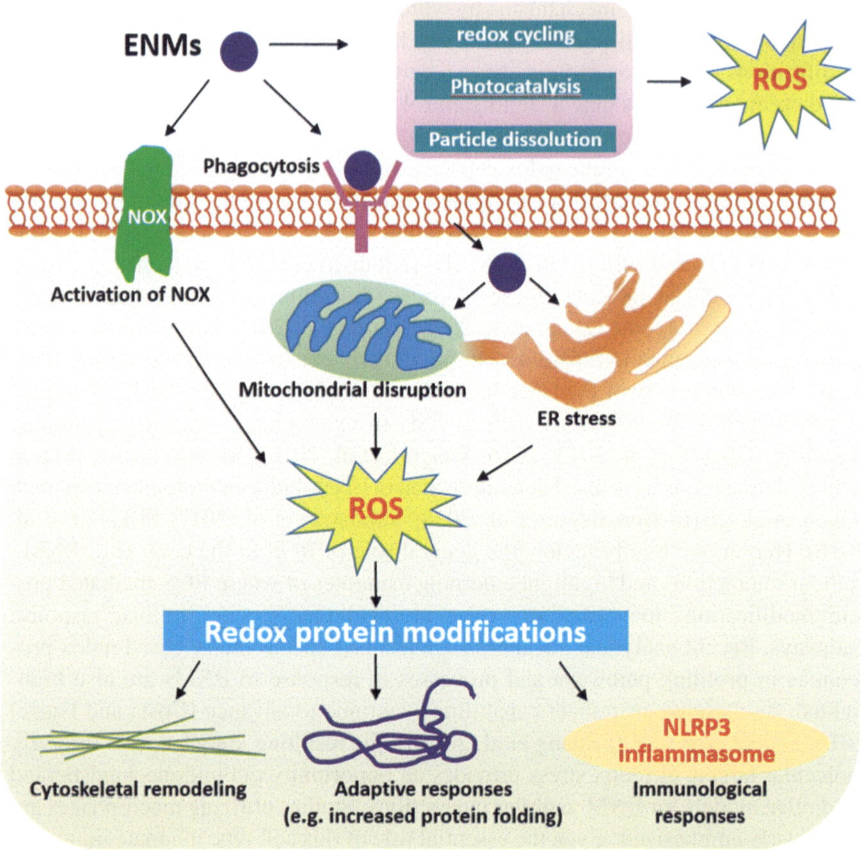

Fig. 8.1 ROS generation and redox modifications in ENM–cellular interactions. ROS can be generated by both direct (examples given in the pink box on the top) and indirect mechanisms (mainly by modulating cellular components such as NOX, mitochondria, and ER) upon ENM exposure. The indirect generation of ROS could also occur after internalization of ENMs, by macrophage phagocytosis, for example. Reversible redox modifications of key cysteine residues on specific proteins represent a leading molecular mechanism by which ROS can function as secondary messengers to transform oxidant signals to biological response

semiconductor-metal hybrid nanostructures bearing ZnO (He et al. 2014), TiO_2 (Xiong et al. 2013), and CdSe/CdS nanorods (Waiskopf et al. 2016). In fact, the localized photocatalysis and formation of $^{\bullet}OH$ and H_2O_2 by these materials are sufficient to modulate protein enzymatic activity through oxidative modification, opening the door to potential therapeutic applications (Waiskopf et al. 2016). Many types of metal/metal oxide ENMs are also prone to dissolution in biological systems, resulting release of soluble free metal ions that are bioactive for ROS generation. For example, the release of silver ions is a likely mediator of pulmonary fibrosis following exposure to silver nanoparticles (Wang et al. 2014). A recent dosimetric analysis of the role of ion dissolution on the toxicity of silver nanoparticles in macrophages revealed that the unbound silver cation is the ultimate toxicant, and ions that form extracellularly drive the cytotoxicity of silver nanoparticles (Munusamy et al. 2015; Smith et al. 2018). Due to the diverse mechanisms for which ROS may be generated, key factors affecting the subsequent oxidative stress may also vary, greatly depending on the physicochemical properties of ENMs. These factors include size, shape, composition, and surface modification, which could ultimately affect the chemical source and degree of cellular ROS (He et al. 2015).

Physiological levels of ROS also play an important role in many normal cellular metabolic functions, such as for the oxidation of NADPH by NOX and other enzymes, electron transfer processes in the mitochondria, peroxisomal enzyme functions, and during proper protein folding in the endoplasmic reticulum (ER). ENM exposure can indirectly promote cellular oxidative stress through either activation or disruption of these cellular functions (Fig. 8.1). For example, silver nanoparticle-induced ROS production may be partially attributed to the upregulation and activation of NOX4 protein (Sun et al. 2017). NOX-derived superoxide is of particular importance in macrophage ROS production, as it has been shown that a large oxidative burst involving localized releases of ROS by NOX plays a critical role in classical macrophage activation in response to bacterial infection. In addition, $O_2^{\bullet-}$ can be converted to H_2O_2, which is permeable to the plasma membrane and can initiate innate immune responses of macrophages (Feduska and Tse 2018).

Incomplete reduction of O_2 in the mitochondria results in the formation of $O_2^{\bullet-}$, representing another important source of endogenous ROS production. Impairment of normal mitochondrial function by ENMs such as hydroxyapatite (Xue et al. 2017), silver (AshaRani et al. 2009), and TiO_2 (Chen et al. 2018) has been reported to cause elevated mitochondrial ROS production. For example, it was reported that silver nanoparticles were localized within the mitochondria of Ag-treated human lung and brain cells, and this was associated with an increased production of ROS, suggesting disruption of mitochondrial integrity and the respiratory chain may be mechanistically responsible (AshaRani et al. 2009). By contrast, TiO_2 ENMs have also been found to enhance mitochondrial ROS generation; yet, transmission electron microscopy showed that in mouse bone marrow-derived macrophages, TiO_2 ENMs were trapped exclusively in multivesicular bodies (Chen et al. 2018). The absence of TiO_2 ENMs within the mitochondria suggests that direct physical interactions are not a prerequisite for disrupting mitochondrial integrity and function. Moreover, inhalation studies in mice suggest that TiO_2 ENMs also induce mitochondrial oxidative stress in vivo (Nichols et al. 2018).

Hydrogen peroxide is also a by-product of normal oxidative protein folding in the ER (Laurindo et al. 2012). Protein folding in the ER is regarded as one of the most error-prone processes in the regulation of gene expression, and is critically dependent on redox-sensitive quality control proteins such as protein disulfide isomerase (PDI) and ER oxidoreductases (Ero) family of enzymes. PDI catalyzes protein disulfide formation in the ER through H_2O_2-generating mechanisms involving

Ero1. A positive interplay between ROS and ER stress proposes that ROS can occur both upstream or downstream of ER stress (Santos et al. 2009). The sources of ROS generation during ER stress include oxidoreductases in the ER, electron transfer in mitochondria, and NOX family members (e.g., NOX4 and NOX2) (Laurindo et al. 2012). PDI has been shown to participate in all three routes of ROS generation (Laurindo et al. 2012), and may be a converging point in regulating ER stress and oxidative stress.

The fact that generation of intracellular ROS is both spatially and temporally tightly controlled and that diverse forms of ROS exist make it challenging to measure the dynamics of cellular redox stress with high sensitivity and specificity. Methodologies for ROS detection and quantification in biological systems have been reviewed elsewhere (Woolley et al. 2013; Rezende et al. 2018), and adaption of these methods to quantify ENM-induced ROS has also been summarized (Tournebize et al. 2013). By far, the most popular methods rely on fluorescent ROS dyes such as 2′,7′-dichlorofluorescin diacetate (H_2DCF) (Han et al. 2017; Nishanth et al. 2011; Zhang et al. 2018b; Chen et al. 2017). Upon oxidation by ROS, this non-fluorescent cell-permeable probe produces green fluorescent 2′,7′-dichlorofluorescein. However, H_2DCF can be oxidized by a variety of ROS, including H_2O_2, nitric oxide (NO) and peroxynitrite anions. Thus, the lack of specificity and low dynamic range of this dye limits its usage to broad screening applications. By virtue of its specificity toward superoxide radicals and permeability to cell membrane, dihydroethidium is another extensively used probe (Libalova et al. 2018). In addition, other dyes designed to target specific cellular compartments to monitor ROS production such as MitoSOX™ Red reagent (ThermoFisher Scientific) have been used to detect mitochondrial ROS production in cells following TiO_2 treatment (Chen et al. 2018).

8.3 Common Redox-Sensitive Cellular Processes Impacted by ENMs

Although a diversity of cellular responses to ENM-induced oxidative stress have been reported, common sets of redox-sensitive cellular processes have emerged across the literature. These include changes in cell morphology and cytoskeleton dynamics, activation of immune signaling cascades, ER stress, mitochondrial dysfunction, and genotoxicity (Choudhury et al. 2017). While co-occurrence does not imply cause–effect relationships, the role of ROS in inducing stress responses and cytotoxic effects is generally inferred by demonstrating that antioxidants diminish the detrimental effects of ENMs to cells (Han et al. 2017), and through positive correlations observed between ENM dose, total ROS generated, and downstream biological effects (Abdal Dayem et al. 2017). Potential molecular mechanisms for ROS-mediated signaling events mediated by ENMs are discussed in more detail in subsequent sections.

8.4 Inflammatory Cascades

It is well established that ROS play important roles in mediating the innate immune response (Blaser et al. 2016) and that ENMs which induce oxidative stress can modulate immune cell functions (Getts et al. 2015). The molecular machinery linking ROS production and inflammatory responses following ENM exposure is not fully understood, although macrophages and other phagocytic cells play essential roles in initiating these processes. Studies employing libraries of structurally diverse metal oxide and rare earth metal ENMs have consistently shown good correlations between the potential for ROS formation after ENM exposure and macrophage activation and pulmonary inflammation (Zhang et al. 2012; Han et al. 2018; Manke et al. 2013). A direct link between ROS-mediated NADPH oxidase activity following phagocytic uptake of silica particles and activation of the NALP3/NLRP3 (NACHT, LRR, and PYD domains-containing protein 3) inflammasome has also been demonstrated (Dostert et al. 2008). Dose-dependent increases in oxidative stress and inflammasome activation have been observed in human THP-1 cells treated with boron nitride nanotubes, and the reported increases in caspase-1, interleukin--1β (IL-1β) and IL-18 support the idea that inflammasome activation by ENMs and apoptotic cascades are coupled (Kodali et al. 2017). These effects can also be attenuated in NLRP3-deficient THP-1 cells, indicating that the NLRP3 inflammasome is centrally involved in mediating these effects. Guglielmo et al. (Guglielmo et al. 2017) showed that needle-like cationic cellulose nanocrystals also induce immunological responses and caspase activation through the NLRP3 inflammasome, and that these effects can be suppressed by co-treatment with curcumin, a natural plant-derived antioxidant. The ability of curcumin to block the inflammatory activity of cellulose nanocrystals is hypothesized to be due to specific decreases in S-glutathionylation of key NLRP3 inflammasome components responsible for sustained inflammasome complex assembly and suppression of IL-1β. Conversely, other ENMs have been shown to possess antiinflammatory and antioxidant activities. For example, lipopolysaccharide (LPS)-induced ROS production in RAW264.7 cells can be inhibited by tetrahedral DNA nanostructures, possibly through upregulation of heme oxygenase-1 (Zhang et al. 2018b).

ENM-mediated inflammasome activation may also be coupled to redox-dependent changes in mitochondrial functions; yet, the underlying mechanisms are not understood. A recent proteomics and metabolomics analysis in primary bone marrow-derived macrophages showed that activation of inflammatory signaling from TiO_2 ENM exposure was associated with an increase in mitochondrial ROS and a decrease in metabolic flux of the TCA cycle as well as decreased ATP production (Chen et al. 2018). It is noteworthy that the NLRP3 directly interacts with mitochondrial membrane components during inflammasome activation. This interaction may provide a mechanism for direct coupling of mitochondrial metabolism, inflammasome activation, and an inflammatory form of cell death known as pyroptosis (Gurung et al. 2015; Elliott et al. 2018), but this relationship has yet to be carefully evaluated for ENM-mediated cell responses.

8.5 ER Stress, Apoptosis, and Autophagy

Many of the essential cellular processes within the ER are redox-regulated, including posttranslational protein folding, lipid biosynthesis, and intracellular storage of calcium. Accordingly, a variety of ENMs including cationic polystyrene nanospheres, metal oxides, quantum dots, and silica nanoparticles have been found to disrupt normal ER homeostasis and induce biochemical markers of ER stress in cells (Duan et al. 2016; Chiu et al. 2015; Cao et al. 2017). Morphological markers of ER stress, such as swelling of the ER and ribosome degranulation, have also been observed in mice and human lung cells exposed to ZnO and TiO_2 nanoparticles (Yang et al. 2015; Yu et al. 2015). These disturbances can also lead to the unfolded protein responses (UPR), a conserved adaptive cellular mechanism designed to restore normal ER function. However, under persistent or severe ER stress conditions where the adaptive efforts and UPR fail to mitigate, proapoptotic pathways can also be activated. The activation of apoptotic pathways due to prolonged ER stress and the accumulation of unfolded proteins may provide a generalized mode of action for toxicity of some ENMs. In support of this hypothesis, pharmacological manipulation of ER stress has also been shown to modulate toxicity of some ENMs. For example, the ER stress inducer thapsigargin enhanced ZnO-induced cytotoxicity in THP-1 macrophages (Chen et al. 2017).

Although the co-occurrence of oxidative stress and activation of the UPR during ENM-cellular interactions have been described (Yang et al. 2015; Chen et al. 2014; Christen et al. 2014), how these processes interact with each other remains unclear. It has been proposed that ER stress and oxidative stress form a positive feed-forward loop to accentuate signaling (Cao and Kaufman 2014). In this model, ENM-induced oxidative stress disturbs the redox homeostatic environment of the ER, resulting in constant activation of the UPR which further potentiates oxidative stress. Mechanistic studies of ER stress induced by ENMs are still limited, but previous work suggests that the excess ROS arising from the UPR response is a necessary signal for apoptosis. In some cases, however, oxidative stress may not be required for ENM-induced ER stress, as reported in the case of Mn_3O_4 nanoparticles in yeast (Yi et al. 2017). Furthermore, the direct role of ROS in other ER stress/UPR-related effects, such as the induction of autophagy, has only begun to be investigated as a mechanism for nanotoxicity (Wang et al. 2018).

The essential role of ROS in activation of autophagy, a highly-regulated process in which cells degrade their own cellular components, has been previously reviewed (Poillet-Perez et al. 2015). A recent study showed that photosensitizing nanoparticles being investigated for treatment of atherosclerosis induce autophagy in THP-1 macrophage-derived foam cells, and these effects are reversible by treatment with N-acetylcysteine (Han et al. 2017). The induction of autophagy by ZnO nanoparticles has also been reported. In this case, autophagy is thought to promote the dissolution of ZnO within the lysosome, resulting in release of free zinc ions that may further stimulate ROS generation (Zhang et al. 2017). The increased ROS generation could also be a result of the impairment of mitochondrial function by free zinc ions (Zhang et al. 2017). While such studies are currently limited to relatively few ENM examples, they highlight important interplays between ER stress, autophagy, and apoptosis that may be critically important in determining toxicity outcomes for some ENMs.

8.5.1 Cytoskeleton Dynamics

Proper regulation of cytoskeleton dynamics is crucial for basic cellular functions including cell adhesion, migration, and phagocytosis. Disruption of this regulation by ENMs leads to aberrant cytoskeletal reorganization that can in turn cause other adverse cellular effects (Liu et al. 2015, 2017). Cytoskeletal proteins such as tubulin and actin can serve as scavengers for metal ions, and dissolution of ENMs such as ZnO triggers actin filament bundling and converts tubulin into rigid microtubules, causing acute cytotoxic effects (Garcia-Hevia et al. 2016).

A link between ROS and cytoskeletal dysregulation has also been observed in a study in which a change in reorganization of actin following superparamagnetic iron oxide ENMs treatment is blocked by ROS scavengers (Buyukhatipoglu and Clyne 2011). Thus, it seems that proper regulation of cellular ROS is vital for normal cytoskeletal dynamics, and this relationship can have important implications for innate immune effects associated with ENMs. For example, exposure to some redox active metal oxide ENMs has been shown to suppress the efficiency by which macrophages phagocytose pathogenic bacteria, such as *Streptococcus pneumonia* (Duan et al. 2016; Kodali et al. 2013; DeLoid et al. 2016). Such observations have important implications to human exposures, given that welders exposed to fumes rich in metal oxide nanoparticles have significantly increased risk of mortality associated with lung infections, including pneumonia (Palmer et al. 2003). In other examples, such as ZnO-exposed human epithelial cells, remodeling of the cytoskeletal components may occur earlier than overall measureable increases in cellular ROS production, suggesting a limited contribution of ROS to cytoskeletal reorganization (Garcia-Hevia et al. 2016). Alternatively, remodeling of the cytoskeleton may in fact serve as a trigger for ROS release from the mitochondria (Garcia-Hevia et al. 2016).

8.6 Protein Thiols as Critical Targets of Cellular Redox Stress

Protein oxidative modifications are a direct product of spatially localized increases in ROS produced by ENMs and other environmental toxicants. Depending on the type and concentration of ROS, protein thiols can be modified to different levels of oxidation state, which further bring about various cellular responses (Fig. 8.2a).

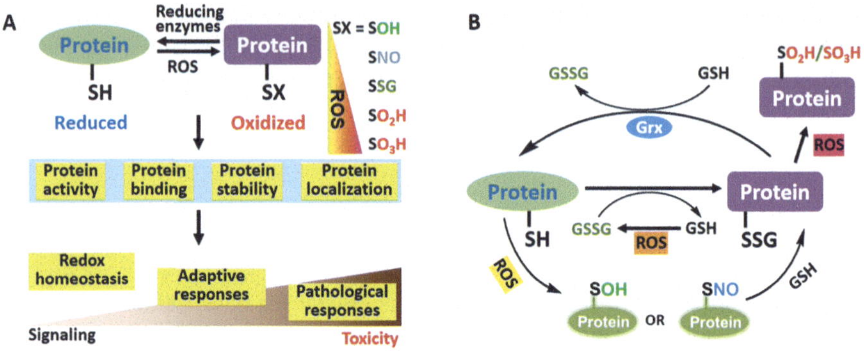

Fig. 8.2 Redox modifications in mediating biological response. (**a**). A general scheme of redox-mediated cellular events. The type and degree of redox modifications are dependent on both the antioxidant systems and ROS. Spatiotemporal control along with the different reactivity of substrate proteins provides specificity in redox-based signaling. These modifications can affect protein functions in a number of ways, ultimately leading to an array of cellular responses ranging from restoring redox homeostasis, to adaptive responses, and ultimately pathology. (**b**). Example of dynamic redox regulation of proteins through protein glutathionylation (SSG) system. Mild level of ROS can result in protein-SSG modification either directly or indirectly through conversion of protein-SOH or protein-SNO modifications. These modifications can readily be reversed by specific enzymes, bringing redox-sensitive proteins back to the reduced status. However, higher level of ROS may result in irreversible modification such as protein-SO_2H and protein-SO_3H. (Figure **a** is adapted with permission from Duan et al. (Duan et al. 2016). Copyright (2016) American Chemical Society). SH free thiol, SOH sulfenylation, SNO S-nitrosylation SSG S-glutathionylation, SO_2H sulfinylation, SO_3H sulfonylation, ROS reactive oxygen species, GSH glutathione, GSSG GSH disulfide, Grx glutaredoxin

Under physiological or moderate levels of ROS, enzymatically reversible protein redox modifications (e.g., SSG, SNO, and SOH) can function in signaling and regulatory mechanisms as well as serve to protect proteins from irreversible hyperoxidations (Duan et al. 2017; Kramer et al. 2015; Hughes et al. 2017). The identification of many proteins as redox switches now enables a much more mechanistic understanding of how oxidative stress can mediate such a diverse array of cellular processes (Moldogazieva et al. 2018; Klomsiri et al. 2011; Grek et al. 2013). Recently, a role of SSG in protecting mitochondria from oxidative stress has also been described in yeast (Gergondey et al. 2017). On the other hand, severe oxidative stress where ROS levels overwhelm the antioxidant system can result in hyperoxidized forms of protein cysteines which are poor substrates for many cellular reductases and thereby largely irreversible, typically resulting in a loss of protein function under pathological conditions (Stadtman and Berlett 1997). In addition, one type of redox modification may be converted into another, depending on the oxidative stress level, as illustrated in Fig. 8.2b. SOH modifications possess the lowest oxidation status for a protein cysteine, but can be further oxidized to form SSG (Chung et al. 2013). Similarly, SNO serves as another important intermediate for SSG formation. Under higher levels of ROS, SSG may undergo further oxidation to hyperoxidized SO_2H and SO_3H modifications, which generally are thought to be poor substrates for enzymatic reversal (Yang et al. 2016; Chung et al. 2013).

Oxidative modification can alter protein function through a variety of mechanisms (Fig. 8.2a), including by altering enzymatic activity. For example, PTEN (phosphatase and tensin homolog) is a redox-sensitive tumor suppressor protein whose phosphatase activity is inhibited upon SNO modification (Gupta et al. 2017). Similarly, inhibition of enzymatic activity via SSG modifications has been reported for GAPDH (glyceraldehyde-3-phosphate dehydrogenase) (Mohr et al. 1999), MKP1 (MAPK phosphate1) (Kim et al. 2012), and eNOS (endothelial nitric oxide synthase) (Chen et al. 2010). Redox modifications may also alter protein stability. This is the case for caveolin-1 (Chanvorachote et al. 2009) and MKP1 (Kim et al. 2012), which showed increased and decreased stability, respectively, upon modification by SNO. Furthermore, redox modifications may result in protein conformational changes that ultimately modulate protein localization and formation of new protein–protein interactions. This is exemplified by Fas (CD95), a member of the tumor necrosis factor receptor superfamily that is important in regulating programmed cell death. SSG modification of Fas promotes aggregation, translocation to lipid rafts, and binding to Fas ligand (Anathy et al. 2009). Interestingly, SNO modifications have been shown to promote the redistribution of Fas to lipid rafts (Leon-Bollotte et al. 2011).

Given the significance of redox modifications on protein function, tight controls must be in place to regulate the extent of these modifications in cells. Accordingly, many types of redox modifications are readily reversible through specific cellular enzyme systems. For example, SNO can be reversed by the S-nitrosoglutathione (GSNO) reductase system, resulting in formation of the intermediate GSNO, which is further reduced to glutathione (GSH) by GSNO reductase and glutaredoxin (Benhar et al. 2009). Alternatively, denitrosylation can be catalyzed by the thioredoxin system consisting of thioredoxin (Trx), Trx reductase (TrxR), and NADPH (Sengupta et al. 2007). SSG modifications can also be reduced through several enzymes including glutaredoxin and Trx (Grek et al. 2013). Thus, the extent of oxidative modifications at a given time is dynamically and rapidly controlled by both forward and the reverse reactions, allowing cells to adjust the overall cellular redox status.

Protein redox modifications are also spatiotemporally constrained. Although exactly how ROS-mediated modifications occur at specific cysteine residues is still not well understood, a couple of hypotheses have emerged. First, gradients of ROS have been observed in both cells and tissues, making spatial regulation of redox signaling possible (Niethammer 2018). In addition, individual cellular compartments maintain different relative redox potentials and thereby have different levels of susceptibility to redox modifications. For instance, while the mitochondrial is among the most reducing organelle environment with a high GSH disulfide ratio (GSH/GSSG), the GSH/GSSG ratio in the ER is much lower ratio, favoring an oxidizing environment for disulfide formation during protein folding (Go and Jones 2008). Thus, redox compartmentalization allows for the concurrence of cellular processes with distinct redox characteristics in different organelles, which has potential implications for understanding ENM-mediated cell responses. For example, the relative susceptibility of ER proteins to redox modification as compared to proteins in the mitochondrial matrix may in part explain how ER stress and mitochondrial metabolism responses are differentially triggered in response to different ENM types, or different levels of ENM-mediated oxidative stress (Duan et al. 2016).

Redox modifications are also site-specific events, occurring non-stochastically on the cysteine residues through the proteome. Generally, exposed cysteine residues with lower pKa values tend to have higher activity toward ROS. Thus, the reaction rates could be enhanced significantly, as much as six orders of magnitude in the case of the antioxidant protein peroxiredoxins (Poole 2015), by factors that lower the pKa of the active cysteine. The high reactivity of specific thiols in redox-responsive proteins, along with selective accessibility, provides one way to achieve redox signaling specificity (Niethammer 2018).

8.6.1 Measurement of Protein Redox Modifications

The overall extent of redox modifications is typically low in cells under most conditions (e.g., an average of 4.5% for SSG modification in mouse muscle (Kramer et al. 2018)). Together with the labile nature of redox modifications such as SNO and SSG, detection and quantification of these oxidative modifications remain analytically challenging. Methods to detect protein redox modifications fall primarily into two categories (for review, see (Duan et al. 2017; Grek et al. 2013)). One approach is to detect the modified proteins directly through methods such as radiolabeling and Western blots. For example, anti-glutathione (Su et al. 2014) and anti-S-nitrosocysteine antibodies (Greco et al. 2006) have been employed to detect SSG and SNO modifications, respectively. Antibodies against irreversible protein modification such as carbonylation have also been used. Another approach involves labeling of protein cysteine residues with thiol-reactive probes such as iodoacetamido fluorescein. The fluorescent-labeled redox proteins are subsequently quantified by gel electrophoresis and image analysis (Bouallegui et al. 2018). Direct labeling of protein thiols has also been developed by taking advantage of the thiol–disulfide exchange reaction (Hill et al. 2010). In this approach, biotinylated GSSG (permeable to the cell membrane) is allowed to react with protein thiols to from protein-SSG-biotin, which can be further purified by biotin–avidin affinity assays.

An alternative approach for detection of protein oxidative modifications involves an initial reduction of the modifications with reducing reagents that are selective for each modification and subsequent detection of their newly formed derivatives. The most popular of these methods is the biotin switch technique, in which SNO groups are selectively reduced by ascorbate and the resulting free thiols are labeled with biotin-conjugates (Jaffrey and Snyder 2001). The newly generated free thiols can also be conjugated to thiol-affinity resins (Guo et al. 2014), making capturing modified proteins less laborious. The resin-assisted approach also enables on-resin trypsin digestion and isobaric labeling for multiplexing and quantification, seamlessly coupling upstream processing with proteomics analysis. Specific enrichment and detection of SNO (Su et al. 2013) and SSG (Duan et al. 2016) modifications have been reported using this resin-assisted enrichment method. Given the high sensitivity, extensive coverage, accurate quantification, and the ability to locate the specific sites of modification, mass spectrometry-based proteomics has become the preferred method for profiling redox modifications.

8.7 ENM-Induced Protein Redox Modifications

Despite the abundant evidence that protein redox modifications have important signaling functions, only a few studies to date have examined protein redox modifications in response to ENM exposure. In one study, significant levels of protein carbonyl were observed in both mouse serum and liver at 24 h following oral administration of CuO ENMs (Kadammattil et al. 2018), suggesting a role of carbonylation in toxicity. Similarly, direct labeling of protein cysteine residues with iodoacetamido fluorescein showed that silver ENMs of different sizes cause protein thiol oxidation in mussels (Bouallegui et al. 2018). Proteomics evidence has also emerged that ENM exposure can modulate proteins that function in redox regulation. A recent study used proteomics to profile the effects of CdSe nanocrystals in diatoms, demonstrating that proteins involved in regulation of S-glutathionylation processes are significantly upregulated (Poirier et al. 2018). Although these studies provide insightful links between oxidative modifications and the biological impacts of ENMs, details of these processes such as the identity of the modified proteins and the exact sites of modification are still lacking.

The first direct profiling of ENM-induced redox modifications was performed in a macrophage cell model, where RAW 264.7 cells were treated with SiO_2, Fe_3O_4, and CoO nanoparticles which were previously shown to induce low, moderate, and high levels of oxidative stress, respectively, resulting in differential effects of macrophage innate immune functions (Duan et al. 2016). Thiol-affinity resin was used to enrich SSG-modified peptides, which were identified and quantified by mass spectrometry. The overall level of SSG modifications was closely correlated with the level of total cellular oxidative stress induced by these ENMs. However, it is important to note that significant increases in protein SSG levels could be quantified by mass spectrometry without detectable changes in total cell glutathione status, demonstrating the greater sensitivity of this approach for assessing redox changes in cells. This analysis identified more than >2000 SSG-modified sites from ~1000 unique proteins. Consistent with the idea that redox modifications are nonstochastic events, it was also found that specific biological pathways were selectively enriched with SSG-modified proteins, depending on the level of overall oxidative stress induced by the ENM treatment. Under low levels of redox stress-induced ENM exposure, the most significantly enriched pathways included those involved in phagosome formation and ER-associated processes involved in protein translation and protein folding. By contrast, at higher levels of ENM-induced oxidative stress, the pattern of SSG modifications displayed reduced specificity across a broader set of cell processes, and included proteins involved in classical stress response pathways and mitochondrial energetics associated with apoptosis.

The site-specific identification of protein redox modifications provides initial evidence that these modifications may act as initiating events involved in recruiting the common ROS-related biological processes observed following ENM exposure. For example, the activation of NLRP3 inflammasome is thought to be dependent on the redox status of NLRP3 (Guglielmo et al. 2017). The suppression of cellulosic nanocrystal-induced NLRP3 inflammasome activation by the natural plant product curcumin is accompanied with a significant shift on the SSG levels on both NLRP3 and caspase-1 (Guglielmo et al. 2017). It is postulated that glutathionylation status may determine the potential of activated caspase-1 to interact with NLRP3 during assembly of the inflammasome complex. Protein redox modifications may be among the early molecular events following ENM exposure involved in the initiation of adaptive responses within the ER. Redox proteomic analyses in macrophages exposed to metal oxide ENMs provide several lines of evidence that support this hypothesis (Duan et al. 2016). First enzymatic pathways within the ER were the most highly enriched in SSG-modifications observed across the proteome. Second, the most susceptible sites of SSG modifications following ENM exposure involved critical cysteine disulfide within active sites (thioredoxin motifs) of PDIs (including PDIA1, PDIA3, and PDIA4) family members, motifs which are essential for disulfide exchange reactions during protein folding. Although additional studies are needed to determine the necessary role of these modifications in ENM-induced ER stress, the inhibition of PDI enzymatic activity after redox modification of these sites has been demonstrated (Nakamura et al. 2013).

Emerging evidence also suggests a role for redox regulation in actin and microtubule dynamics (Wilson and González-Billault 2015). A notable example is the dysregulation of actin polymerization and phagosome assembly by SSG modifications. Macrophage phagocytic function can be impaired following exposure to some redox-active ENMs (Duan et al. 2016; Kodali et al. 2013; DeLoid et al. 2016; Dalle-Donne et al. 2003; Sakai et al. 2012). The specific sites of protein glutathionylation after ENM exposure identified by mass spectrometry coincide with critical actin cysteine residues which have been shown to be necessary for efficient phagosome formation in neutrophils (Sakai et al. 2012). Additional cytoskeletal proteins, including 14–3–3 proteins necessary for phagocytic function, are also susceptible to ENM-induced SSG modifications (Duan et al. 2016). Furthermore, macrophages or neutrophils that are deficient in glutaredoxin-1 (Grx1), the major cytosolic enzyme responsible for reversal of SSG modifications, have impaired phagocytic function toward bacterial pathogens (Duan et al. 2016; Sakai et al. 2012), mimicking the phenotypic effect observed following treatment with some redox-active ENMs.

8.8 Conclusions and Perspectives

While oxidative stress should not be viewed as a universal mechanism of nanotoxicity, over the last decade, it has clearly emerged as one of the dominant mechanisms in the literature. It is important to recognize that ENMs such as multiwalled carbon nanotubes may cause toxicity through alternative mechanisms (Tabet et al. 2009). At the same time, it is important to note that many commonly employed measurements of cellular oxidative stress, while useful in routine screening assays, often lack sensitivity and specificity to firmly rule out contributing roles of oxidative stress and targeted oxidative modifications in mediating cellular responses to ENMs. The advancement of sensitive measurement techniques such as mass spectrometry, which provide site-specific quantitation of redox modifications induced by ENMs, will undoubtedly shed new light into these underlying mechanisms. An important future goal for nanosafety and predictive nanotoxicology is to understand whether (and how) the early molecular targets of ENM-induced oxidative stress are dependent on the physical and chemical attributes of an ENM. Emerging evidence demonstrating both the nonstochastic nature and signaling functions of protein redox modifications suggests that it may be feasible to mechanistically link specific biological responses pathways to different classes of ENMs based on the pattern of redox modifications in a more predictive manner. Ultimate validation of these concepts through future studies with emerging classes of next-generation ENMs will provide valuable insights on the early molecular events that distinguish cellular adaptive responses from those that are responsible for adverse outcomes from ENM exposures, potentially leading to new classes of biomarkers that can be used for rapid hazard screening.

Acknowledgments The authors acknowledge support for this work provided by the National Institutes for Environmental Health Sciences (NIEHS) under grant ES027292 (BDT).

References

Abdal Dayem A, Hossain MK, Lee SB, Kim K, Saha SK, Yang G-M, Choi HY, Cho S-G. The role of Reactive Oxygen Species (ROS) in the biological activities of metallic nanoparticles. Int J Mol Sci. 2017;18(1):120. https://doi.org/10.3390/ijms18010120.

Anathy V, Aesif SW, Guala AS, Havermans M, Reynaert NL, Ho YS, Budd RC, Janssen-Heininger YM. Redox amplification of apoptosis by caspase-dependent cleavage of glutaredoxin 1 and S-glutathionylation of Fas. J Cell Biol. 2009;184(2):241–52.

Angelé-Martínez C, Nguyen KVT, Ameer FS, Anker JN, Brumaghim JL. Reactive oxygen species generation by copper(II) oxide nanoparticles determined by DNA damage assays and EPR spectroscopy. Nanotoxicology. 2017;11(2):278–88. https://doi.org/10.1080/17435390.2017.1293750.

AshaRani PV, Low Kah Mun G, Hande MP, Valiyaveettil S. Cytotoxicity and genotoxicity of silver nanoparticles in human cells. ACS Nano. 2009;3(2):279–90. https://doi.org/10.1021/nn800596w.

Benhar M, Forrester MT, Stamler JS. Protein denitrosylation: enzymatic mechanisms and cellular functions. Nat Rev Mol Cell Biol. 2009;10(10):721–32. https://doi.org/10.1038/nrm2764.

Blaser H, Dostert C, Mak TW, Brenner D. TNF and ROS crosstalk in inflammation. Trends Cell Biol. 2016;26(4):249–61. https://doi.org/10.1016/j.tcb.2015.12.002.

Boraschi D, Italiani P, Palomba R, Decuzzi P, Duschl A, Fadeel B, Moghimi SM. Nanoparticles and innate immunity: new perspectives on host defence. Semin Immunol. 2017;34:33–51. https://doi.org/10.1016/j.smim.2017.08.013.

Bouallegui Y, Ben Younes R, Oueslati R, Sheehan D. Role of endocytotic uptake routes in impacting the ROS-related toxicity of silver nanoparticles to Mytilus galloprovincialis: a redox proteomic investigation. Aquat Toxicol. 2018;200:21–7. https://doi.org/10.1016/j.aquatox.2018.04.013.

Burello E, Worth AP. A theoretical framework for predicting the oxidative stress potential of oxide nanoparticles. Nanotoxicology. 2011;5(2):228–35. https://doi.org/10.3109/17435390.2010.502980.

Buyukhatipoglu K, Clyne AM. Superparamagnetic iron oxide nanoparticles change endothelial cell morphology and mechanics via reactive oxygen species formation. J Biomed Mater Res A. 2011;96(1):186–95. https://doi.org/10.1002/jbm.a.32972.

Cao SS, Kaufman RJ. Endoplasmic reticulum stress and oxidative stress in cell fate decision and human disease. Antioxid Redox Signal. 2014;21(3):396–413. https://doi.org/10.1089/ars.2014.5851.

Cao Y, Long J, Liu L, He T, Jiang L, Zhao C, Li Z. A review of endoplasmic reticulum (ER) stress and nanoparticle (NP) exposure. Life Sci. 2017;186:33–42. https://doi.org/10.1016/j.lfs.2017.08.003.

Chanvorachote P, Nimmannit U, Lu Y, Talbott S, Jiang BH, Rojanasakul Y. Nitric oxide regulates lung carcinoma cell anoikis through inhibition of ubiquitin-proteasomal degradation of caveolin-1. J Biol Chem. 2009;284(41):28476–84.

Chen CA, Wang TY, Varadharaj S, Reyes LA, Hemann C, Talukder MA, Chen YR, Druhan LJ, Zweier JL. S-glutathionylation uncouples eNOS and regulates its cellular and vascular function. Nature. 2010;468(7327):1115–8. https://doi.org/10.1038/nature09599.

Chen R, Huo L, Shi X, Bai R, Zhang Z, Zhao Y, Chang Y, Chen C. Endoplasmic reticulum stress induced by zinc oxide nanoparticles is an earlier biomarker for nanotoxicological evaluation. ACS Nano. 2014;8(3):2562–74. https://doi.org/10.1021/nn406184r.

Chen G, Shen Y, Li X, Jiang Q, Cheng S, Gu Y, Liu L, Cao Y. The endoplasmic reticulum stress inducer thapsigargin enhances the toxicity of ZnO nanoparticles to macrophages and macrophage-endothelial co-culture. Environ Toxicol Pharmacol. 2017;50:103–10. https://doi.org/10.1016/j.etap.2017.01.020.

Chen Q, Wang N, Zhu M, Lu J, Zhong H, Xue X, Guo S, Li M, Wei X, Tao Y, Yin H. TiO(2) nanoparticles cause mitochondrial dysfunction, activate inflammatory responses, and attenuate phagocytosis in macrophages: a proteomic and metabolomic insight. Redox Biol. 2018;15:266–76. https://doi.org/10.1016/j.redox.2017.12.011.

Chiu HW, Xia T, Lee YH, Chen CW, Tsai JC, Wang YJ. Cationic polystyrene nanospheres induce autophagic cell death through the induction of endoplasmic reticulum stress. Nanoscale. 2015;7(2):736–46. https://doi.org/10.1039/c4nr05509h.

Choudhury SR, Ordaz J, Lo CL, Damayanti NP, Zhou F, Irudayaraj J. From the cover: zinc oxide nanoparticles-induced reactive oxygen species promotes multimodal cyto- and epigenetic toxicity. Toxicol Sci. 2017;156(1):261–74. https://doi.org/10.1093/toxsci/kfw252.

Christen V, Camenzind M, Fent K. Silica nanoparticles induce endoplasmic reticulum stress response, oxidative stress and activate the mitogen-activated protein kinase (MAPK) signaling pathway. Toxicol Rep. 2014;1:1143–51.

Chung HS, Wang S-B, Venkatraman V, Murray CI, Van Eyk JE. Cysteine oxidative posttranslational modifications: emerging regulation in the cardiovascular system. Circ Res. 2013;112(2):382–92. https://doi.org/10.1161/CIRCRESAHA.112.268680.

Costa PM, Fadeel B. Emerging systems biology approaches in nanotoxicology: towards a mechanism-based understanding of nanomaterial hazard and risk. Toxicol Appl Pharmacol. 2016;299:101–11. https://doi.org/10.1016/j.taap.2015.12.014.

Dalle-Donne I, Giustarini D, Rossi R, Colombo R, Milzani A. Reversible S-glutathionylation of Cys 374 regulates actin filament formation by inducing structural changes in the actin molecule. Free Radic Biol Med. 2003;34(1):23–32.

DeLoid G, Casella B, Pirela S, Filoramo R, Pyrgiotakis G, Demokritou P, Kobzik L. Effects of engineered nanomaterial exposure on macrophage innate immune function. NanoImpact. 2016;2:70–81.

Donaldson K, Borm PJ, Castranova V, Gulumian M. The limits of testing particle-mediated oxidative stress in vitro in predicting diverse pathologies; relevance for testing of nanoparticles. Part Fibre Toxicol. 2009;6:13.

Dostert C, Petrilli V, Van Bruggen R, Steele C, Mossman BT, Tschopp J. Innate immune activation through Nalp3 inflammasome sensing of asbestos and silica. Science. 2008;320(5876):674–7.

Duan J, Kodali VK, Gaffrey MJ, Guo J, Chu RK, Camp DG, Smith RD, Thrall BD, Qian WJ. Quantitative profiling of protein S-glutathionylation reveals redox-dependent regulation of macrophage function during nanoparticle-induced oxidative stress. ACS Nano. 2016;10(1):524–38. https://doi.org/10.1021/acsnano.5b05524.

Duan J, Gaffrey MJ, Qian WJ. Quantitative proteomic characterization of redox-dependent post-translational modifications on protein cysteines. Mol BioSyst. 2017;13(5):816–29. https://doi.org/10.1039/c6mb00861e.

Elliott EI, Miller AN, Banoth B, Iyer SS, Stotland A, Weiss JP, Gottlieb RA, Sutterwala FS, Cassel SL. Cutting edge: mitochondrial assembly of the NLRP3 inflammasome complex is initiated at priming. J Immunol. 2018;200(9):3047–52.

Fadeel B, Farcal L, Hardy B, Vazquez-Campos S, Hristozov D, Marcomini A, Lynch I, Valsami-Jones E, Alenius H, Savolainen K. Advanced tools for the safety assessment of nanomaterials. Nat Nanotechnol. 2018;13(7):537–43. https://doi.org/10.1038/s41565-018-0185-0.

Feduska JM, Tse HM. The proinflammatory effects of macrophage-derived NADPH oxidase function in autoimmune diabetes. Free Radic Biol Med. 2018;125:81–9. https://doi.org/10.1016/j.freeradbiomed.2018.04.581.

Fu PP, Xia Q, Hwang HM, Ray PC, Yu H. Mechanisms of nanotoxicity: generation of reactive oxygen species. J Food Drug Anal. 2014;22(1):64–75. https://doi.org/10.1016/j.jfda.2014.01.005.

Garcia-Hevia L, Valiente R, Martin-Rodriguez R, Renero-Lecuna C, Gonzalez J, Rodriguez-Fernandez L, Aguado F, Villegas JC, Fanarraga ML. Nano-ZnO leads to tubulin macrotube assembly and actin bundling, triggering cytoskeletal catastrophe and cell necrosis. Nanoscale. 2016;8(21):10963–73. https://doi.org/10.1039/c6nr00391e.

Gergondey R, Garcia C, Marchand CH, Lemaire SD, Camadro JM, Auchere F. Modulation of the specific glutathionylation of mitochondrial proteins in the yeast Saccharomyces cerevisiae under basal and stress conditions. Biochem J. 2017;474(7):1175–93. https://doi.org/10.1042/bcj20160927.

Getts DR, Shea LD, Miller SD, King NJ. Harnessing nanoparticles for immune modulation. Trends Immunol. 2015;36(7):419–27. https://doi.org/10.1016/j.it.2015.05.007.

Go YM, Jones DP. Redox compartmentalization in eukaryotic cells. Biochim Biophys Acta. 2008;1780(11):1273–90.

Greco TM, Hodara R, Parastatidis I, Heijnen HF, Dennehy MK, Liebler DC, Ischiropoulos H. Identification of S-nitrosylation motifs by site-specific mapping of the S-nitrosocysteine proteome in human vascular smooth muscle cells. Proc Natl Acad Sci U S A. 2006;103(19):7420–5. https://doi.org/10.1073/pnas.0600729103.

Grek CL, Zhang J, Manevich Y, Townsend DM, Tew KD. Causes and consequences of cysteine S-glutathionylation. J Biol Chem. 2013;288(37):26497–504.

Grimsrud PA, Xie H, Griffin TJ, Bernlohr DA. Oxidative stress and covalent modification of protein with bioactive aldehydes. J Biol Chem. 2008;283(32):21837–41. https://doi.org/10.1074/jbc.R700019200.

Guglielmo A, Sabra A, Elbery M, Cerveira MM, Ghenov F, Sunasee R, Ckless K. A mechanistic insight into curcumin modulation of the IL-1beta secretion and NLRP3 S-glutathionylation induced by needle-like cationic cellulose nanocrystals in myeloid cells. Chem Biol Interact. 2017;274:1–12. https://doi.org/10.1016/j.cbi.2017.06.028.

Guo J, Gaffrey MJ, Su D, Liu T, Camp DG, Smith RD, Qian W-J. Resin-assisted enrichment of thiols as a general strategy for proteomic profiling of cysteine-based reversible modifications. Nat Protoc. 2014;9(1):64–75. https://doi.org/10.1038/nprot.2013.161.

Gupta A, Anjomani-Virmouni S, Koundouros N, Dimitriadi M, Choo-Wing R, Valle A, Zheng Y, Chiu YH, Agnihotri S, Zadeh G, Asara JM, Anastasiou D, Arends MJ, Cantley LC,

Poulogiannis G. PARK2 depletion connects energy and oxidative stress to PI3K/Akt activation via PTEN S-nitrosylation. Mol Cell. 2017;65(6):999–1013 e1017. https://doi.org/10.1016/j.molcel.2017.02.019.

Gurung P, Lukens JR, Kanneganti TD. Mitochondria: diversity in the regulation of the NLRP3 inflammasome. Trends Mol Med. 2015;21(3):193–201.

Halamoda-Kenzaoui B, Bremer-Hoffmann S. Main trends of immune effects triggered by nanomedicines in preclinical studies. Int J Nanomedicine. 2018;13:5419–31.

Han XB, Li HX, Jiang YQ, Wang H, Li XS, Kou JY, Zheng YH, Liu ZN, Li H, Li J, Dou D, Wang Y, Tian Y, Yang LM. Upconversion nanoparticle-mediated photodynamic therapy induces autophagy and cholesterol efflux of macrophage-derived foam cells via ROS generation. Cell Death Dis. 2017;8(6):e2864.

Han Y, Lee DK, Kim SH, Lee S, Jeon S, Cho WS. High inflammogenic potential of rare earth oxide nanoparticles: the New Hazardous Entity. Nanotoxicology. 2018;12(7):712–28. https://doi.org/10.1080/17435390.2018.1472311.

He W, Zhou YT, Wamer WG, Boudreau MD, Yin JJ. Mechanisms of the pH dependent generation of hydroxyl radicals and oxygen induced by Ag nanoparticles. Biomaterials. 2012;33(30):7547–55. https://doi.org/10.1016/j.biomaterials.2012.06.076.

He W, Wu H, Wamer WG, Kim HK, Zheng J, Jia H, Zheng Z, Yin JJ. Unraveling the enhanced photocatalytic activity and phototoxicity of ZnO/metal hybrid nanostructures from generation of reactive oxygen species and charge carriers. ACS Appl Mater Interfaces. 2014;6(17):15527–35. https://doi.org/10.1021/am5043005.

He W, Jia H, Yang D, Xiao P, Fan X, Zheng Z, Kim HK, Wamer WG, Yin JJ. Composition directed generation of reactive oxygen species in irradiated mixed metal sulfides correlated with their photocatalytic activities. ACS Appl Mater Interfaces. 2015;7(30):16440–9. https://doi.org/10.1021/acsami.5b03626.

Hill BG, Ramana KV, Cai J, Bhatnagar A, Srivastava SK. Measurement and identification of S-glutathiolated proteins. Methods Enzymol. 2010;473:179–97. https://doi.org/10.1016/S0076-6879(10)73009-3.

Hughes MM, McGettrick AF, O'Neill LAJ. Glutathione and glutathione transferase omega 1 as key posttranslational regulators in macrophages. Microbiol Spectr. 2017;5(1). doi:https://doi.org/10.1128/microbiolspec.MCHD-0044-2016.

Jaffrey SR, Snyder SH. The biotin switch method for the detection of S-nitrosylated proteins. Sci STKE. 2001;2001(86):pl1. https://doi.org/10.1126/stke.2001.86.pl1.

Kadammattil AV, Sajankila SP, Prabhu S, Rao BN, Rao BSS. Systemic toxicity and teratogenicity of copper oxide nanoparticles and copper sulfate. J Nanosci Nanotechnol. 2018;18(4):2394–404. https://doi.org/10.1166/jnn.2018.14542.

Khanna P, Ong C, Bay BH, Baeg GH. Nanotoxicity: an interplay of oxidative stress, inflammation and cell death. Nanomaterials (Basel). 2015;5(3):1163–80. https://doi.org/10.3390/nano5031163.

Kim HS, Ullevig SL, Zamora D, Lee CF, Asmis R. Redox regulation of MAPK phosphatase 1 controls monocyte migration and macrophage recruitment. Proc Natl Acad Sci U S A. 2012;109(41):E2803–12.

Klomsiri C, Karplus PA, Poole LB. Cysteine-based redox switches in enzymes. Antioxid Redox Signal. 2011;14(6):1065–77.

Kodali V, Thrall BD. Oxidative stress and nanomaterial-cellular interactions. Studies on experimental toxicology and pharmacology. Oxidative stress in applied basic research and clinical practice: Springer International Publishing Switzerland Humana Press; 2015.

Kodali V, Littke MH, Tilton SC, Teeguarden JG, Shi L, Frevert CW, Wang W, Pounds JG, Thrall BD. Dysregulation of macrophage activation profiles by engineered nanoparticles. ACS Nano. 2013;7(8):6997–7010.

Kodali VK, Roberts JR, Shoeb M, Wolfarth MG, Bishop L, Eye T, Barger M, Roach KA, Friend S, Schwegler-Berry D, Chen BT, Stefaniak A, Jordan KC, Whitney RR, Porter DW, Erdely AD. Acute in vitro and in vivo toxicity of a commercial grade boron nitride nanotube mixture. Nanotoxicology. 2017;11(8):1040–58. https://doi.org/10.1080/17435390.2017.1390177.

Kramer PA, Duan J, Qian WJ, Marcinek DJ. The measurement of reversible redox dependent post-translational modifications and their regulation of mitochondrial and skeletal muscle function. Front Physiol. 2015;6:347. https://doi.org/10.3389/fphys.2015.00347.

Kramer PA, Duan J, Gaffrey MJ, Shukla AK, Wang L, Bammler TK, Qian W-J, Marcinek DJ. Fatiguing contractions increase protein S-glutathionylation occupancy in mouse skeletal muscle. Redox Biol. 2018;17:367–76. https://doi.org/10.1016/j.redox.2018.05.011.

Laurindo FR, Pescatore LA, Fernandes Dde C. Protein disulfide isomerase in redox cell signaling and homeostasis. Free Radic Biol Med. 2012;52(9):1954–69. https://doi.org/10.1016/j.freeradbiomed.2012.02.037.

Leon-Bollotte L, Subramaniam S, Cauvard O, Plenchette-Colas S, Paul C, Godard C, Martinez-Ruiz A, Legembre P, Jeannin JF, Bettaieb A. S-nitrosylation of the death receptor fas promotes fas ligand-mediated apoptosis in cancer cells. Gastroenterology. 2011;140(7):2009–18, 2018 e2001–2004. https://doi.org/10.1053/j.gastro.2011.02.053.

Libalova H, Costa PM, Olsson M, Farcal L, Ortelli S, Blosi M, Topinka J, Costa AL, Fadeel B. Toxicity of surface-modified copper oxide nanoparticles in a mouse macrophage cell line: interplay of particles, surface coating and particle dissolution. Chemosphere. 2018;196:482–93. https://doi.org/10.1016/j.chemosphere.2017.12.182.

Liu YX, Karsai A, Anderson DS, Silva RM, Uyeminami DL, Van Winkle LS, Pinkerton KE, Liu GY. Single-cell mechanics provides an effective means to probe in vivo interactions between alveolar macrophages and silver nanoparticles. J Phys Chem B. 2015;119(49):15118–29. https://doi.org/10.1021/acs.jpcb.5b07656.

Liu Y, Rogel N, Harada K, Jarett L, Maiorana CH, German GK, Mahler GJ, Doiron AL. Nanoparticle size-specific actin rearrangement and barrier dysfunction of endothelial cells. Nanotoxicology. 2017;11(7):846–56. https://doi.org/10.1080/17435390.2017.1371349.

Manke A, Wang L, Rojanasakul Y. Mechanisms of nanoparticle-induced oxidative stress and toxicity. Biomed Res Int. 2013;2013:942916. https://doi.org/10.1155/2013/942916.

Meng H, Xia T, George S, Nel AE. A predictive toxicological paradigm for the safety assessment of nanomaterials. ACS Nano. 2009;3(7):1620–7. https://doi.org/10.1021/nn9005973.

Mohr S, Hallak H, de Boitte A, Lapetina EG, Brune B. Nitric oxide-induced S-glutathionylation and inactivation of glyceraldehyde-3-phosphate dehydrogenase. J Biol Chem. 1999;274(14):9427–30.

Moldogazieva NT, Mokhosoev IM, Feldman NB, Lutsenko SV. ROS and RNS signalling: adaptive redox switches through oxidative/nitrosative protein modifications. Free Radic Res. 2018;52(5):507–43. https://doi.org/10.1080/10715762.2018.1457217.

Munusamy P, Wang C, Engelhard MH, Baer DR, Smith JN, Liu C, Kodali V, Thrall BD, Chen S, Porter AE, Ryan MP. Comparison of 20 nm silver nanoparticles synthesized with and without a gold core: structure, dissolution in cell culture media, and biological impact on macrophages. Biointerphases. 2015;10(3):031003. https://doi.org/10.1116/1.4926547.

Nakamura T, Tu S, Akhtar MW, Sunico CR, Okamoto S, Lipton SA. Aberrant protein s-nitrosylation in neurodegenerative diseases. Neuron. 2013;78(4):596–614.

Nakayama M. Macrophage recognition of crystals and nanoparticles. Front Immunol. 2018;9:103. https://doi.org/10.3389/fimmu.2018.00103.

Nichols CE, Shepherd DL, Hathaway QA, Durr AJ, Thapa D, Abukabda A, Yi J, Nurkiewicz TR, Hollander JM. Reactive oxygen species damage drives cardiac and mitochondrial dysfunction following acute nano-titanium dioxide inhalation exposure. Nanotoxicology. 2018;12(1):32–48.

Niethammer P. Wound redox gradients revisited. Semin Cell Dev Biol. 2018;80:13–6.

Nishanth RP, Jyotsna RG, Schlager JJ, Hussain SM, Reddanna P. Inflammatory responses of RAW 264.7 macrophages upon exposure to nanoparticles: role of ROS-NFkappaB signaling pathway. Nanotoxicology. 2011;5(4):502–16. https://doi.org/10.3109/17435390.2010.541604.

Palmer KT, Poole J, Ayres JG, Mann J, Burge PS, Coggon D. Exposure to metal fume and infectious pneumonia. Am J Epidemiol. 2003;157(3):227–33.

Poillet-Perez L, Despouy G, Delage-Mourroux R, Boyer-Guittaut M. Interplay between ROS and autophagy in cancer cells, from tumor initiation to cancer therapy. Redox Biol. 2015;4:184–92. https://doi.org/10.1016/j.redox.2014.12.003.

Poirier I, Pallud M, Kuhn L, Hammann P, Demortiere A, Jamali A, Chicher J, Caplat C, Gallon RK, Bertrand M. Toxicological effects of CdSe nanocrystals on the marine diatom Phaeodactylum tricornutum: the first mass spectrometry-based proteomic approach. Ecotoxicol Environ Saf. 2018;152:78–90. https://doi.org/10.1016/j.ecoenv.2018.01.043.

Poole LB. The basics of thiols and cysteines in redox biology and chemistry. Free Radic Biol Med. 2015;80:148–57. https://doi.org/10.1016/j.freeradbiomed.2014.11.013.

Rezende F, Brandes RP, Schroder K. Detection of hydrogen peroxide with fluorescent dyes. Antioxid Redox Signal. 2018;29(6):585–602. https://doi.org/10.1089/ars.2017.7401.

Sakai J, Li J, Subramanian KK, Mondal S, Bajrami B, Hattori H, Jia Y, Dickinson BC, Zhong J, Ye K, Chang CJ, Ho YS, Zhou J, Luo HR. Reactive oxygen species-induced actin glutathionylation controls actin dynamics in neutrophils. Immunity. 2012;37(6):1037–49. https://doi.org/10.1016/j.immuni.2012.08.017.

Santos CX, Tanaka LY, Wosniak J, Laurindo FR. Mechanisms and implications of reactive oxygen species generation during the unfolded protein response: roles of endoplasmic reticulum oxidoreductases, mitochondrial electron transport, and NADPH oxidase. Antioxid Redox Signal. 2009;11(10):2409–27. https://doi.org/10.1089/ars.2009.2625.

Sengupta R, Ryter SW, Zuckerbraun BS, Tzeng E, Billiar TR, Stoyanovsky DA. Thioredoxin catalyzes the denitrosation of low-molecular mass and protein S-nitrosothiols. Biochemistry. 2007;46(28):8472–83. https://doi.org/10.1021/bi700449x.

Sharma G, Kodali V, Gaffrey M, Wang W, Minard KR, Karin NJ, Teeguarden JG, Thrall BD. Iron oxide nanoparticle agglomeration influences dose rates and modulates oxidative stress-mediated dose-response profiles in vitro. Nanotoxicology. 2014;8(6):663–75. https://doi.org/10.3109/17435390.2013.822115.

Smith JN, Thomas DG, Jolley H, Kodali VK, Littke MH, Munusamy P, Baer DR, Gaffrey MJ, Thrall BD, Teeguarden JG. All that is silver is not toxic: silver ion and particle kinetics reveals the role of silver ion aging and dosimetry on the toxicity of silver nanoparticles. Part Fibre Toxicol. 2018;15(1):47.

Srikanth K, Pereira E, Duarte AC, Rao JV. Evaluation of cytotoxicity, morphological alterations and oxidative stress in Chinook salmon cells exposed to copper oxide nanoparticles. Protoplasma. 2016;253(3):873–84. https://doi.org/10.1007/s00709-015-0849-7.

Stadtman ER, Berlett BS. Reactive oxygen-mediated protein oxidation in aging and disease. Chem Res Toxicol. 1997;10(5):485–94. https://doi.org/10.1021/tx960133r.

Su D, Shukla AK, Chen B, Kim J-S, Nakayasu E, Qu Y, Aryal U, Weitz K, Clauss TRW, Monroe ME, Camp DG, Bigelow DJ, Smith RD, Kulkarni RN, Qian W-J. Quantitative site-specific reactivity profiling of S-nitrosylation in mouse skeletal muscle using cysteinyl peptide enrichment coupled with mass spectrometry. Free Radic Biol Med. 2013;57:68–78. https://doi.org/10.1016/j.freeradbiomed.2012.12.010.

Su D, Gaffrey MJ, Guo J, Hatchell KE, Chu RK, Clauss TRW, Aldrich JT, Wu S, Purvine S, Camp DG, Smith RD, Thrall BD, Qian W-J. Proteomic identification and quantification of S-glutathionylation in mouse macrophages using resin-assisted enrichment and isobaric labeling. Free Radic Biol Med. 2014;67:460–70. https://doi.org/10.1016/j.freeradbiomed.2013.12.004.

Sun X, Yang Y, Shi J, Wang C, Yu Z, Zhang H. NOX4- and Nrf2-mediated oxidative stress induced by silver nanoparticles in vascular endothelial cells. J Appl Toxicol. 2017;37(12):1428–37. https://doi.org/10.1002/jat.3511.

Sundarraj K, Manickam V, Raghunath A, Periyasamy M, Viswanathan MP, Perumal E. Repeated exposure to iron oxide nanoparticles causes testicular toxicity in mice. Environ Toxicol. 2017;32(2):594–608. https://doi.org/10.1002/tox.22262.

Tabet L, Bussy C, Amara N, Setyan A, Grodet A, Rossi MJ, Pairon JC, Boczkowski J, Lanone S. Adverse effects of industrial multiwalled carbon nanotubes on human pulmonary cells. J Toxicol Environ Health A. 2009;72(2):60–73.

Tournebize J, Sapin-Minet A, Bartosz G, Leroy P, Boudier A. Pitfalls of assays devoted to evaluation of oxidative stress induced by inorganic nanoparticles. Talanta. 2013;116:753–63. https://doi.org/10.1016/j.talanta.2013.07.077.

Ullevig S, Kim HS, Asmis R. S-glutathionylation in monocyte and macrophage (dys)function. Int J Mol Sci. 2013;14(8):15212–32.

Ullevig SL, Kim HS, Short JD, Tavakoli S, Weintraub ST, Downs K, Asmis R. Protein S-glutathionylation mediates macrophage responses to metabolic cues from the extracellular environment. Antioxid Redox Signal. 2016;25(15):836–51. https://doi.org/10.1089/ars.2015.6531.

Waiskopf N, Ben-Shahar Y, Galchenko M, Carmel I, Moshitzky G, Soreq H, Banin U. Photocatalytic reactive oxygen species formation by semiconductor-metal hybrid nanoparticles. Toward light-induced modulation of biological processes. Nano Lett. 2016;16(7):4266–73. https://doi.org/10.1021/acs.nanolett.6b01298.

Wang X, Ji Z, Chang CH, Zhang H, Wang M, Liao YP, Lin S, Meng H, Li R, Sun B, Winkle LV, Pinkerton KE, Zink JI, Xia T, Nel AE. Use of coated silver nanoparticles to understand the relationship of particle dissolution and bioavailability to cell and lung toxicological potential. Small. 2014;10(2):385–98.

Wang J, Li Y, Duan J, Yang M, Yu Y, Feng L, Yang X, Zhou X, Zhao Z, Sun Z. Silica nanoparticles induce autophagosome accumulation via activation of the EIF2AK3 and ATF6 UPR pathways in hepatocytes. Autophagy. 2018;14(7):1185–200.

Weigert A, von Knethen A, Fuhrmann D, Dehne N, Brune B. Redox-signals and macrophage biology. Mol Asp Med. 2018;63:70–87. https://doi.org/10.1016/j.mam.2018.01.003.

Wilson C, González-Billault C. Regulation of cytoskeletal dynamics by redox signaling and oxidative stress: implications for neuronal development and trafficking. Front Cell Neurosci. 2015;9:381. https://doi.org/10.3389/fncel.2015.00381.

Woolley JF, Stanicka J, Cotter TG. Recent advances in reactive oxygen species measurement in biological systems. Trends Biochem Sci. 2013;38(11):556–65. https://doi.org/10.1016/j.tibs.2013.08.009.

Xia T, Kovochich M, Brant J, Hotze M, Sempf J, Oberley T, Sioutas C, Yeh JI, Wiesner MR, Nel AE. Comparison of the abilities of ambient and manufactured nanoparticles to induce cellular toxicity according to an oxidative stress paradigm. Nano Lett. 2006;6(8):1794–807. https://doi.org/10.1021/nl061025k.

Xiong S, George S, Ji Z, Lin S, Yu H, Damoiseaux R, France B, Ng KW, Loo SCJ. Size of TiO(2) nanoparticles influences their phototoxicity: an in vitro investigation. Arch Toxicol. 2013;87(1):99–109. https://doi.org/10.1007/s00204-012-0912-5.

Xue Y, Chen Q, Sun J. Hydroxyapatite nanoparticle-induced mitochondrial energy metabolism impairment in liver cells: in vitro and in vivo studies. J Appl Toxicol. 2017;37(8):1004–16. https://doi.org/10.1002/jat.3450.

Yang X, Shao H, Liu W, Gu W, Shu X, Mo Y, Chen X, Zhang Q, Jiang M. Endoplasmic reticulum stress and oxidative stress are involved in ZnO nanoparticle-induced hepatotoxicity. Toxicol Lett. 2015;234(1):40–9. https://doi.org/10.1016/j.toxlet.2015.02.004.

Yang J, Carroll KS, Liebler DC. The expanding landscape of the thiol redox proteome. Mol Cell Proteomics. 2016;15(1):1–11.

Yi X, Zhao WL, Li JR, Zhang B, Yu QL, Li MC. Mn3O4 nanoparticles cause endoplasmic reticulum stress-dependent toxicity to Saccharomyces cerevisiae. RSC Adv. 2017;7(73):46028–35. https://doi.org/10.1039/c7ra07458a.

Yu KN, Chang SH, Park SJ, Lim J, Lee J, Yoon TJ, Kim JS, Cho MH. Titanium dioxide nanoparticles induce endoplasmic reticulum stress-mediated Autophagic cell death via mitochondria-associated endoplasmic reticulum membrane disruption in normal lung cells. PLoS One. 2015;10(6):e0131208.

Zhang H, Ji Z, Xia T, Meng H, Low-Kam C, Liu R, Pokhrel S, Lin S, Wang X, Liao YP, Wang M, Li L, Rallo R, Damoiseaux R, Telesca D, Madler L, Cohen Y, Zink JI, Nel AE. Use of metal oxide nanoparticle band gap to develop a predictive paradigm for oxidative stress and acute pulmonary inflammation. ACS Nano. 2012;6(5):4349–68. https://doi.org/10.1021/nn3010087.

Zhang J, Qin X, Wang B, Xu G, Qin Z, Wang J, Wu L, Ju X, Bose DD, Qiu F, Zhou H, Zou Z. Zinc oxide nanoparticles harness autophagy to induce cell death in lung epithelial cells. Cell Death Dis. 2017;8(7):e2954. https://doi.org/10.1038/cddis.2017.337.

Zhang T, Gaffrey MJ, Thrall BD, Qian WJ. Mass spectrometry-based proteomics for system-level characterization of biological responses to engineered nanomaterials. Anal Bioanal Chem. 2018a;410(24):6067–77. https://doi.org/10.1007/s00216-018-1168-6.

Zhang Q, Lin S, Shi S, Zhang T, Ma Q, Tian T, Zhou T, Cai X, Lin Y. Anti-inflammatory and antioxidative effects of tetrahedral DNA nanostructures via the modulation of macrophage responses. ACS Appl Mater Interfaces. 2018b;10(4):3421–30. https://doi.org/10.1021/acsami.7b17928.

Chapter 9
Allergy and Immunity Induced by Nanomaterials

Harri Alenius and Kai Savolainen

Abstract Specific types of engineered nanomaterials have been shown to induce allergic inflammation in the lungs or in the skin. For example, rigid or rod-like carbon nanotubes delivered to the lungs of mice are capable of inducing direct asthmatic-like effects, while tangled carbon nanotubes exacerbate existing allergic lung inflammation in various mouse models of asthma. The skin is also an important target organ of allergic effects of engineered nanomaterials. For example, zinc or silver nanoparticles have been shown to modify atopic dermatitis in mice or humans, respectively. These findings indicate that engineered nanomaterials may have direct or modifying effects on allergic immune reactions in important target organs, notably the lungs and the skin.

Keywords Engineered nanomaterials · Allergy · Asthma · Dermatitis

9.1 Immunity Behind Allergic Inflammation

The prevalence of allergic diseases has dramatically increased especially in the developed countries during the recent 50 years. It has been estimated that up to 40% of the population worldwide is sensitized (i.e., exhibits IgE antibodies to foreign proteins in the environment) and has an increased risk for developing allergic diseases. Allergies are the most frequent chronic disease in the EU with 44–76 million individuals suffering from allergic diseases. The U.S. Global Change Research

H. Alenius (✉)
Institute of Environmental Medicine, Karolinska Institute, Stockholm, Sweden

Human Microbiome Research (HUMI), Medical Faculty, University of Helsinki, Helsinki, Finland
e-mail: harri.alenius@ki.se

K. Savolainen
Work Environment, Finnish Institute of Occupational Health, Helsinki, Finland

J. C. Bonner, J. M. Brown (eds.), *Interaction of Nanomaterials with the Immune System*, Molecular and Integrative Toxicology,
https://doi.org/10.1007/978-3-030-33962-3_9

Program predicted worsened allergy and asthma conditions due to changes in climate and biodiversity loss leading to poor and allergy promoting outdoor and indoor air quality (Crimmins et al. 2016). By today, allergy has become a major healthcare problem (www.worldallergy.org).

Allergic reactions are the result of inappropriate immune reactions to substances (typically proteins) that are normally harmless to the host. Allergic reactions are most commonly classified to immediate hypersensitivity reactions that involve IgE-mediated release of bioactive mediators from the sensitized mast cells and to delayed reactions which are mediated by T cells and monocytes/macrophages. Allergic immune response is initiated by sensitization phase in which allergen exposure together with epithelial insults results in internalization of allergens by immature dendritic cells (DC). Secretion of proinflammatory cytokines from the activated airway epithelium induces DC maturation and expression of homing molecules, resulting in the recruitment of DC to the draining lymph nodes. Generation of allergen-specific Th2 cells and the production of allergen-specific IgE antibodies by the B cells will take a place in draining lymph nodes. IgE antibodies are released into the circulation, and finally they bind the high-affinity IgE receptors on the surface of mast cells in the tissue causing mast cell sensitization.

Immediate hypersensitivity reactions (Type I reactions – appear within the minutes after the allergen encounter) manifest clinically as allergic asthma, allergic rhinitis, food allergy, and atopic dermatitis. Immediate reactions are mediated by the release of inflammatory mediators such as histamine and tryptase from mast cells, causing clinical symptoms like vasodilatation, mucus secretion, increased vascular permeability, and increased smooth muscle contraction. The delayed type reactions (Type IV reactions) occur about 48–72 hours later, after the encounter with allergen, and are mediated by antigen-specific effector T cells. Type IV reactions involve especially activated T cells, which release cytokines and chemokines, macrophages, and cytotoxic CD8+ T cells that are attracted by these biomolecules. Type IV hypersensitivity is the main mechanism of chemical-induced allergic contact dermatitis.

9.1.1 Allergic Airway Inflammation and Asthma

Asthma is a chronic inflammatory disease of the airways affecting about 6.8% of the working adults in the United States during 2011–2016 (Mazurek and Syamlal 2018). Hallmarks of asthma include bronchial hyperreactivity, mucus overproduction, and remodeling of airway walls. Typical clinical symptoms of asthmatics are wheezing, shortness of breath, and chest tightness (Lambrecht and Hammad 2015). Asthma can be divided roughly into nonallergic (intrinsic) and allergic (extrinsic) forms. Allergic asthma usually starts early in life and is triggered by environmental allergens such as pet dander and pollen. In contrast, nonallergic asthma is more commonly diagnosed in adults and is induced by exercise, respiratory infections, or cold air (Kim et al. 2010).

Allergic asthma is the most common form of asthma. Sensitization occurs in the first contact with an allergen (typically protein), resulting in the production of allergen-specific IgE antibodies which attach on FcεRI receptor on the surface of mast cells and basophils. Reencountering with the same allergen induces full-scale allergic response (Galli and Tsai 2012). The allergen binds to IgE antibodies on mast cells and basophils initiating early-phase reactions in which bioactive mediators (e.g., histamine, lipid mediators, cytokines, and chemokines) will be released that cause increased vascular permeability, mucus over production, and bronchoconstriction within minutes after exposure. Mast cell mediators contribute to the transition to the late-phase reaction (hours after allergen exposure) by promoting an influx of circulating leukocytes. Late-phase reaction includes activation and migration of CD4+ Th2 cells, eosinophils, and other leukocytes with continuous mediator release from local cells (Galli and Tsai 2012).

In addition to the acute responses, asthma is often characterized by chronic airway inflammation and airway wall remodeling. Repetitive or persistent exposure to allergens may occur for several years finally resulting in structural changes of the airway wall – that is, remodeling. Typical features of airway remodeling include increased numbers of mucus-producing goblet cells, epithelial cell shedding, thickening of the smooth muscle layer and basement membrane, and a pathological increase of vasculature (Galli and Tsai 2012).

9.1.2 Allergic Skin Inflammation

Skin is an important organ of our body – it is a buffer between internal tissues and the environment, as well as a protection from pathogens and hazardous substances. It is composed of stratified epidermis and underlying dermis. Epidermis, the outermost layer of the skin, consists mainly of keratinocytes, but it also contains melanocytes, Merkel cells, and Langerhans' cells. The most superficial cornified layer of the epidermis (*stratum corneum*) provides an efficient protection against the external environment by continuously shedding from the surface to clean and thereby renewing the skin. Dermis – a layer between the epidermis and the subcutaneous tissue – consists of a net of elastin and collagen fibers, in which mast cells, macrophages, and dendritic cells reside. Dermis contains also hair follicles and sweat and sebaceous glands.

9.1.2.1 Atopic Dermatitis

Atopic dermatitis (AD) is a most common chronic inflammatory skin disease that usually starts in an early age. It affects about 15–20% of the population in developed countries (Weidinger et al. 2018). The prevalence of AD is considerably high among children (up to 25%), but it is frequent also in adults (about 7–10%). The most common comorbidities of AD are allergic rhinitis and asthma, which co-occur in 25–30% of patients (Weidinger et al. 2018).

The strongest known risk factor for AD is a family history of atopic diseases, in particular AD. Twin studies suggest a heritability of about 75% (Thomsen et al. 2007). Also, multiple genetic risk factors for AD have been identified, including factors associated with innate and adaptive immune pathways and skin barrier function. The strongest genetic risk factor for AD is a loss-of-function mutations of the filaggrin gene, which is a key component in the skin barrier function (Sandilands et al. 2009). Th2 cytokines (IL-4, IL-13, and IL-33), which are characteristic to the immune response in atopic skin inflammation, can perturb skin barrier function via multiple mechanisms, and this complex interplay between the immune system and the barrier function can create a microenvironment where skin sensitization can lead to atopic disease (Seltmann et al. 2015).

The skin of AD patients is highly colonized with *Staphylococcus aureus*, which is likely partially due to the Th2-type environment in the AD causing the suppression of antimicrobial peptide production. However, *S. aureus* produces enterotoxins that act as superantigens – they bind MHC and TCR molecules nonspecifically, causing an uncontrollable activation of T cells and overproduction of cytokines that results in aggravation of AD (Bieber 2008).

9.1.2.2 Allergic Contact Dermatitis

Allergic contact dermatitis (ACD) is developed in response to a hapten in sensitized individuals. Hapten is a low-molecular-weight molecule, such as chemical or divalent cation, for example, nickel, unable to provoke an immune response by itself, but once bound to self-proteins, it generates antigenic determinants. ACD is a common disease affecting up to 20% of the general population (Diepgen et al. 2016), resulting in considerable physical handicap and reduced quality of life.

ACD is the result of a delayed contact hypersensitivity reaction elicited by allergens, and the inflammatory response is orchestrated by allergen-specific memory T cells, including a sensitization phase and an elicitation phase. During the sensitization phase, low-molecular-weight haptens induce an inflammatory response, activating hapten-specific T cells which are recruited to the skin. The haptens form complexes with self-proteins (such as skin proteins or serum albumin), which are internalized and processed by DCs, which present them to specific T cells in draining lymph nodes. Allergen-specific, clonally expanded T cells then leave the lymph nodes and circulate in the blood and secondary lymphatic organs and tissues, including the skin. In the elicitation phase, following repeated contact with the same allergen, the allergen-specific T cells already present in the skin will orchestrate an inflammatory response, manifesting in clinical dermatitis (Fyhrquist et al. 2014).

9.2 Effect of Engineered Nanomaterials on Allergic Pulmonary Inflammation

Nanomaterials may cause risks for immune-mediated lung diseases through occupational, consumer, or environmental exposures (Bonner 2010; Card et al. 2008; Savolainen et al. 2010). Immunotoxicity of nanomaterials can be defined as any adverse effect on the immune system following exposure to these materials leading to immune stimulation or immune suppression (Hussain et al. 2012). Engineered nanomaterials may also exhibit either immunostimulatory or immunosuppressive effects in the lungs. The types of effects that predominate largely depend on the specific type of nanomaterial. Effects of nanomaterials on the immune system may also depend on the type of exposure; for example, repeated exposures versus a single nanomaterial exposure after the establishment of an allergic inflammation.

Alveolar macrophages serve the central immune defense against inhaled particles and fibers. Most inhaled engineered nanomaterials, deposited in the distal regions of the lung, are readily engulfed by alveolar macrophages. However, the uptake of engineered nanomaterials may have a variety of consequences related to macrophage biology and function. The high aspect ratio nanomaterials (carbon nanotubes) cause inflammasome activation in macrophages, which results in the processing and secretion of the pro-inflammatory cytokines IL-1β and IL-18 (Meunier et al. 2012; Palomaki et al. 2011; Sun et al. 2013). Inflammasome activation by multiwalled carbon nanotubes (MWCNT) and other high aspect ratio materials (e.g., nanofibers, TiO_2 nanobelts) is mediated by lysosomal disruption and ROS production (Sun et al. 2013). Exposure to certain nanomaterials (TiO_2, ZnO, CNTs) results in a Th1 immune cell microenvironment that promotes polarization of classically activated macrophages (M1) capable of inflammasome activation. In allergic asthma or fibrosis, however, macrophages are polarized to alternatively activated macrophages (M2) in the presence of Th2 cytokines IL-4 or IL-13 (Scotton et al. 2005). Human THP-1 macrophages stimulated with IL-4 or IL-13 in vitro have reduced IL-1β production, and mice with allergic lung inflammation induced by house dust mite have reduced IL-1β levels in bronchoalveolar lavage fluid, indicating that a Th2 microenvironment suppresses the inflammasome mechanism (Shipkowski et al. 2015). Eosinophils are a primary infiltrating cell type into the lungs following initiation of a Th2 immune response and are the hallmark of asthma. Some engineered nanomaterials have also caused eosinophilic inflammation in the lungs of mice, including certain types of MWCNT (rod-like) (Rydman et al. 2014). Mast cells have been shown to participate in nanomaterial-induced immune and inflammatory responses after pulmonary exposure to influence physiological responses in other organ systems such as the heart (Katwa et al. 2012). In addition, anatase and rutile TiO_2 nanoparticles induce histamine secretion in mast cells (Chen et al. 2012). Hence, there is convincing evidence that some engineered nanomaterials are capable of either directly inducing allergic inflammation or exacerbating preexisting allergic lung disease (Bonner 2010).

9.2.1 Carbon Nanomaterials

Carbon nanotubes (CNT) exhibit immunotoxic responses such as allergic sensitization and immunosuppression in experimental animals (primarily rodents) following inhalation exposure or other pulmonary delivery routes such as oropharyngeal aspiration or intranasal aspiration (Thompson et al. 2015). However, no epidemiological evidence is available to support immunotoxicity or any adverse outcome in humans exposed occupationally, through unanticipated environmental contact or through exposure to consumer products containing CNTs. This may be due to the novelty of nanotechnology and small quantities of CNT produced globally rendering the existence of CNT at large amounts unlikely in the workplace air or surfaces. Based on data from experimental animals, there has been guidance for occupational exposure limits that consider CNTs as a single type of material (NIOSH 2013). However, CNTs represent a diverse class of nanomaterials that have variable physicochemical characteristics; for example, single-walled carbon nanotubes (SWCNTs) vs multi-walled carbon nanotubes (MWCNTs), different aspect ratios, rigid vs flexible CNTs, or modified forms after CNT synthesis by surface functionalization. There are, however, data to document immunotoxicity in response to well-characterized CNTs in experimental animals (Kuempel et al. 2017).

CNTs primarily impact innate immune function by affecting the function of macrophages, that is, professional phagocytic cells. Long, rigid, and biopersistent MWCNTs behave like asbestos fibers to cause frustrated phagocytosis and the rupture of lysosomal or cell membranes to cause leakage of proteases and cytokines that cause inflammation (Donaldson et al. 2010). Lysosomal membrane damage by MWCNTs is a key factor in releasing cathepsin B into the cytoplasm that plays a role in inflammasome activation, resulting in the secretion of interleukin (IL)-1beta and IL-18 as a primary innate immune response (Hamilton et al. 2012; Palomaki et al. 2011).

The acquired immune response is also affected by pulmonary CNT exposure. For example, a challenge exposure to either SWCNT or MWCNT in mouse models of allergic asthma after allergen sensitization exacerbates airway remodeling (fibrosis), mucous cell metaplasia and mucus hypersecretion, eosinophilic inflammation, increased levels of T-helper 2 (Th2) cytokines, such as IL-13, and elevates levels of allergen-induced serum IgE levels (Inoue 2011; Ryman-Rasmussen et al. 2009). Exacerbation of allergen-induced airway disease has taken place with SWCNT or tangled flexible MWCNTs, but most of these CNTs do not directly elicit allergic Th2 responses in the lungs of mice upon a single inhalation or oropharyngeal exposure. Instead, the typical pulmonary immune response to SWCNT or tangled MWCNTs is a Th1 immune response characterized by acute neutrophilic inflammation that progresses to pulmonary fibrosis or granuloma formation over the course of several weeks (Bonner et al. 2013; Porter et al. 2010). However, more recent work demonstrated that rigid, rod-like MWCNTs can directly mediate asthma-like immunologic effects (Th2 cytokines), pathologic effects (eosinophilia, mucus hypersecretion, fibrosis), and physiological effects (airway hyperresponsiveness) after inhalation exposure in the absence of any allergen co-exposure (Rydman et al. 2014). Mast cells were found to partly regulate allergic inflammation caused by rod-like MWCNTs. However, mast cells are not solely linked to allergic inflammatory responses, as these inflammatory cells have been demonstrated to play a key role in pulmonary and cardiovascular response to tangled MWCNTs (Katwa et al. 2012). An emerging concept is that different types of MWCNTs with different physical and chemical characteristics can cause widely different acquired immune responses (e.g., Th1 vs Th2 lymphocyte polarization). Even when different types of CNTs produce seemingly similar inflammatory responses in the lungs of mice, they may elicit differences in transcriptional and protein biomarkers of fibrosis or innate immune responses of macrophages (Palomaki et al. 2011; Poulsen et al. 2015).

Whether or not CNTs directly cause allergic lung inflammation or exacerbate lung inflammation may depend in mouse strain differences. However, the ability of some CNTs to directly promote allergic lung inflammation is also influenced by physical and/or chemical characteristics. For example, a comparison of tangled, flexible MWCNTs and rigid, rod-like MWCNTs delivered by occupationally relevant doses via inhalation exposure showed that rod-like MWCNTs caused allergic airway inflammation that was characterized by increased numbers of eosinophils and mucus hypersecretion, along with Th2 cytokines and chemokines (Rydman et al. 2014). In contrast, tangled MWCNTs caused only a very weak lung inflammation without Th2-type characteristics.

There is also evidence that specific genes play a role in susceptibility to MWCNT-induced exacerbation of allergic lung disease. For example, COX-2 KO mice have exaggerated lung inflammation when exposed sequentially to ovalbumin and then MWCNTs. These mice also exhibit elevated levels of serum IgE and increased IL-13 mRNA (Sayers et al. 2013). STAT1 KO mice are also susceptible to lung inflammation and fibrosis induced by ovalbumin presensitization followed by MWCNT exposure by oropharyngeal aspiration and have increased biomarkers of Th2 inflammation (Thompson et al. 2015). In addition, elevated levels of IgE antibodies are evoked by MWCNT in ICR mice along with increased levels of Th2 cytokines, such as IL-4 and IL-5 (Park et al. 2009). Similarly, allergic lung inflammation and elevated serum IgE levels were induced after oropharyngeal aspiration exposure to MWCNTs in ICR mice (Inoue et al. 2009; Inoue et al. 2010). Furthermore, exposure to tangled MWCNTs exacerbated ovalbumin allergen-induced airway inflammation and Th2 cytokine levels in C57BL6 mice (Ryman-Rasmussen et al. 2009). There is, hence, convincing evidence that some types of CNT may induce allergic asthma alone or exacerbate existing asthma.

9.2.2 Nanocelluloses

The use of different types of nanocellulose molecules, crystals and fibers, has widely increased due to their nanotechnological advantages that have made such applications available for a number of new industrial and consumer applications.

Nanocelluloses have become important in applications such as nanofillers in composite polymers, in building materials such as insulation, in the construction industry, and in the cosmetics, food, and drug industries. Possible health effects of cellulose nanomaterials have caused concerns because in some cases crystalline or fibrous structure of nanocellulose resembles carbon nanotubes or even asbestos fibers – raising worries that their use may lead to adverse effects similar to those caused by exposure to asbestos (Yanamala et al. 2014). The wide use of cellulose, in general, and the likelihood of the increasing use of nanocellulose have prompted much research on the potentially harmful effects of nanocellulose, especially in the lungs.

Cellulose nanocrystal powder and cellulose nanocrystal suspension given to mice via the lungs by using pharyngeal aspiration at low doses facilitated innate inflammatory response as shown by increases of leucocytes and eosinophils in the bronchoalveolar lavage (BAL). There was a marked elevation in the levels of oxidatively modified proteins after exposure to nanocellulose dispersions. The powder form of the cellulose nanocrystals was, in turn, potent in increasing the levels of pro-inflammatory cytokines in the BAL of exposed mice (Yanamala et al. 2014).

Nanocellulose fibers were more cytotoxic than nanocellulose crystals when human A549 cells were exposed for 24 or 72 h to low-graded doses of these materials (Menas et al. 2017). The nanocellulose fibers induced cytotoxicity of the cells, whereas the crystals induced a more intense innate immunological reaction in the exposed cells. The reaction was characterized by markedly elevated levels of pro-inflammatory cytokines. The nanocellulose nanocrystals were engulfed by the A549 cells, whereas the fibers were not. In summary, nanocellulose fibers seemed to be more cytotoxic than nanocellulose crystals, which in turn readily induced an innate inflammatory reaction in the cells.

Nanofibrillated cellulose (NFC) oxidized with TEMPO-(2,2,6,6-tetramethyl-piperidin-1-oxyl) (Catalan et al. 2016) was administered to mice at doses of 10, 40, 80, or 200 μg/animal via single pharyngeal aspiration. The two lower doses induced local DNA damage in lung cells, but not systemic genotoxicity in bone marrow cells. NFC increased the recruitment of inflammatory cells to the lungs, together with a dose-dependent increase in mRNA expression of several pro-inflammatory cytokines. In addition, dose-related accumulation of NFC occurred in the bronchi, the alveoli, and macrophages. There was also a dose-dependent neutrophilic accumulation in the alveolar lung space. Hence, NFC administered by pharyngeal aspiration caused an acute inflammatory response and DNA damage in the lungs, but neither a systemic genotoxic effect in the bone marrow nor clear signs of allergic inflammation in the lungs.

Cultured human THP-1 cells or mice, through oropharyngeal aspiration, were exposed to two types of nano-fibrillated celluloses or cellulose fibers modified by carboxymethylation or carboxylation, or unmodified cellulose fibers (Ilves et al. 2018). One of the nonfunctionalized cellulose nanofibers induced cytotoxicity in vitro and triggered an innate immune reaction and increased the levels of pro-inflammatory cytokines. When delivered to the lungs of mice in vivo, the functionalized nanocellulose fibers induced a strong innate immunological reaction 24 hours post exposure associated with markedly increased levels of pro-inflammatory cytokines. The nonfunctionalized cellulose nanofibers, in turn, induced an immune reaction with Th2-type features. Furthermore, all types of nanocellulose were present in the lungs of mice 28 days post exposure and associated with a mild immunological response. Hence, the nanocellulose fibers, depending on the type of the fiber, induced either a Th1- or a Th2-type reaction and remained in the lungs for more than 28 days, indicating that this exposure could modify the immune status of mice for more than a month, partly due to a slow clearance of the fibers from the lungs.

9.2.3 Metal Oxides

Metal oxide nanoparticles are used in a number of consumer and industrial applications varying from cosmetic and sports products to windows in the construction industry, car tires, and semiconductors in printed electronics. Because of their wide use, especially workers can be exposed to them in a great number of occupational settings. Furthermore, releases of these materials into the environment may cause burden to environmental species. Hence, there is a good reason to assess their potential to induce hazards to human health and the environment.

When titanium dioxide (TiO_2) nanoparticles were given intraperitoneally to mice, exposed to ovalbumin (OVA) alone or together with TiO_2 or aluminum hydroxide and then challenged with OVA inhalationally, TiO_2 aggravated the Th2 immune response in the mice as reflected by increased levels in serum of OVA-specific IgE and IgG1. This exposure also increased the recruitment of eosinophils, lymphocytes, and neutrophils and mucus in the bronchoalveolar lavage fluid (BALF) (Larsen et al. 2010). TiO_2 induced markedly higher levels of OVA-specific IgE and IgG1 levels in the serum than aluminum hydroxide, but both adjuvants were equally effective in inducing eosinophilia and increased numbers of lymphocytes in the BALF. Hence, exposure to TiO_2 seemed to aggravate OVA-induced allergic inflammation in mice.

In another study, healthy (nonasthmatic) and asthmatic (OVA-sensitized) mice were exposed for 2 hours a day, 3 days for a week, and for weeks to 10 mg/m^3 of titanium dioxide (TiO_2) nanoparticles, and lungs of the animals were studied for pulmonary effects. In healthy mice, TiO_2 nanoparticles strongly induced the levels of the chemokines CXCL5 and induced a strong increase of neutrophils, lymphocytes, and eosinophils in the lungs. In the asthmatic mice, TiO_2 nanoparticles dramatically reduced all asthmatic signs, including levels of chemokines, cytokines, leukocytes, and antibodies. The results strongly suggest that TiO_2 can markedly modify the immunologic status of lungs of mice (Rossi et al. 2010).

TiO_2 nanoparticles also augmented allergic airway inflammation and Socs3 expression via NF-kappaB pathway in a murine model of asthma (Mishra et al. 2016). Nanosized TiO_2 nanoparticles, when administered with OVA, during sensitization phase augmented airway hyperresponsiveness (AHR), biochemical markers of lung damage, and mixed Th2/Th1-dependent immune response. Many signs of the responses were blocked by blocking the NF-kappaB pathway. The results tally with the interpretation that TiO_2 nanoparticles exacerbate the inflammatory response in the lungs or presensitized allergic individuals via NF-kappaB-regulated pathway. Taken together, there is quite a convincing evidence that exposure to TiO_2 nanoparticles may contribute to allergic inflammation in the lungs.

Potential health risks of a panel of well-characterized metal oxide nanoparticles were explored in the lungs of rats; this panel consisted of cerium dioxide (CeO_2), titanium dioxide (TiO_2), carbon black (CB), silicon dioxide (SiO_2), nickel dioxide (NiO_2), zinc oxide (ZnO), copper oxide (CuO), and amine-modified polystyrene beads (Cho et al. 2010). The inflammatory potency of the nanoparticles was evaluated 24 hours and 4 weeks post exposure. Only the metal oxide nanoparticles CeO_2, NiO, ZnO, and CuO were inflammogenic to lungs of rats at the high doses used in the study, whereas carbon black and the amine-modified polystyrene beads were without any effects. Each of these inflammogenic nanoparticles induced a unique inflammatory footprint both at 24 hours and 4 weeks post exposure. At 24 hours, infiltrates of neutrophils and eosinophils differed after the short-term exposure to CeO_2, NiO, ZnO, and CuO nanoparticles. The inflammatory responses of the same nanoparticles also differed after 4 weeks of exposure with regard to infiltration of neutrophils, lymphocytes, and eosinophils, and the induction of granulomas and fibrotic lesions. The different inflammatory responses induced by different metal oxide nanoparticles emphasize the importance to identify engineered nanoparticle-induced hazards individually until a unifying paradigm can be used. These results also emphasize the value of in vivo studies as compared with in vitro studies because the inflammatory responses identified here do not take place in vitro.

Copper oxide nanoparticles administered via the lungs to Balb/c mice induced airway hyper responsiveness (AHR) and increased inflammatory cell counts, mucus secretion, the levels of pro-inflammatory cytokines in the lungs, and the production of reactive oxygen species. CuO-induced inflammation was associated with phosphorylation of MAPKs (ERK, JNK, p38). In the OVA-induced asthma model, CuO increased AHR, inflammatory cell counts, pro-inflammatory cytokines, ROS, and the levels of IgE induced by the OVA exposure. In the asthma model used, CuO nanoparticles also induced infiltration of phagocytic cells into the lung parenchyma, phosphorylation of MAPK, and mucus secretion. Taken together, these results suggest that CuO nanoparticles exert pulmonary immunotoxicity that is associated with MAPK phosphorylation. In addition, CuO nanoparticle exposure aggravated the development of allergic lung inflammation in a mouse model of asthma (Park et al. 2016).

Mice were administered TiO_2, SiO_2, and ZnO nanoparticles via pharyngeal aspiration into the lungs and then subsequently were given OVA via inhalation for eight times during three following weeks. After OVA administration, ZnO-treated mice expressed elevated total IgE, OVA-specific Ige, and IgG levels. TiO_2 or SiO_2 did not have any effects on serum Ig levels. The authors concluded that the release of Zn^{2+} ions is necessary for the aggravation of asthmatic symptoms. Mice that were given ZnCl2 did not exhibit any signs of asthma. The authors concluded, therefore, that a continuous release of Zn^{2+} ions from ZnO nanoparticles is necessary for the aggravation of asthma (Horie et al. 2015).

Available literature on metal oxide-induced pulmonary inflammation suggests that several metal oxide nanoparticles readily evoke innate pulmonary inflammation in vitro and in vivo. There are also data to suggest that several metal oxide nanoparticles can induce allergic pulmonary inflammation as indicated increased numbers of Th2 lymphocytes, increased AHR, mucus secretion and numbers of eosinophils, all hallmarks of asthma.

9.3 Engineered Nanomaterials and Allergic Skin Inflammation

Dermal exposure to engineered nanomaterials has generally been considered insignificant due to the view that healthy skin is an effective barrier for nanomaterial entry. Most studies that have investigated engineered nanomaterials' transdermal penetration in vitro or in vivo have found that the particles do not pass through the intact skin (Lin et al. 2011; Monteiro-Riviere et al. 2011; Senzui et al. 2010). In contrast to this view, Gulson et al. reported in human volunteers that after repeated application of isotope-labeled ZnO containing sunscreen, small amounts of exogenous zinc were found in blood and urine (Gulson et al. 2010). Information about the penetration of engineered nanomaterials into irritated, injured, or diseased skin is very limited, and only few studies exist in this area (Lin et al. 2011; Monteiro-Riviere et al. 2011; Senzui et al. 2010).

9.3.1 Effects of Engineered Nanomaterials on Atopic Dermatitis

Metal oxide nanoparticles such as ZnO are used in sunscreens as they improve their optical properties against UV light. However, virtually nothing is known about the immunomodulatory properties of the topically applied ZnO nanoparticles in the context of the diseased skin. Ilves et al. investigated whether the ZnO particles were able to penetrate injured and allergen-exposed skin in the mouse atopic dermatitis (AD) model. Results demonstrated that only nanosized ZnO (nZnO) penetrated into the deep layers of the allergic skin after repeated topical application of ZnO particles, whereas bulk size (bZnO) stayed only in the upper skin layers. In addition, both types of particles diminished allergen-induced skin inflammation in mice; however, nZnO had a higher potential to suppress the local effects. On the other hand, especially nZnO induced vigorous production of total and allergen-specific IgE antibodies, evidence of allergy promoting adjuvant properties for topically applied nZnO. These results provide new information about immunomodulatory effects of metal oxide nanoparticles used in skin-care products (Ilves et al. 2014)

Silver is increasingly being used in garments to exploit its antibacterial properties. Bianco et al. investigated the presence of AgNPs in a silver containing garment and in the stratum corneum (SC) of healthy subjects (CTRLs) and individuals with atopic dermatitis (AD). Seven CTRLs and seven AD patients wore a silver sleeve 8 h/day for 5 days on a forearm and a placebo sleeve on the other forearm. After 5 days, the layers of the SC were collected by adhesive tapes. Atomic force microscopy (AFM) and scanning electron microscopy (SEM) revealed the presence of submicrometer particles on the surface of the garment that were identified as silver. On the SC tapes collected from different depths, aggregates of AgNPs with a wide range of sizes were found, but there was no difference in the number or size of the aggregates observed in SC between healthy and AD subjects (Bianco et al. 2016).

Shershakova et al. (2016) investigated anti-inflammatory effects of aqueous dispersion of fullerene C60 (nC60) in the mouse models of atopic dermatitis. To induce skin inflammation, female BALB/c mice were epicutaneously (EC) sensitized with ovalbumin and the nC60 solution was administrated in mice subcutaneously (SC) and EC. A significant suppression of IgE and Th2 cytokine production and a significant increase in the levels of Foxp3(+) and filaggrin mRNA expression were observed in nC60-treated groups at EC application. Reduction of the eosinophil and leukocyte infiltration in the ovalbumin-treated skin sites was also observed. The authors suggest that nC60 treatment can be a good alternative for the treatment of allergic skin diseases since it shifts immune response from Th2 to Th1 and restores to some extent the function of the skin barrier (Shershakova et al. 2016).

Izumi et al. (2016) evaluated the effect of chitin nanofibril (CNF) application via skin on an experimental atopic dermatitis (AD) model. Topical CNF treatment reduced skin hypertrophy and hyperkeratosis of the epidermis and inhibited also inflammatory cell infiltration in the skin. It was of interest that suppressive effects of CNFs were equal to those of corticosteroid application; however, bulk size chitin did not show these effects. It was suggested that CNT induced suppression of nuclear factor kappa B, cyclooxygenase-2, and inducible nitric oxide synthase were possible mechanisms underlying CNF-induced anti-inflammatory effects. The authors suggest that CNF application could be potentially used in the prevention or treatment of AD skin lesions (Izumi et al. 2016)

9.3.2 ENM and Contact Dermatitis

Nanocrystalline silver has both antimicrobial and anti-inflammatory properties, but the mechanisms underlying these activities are not well known. Bhol and Schechter (Bhol and Schechter, 2005) investigated the immunomodulatory effects of nanocrystalline silver using a murine model of allergic contact dermatitis and compared the effects with those induced by tacrolimus and a high potency steroid. Dermatitis was induced on the ears of BALB/c mice using dinitrofluorobenzene. Topical treatment, including vehicles, 1% nanocrystalline silver cream, tacrolimus ointment, and a high potency steroid was applied once a day for 4 days. Significant reductions of ear swelling, erythema, and histopathological inflammation in mice ears were observed after 4 days of treatment with 1% nanocrystalline silver cream, tacrolimus ointment, or a high potency steroid with no significant difference among them. Nanocrystalline silver, tacrolimus, and steroid significantly suppressed the expression of proinflammatory cytokines TNF-alpha and IL-12 in the ear samples. Moreover, a significant increase in the numbers of apoptotic cells was found in the nanocrystalline silver-treated groups compared with groups treated with tacrolimus or steroid. Results demonstrated that topical exposure to nanocrystalline silver inhibits allergen-induced contact dermatitis in mice, similar to the steroid and tacrolimus. Suppression of proinflammatory cytokines and induction of apoptosis may be the underlying reason for nanocrystalline silver-induced anti-inflammatory effects (Bhol and Schechter 2005).

In line with the study by Bhol and Schechter (Bhol and Schechter, 2005), the anti-inflammatory activity of nanocrystalline silver was examined also using a porcine model of contact dermatitis. Inflammation was induced with dinitrochlorobenzene and then treated daily with nanocrystalline silver dressings, 0.5% silver nitrate, or saline. Nanocrystalline silver-treated pigs had near-normal skin 72 hours after the allergen exposure, while other treatment groups remained inflamed exhibiting significant erythema and edema. The decreased inflammation by the nanocrystalline silver was accompanied with increased apoptosis and a decreased expression of proinflammatory cytokines. These results demonstrate that nanocrystalline silver had a direct anti-inflammatory effect in the porcine contact dermatitis model that improved the overall outcome of the healing process (Nadworny et al. 2008).

9.4 Conclusions and Take-Home Message

Several types of engineered nanomaterials have been shown to influence or induce allergic inflammation in the lungs or in the skin. For example, rigid or rod-like carbon nanotubes are capable of inducing directly asthmatic-like effects in mice exposed oropharyngeal aspiration. Furthermore, it has been clearly shown that exposure to various carbon nanotubes exacerbates existing asthma in the different mouse models in which pulmonary effects of carbon nanotubes have been investigated. Another group of fibrous – and crystalline – nanomaterials, notably cellulose nanofibers and crystals, has been shown to evoke inflammation of the lungs, and there is evidence that in addition to an innate immune response, these fibers may also in some cases induce an allergic immune reaction in the lungs. Metal oxide nanoparticles also induce pulmonary inflammation; these inflammogenic metal oxide nanoparticles include TiO_2, CoO, CeO, ZnO, and SiO_2.

In addition to the lungs, also skin is an important target organ of allergic effects of engineered nanomaterials. Engineered nanomaterials as different as Zn, Ag, fullerenes, and chitin nanofibrils have been shown to modify atopic dermatitis, demonstrating that very different nanomaterials properties may be operative in inducing this effect in the skin. It has also been shown that nanocrystalline silver may alleviate contact dermatitis induced by dinitrofluorobenzene. This effect resembled that induced by tacrolimus ointment and a potent steroid.

All the effects described above clearly indicate that engineered nanomaterials may have marked direct of modifying effects on allergic immune reactions in important target organs, notably the lungs and the skin. Hence, when producing novel engineered nanomaterials or evaluating the safety of existing ones, their impact on the allergic immune response is important to assess.

References

Bhol KC, Schechter PJ. Topical nanocrystalline silver cream suppresses inflammatory cytokines and induces apoptosis of inflammatory cells in a murine model of allergic contact dermatitis. Br J Dermatol. 2005;152:1235–42.

Bianco C, Visser MJ, Pluut OA, Svetlicic V, Pletikapic G, Jakasa I, Riethmuller C, Adami G, Larese Filon F, Schwegler-Berry D, et al. Characterization of silver particles in the stratum corneum of healthy subjects and atopic dermatitis patients dermally exposed to a silver-containing garment. Nanotoxicology. 2016;10:1480–91.

Bieber T. Atopic dermatitis. N Engl J Med. 2008;358:1483–94.

Bonner JC. Mesenchymal cell survival in airway and interstitial pulmonary fibrosis. Fibrogenesis Tissue Repair. 2010;3:15.

Bonner JC, Silva RM, Taylor AJ, Brown JM, Hilderbrand SC, Castranova V, Porter D, Elder A, Oberdorster G, Harkema JR, et al. Interlaboratory evaluation of rodent pulmonary responses to engineered nanomaterials: the NIEHS Nano GO Consortium. Environ Health Perspect. 2013;121:676–82.

Card JW, Zeldin DC, Bonner JC, Nestmann ER. Pulmonary applications and toxicity of engineered nanoparticles. Am J Physiol Lung Cell Mol Physiol. 2008;295:L400–11.

Catalan J, Siivola KM, Nymark P, Lindberg H, Suhonen S, Jarventaus H, Koivisto AJ, Moreno C, Vanhala E, Wolff H, et al. In vitro and in vivo genotoxic effects of straight versus tangled multi-walled carbon nanotubes. Nanotoxicology. 2016;10:794–806.

Chen EY, Garnica M, Wang YC, Mintz AJ, Chen CS, Chin WC. A mixture of anatase and rutile TiO2 nanoparticles induces histamine secretion in mast cells. Part Fibre Toxicol. 2012;9:2.

Cho WS, Duffin R, Poland CA, Howie SE, MacNee W, Bradley M, Megson IL, Donaldson K. Metal oxide nanoparticles induce unique inflammatory footprints in the lung: important implications for nanoparticle testing. Environ Health Perspect. 2010;118:1699–706.

USGCRP, 2016: The Impacts of Climate Change on Human Health in the United States: A Scientific Assessment. Crimmins, A., J. Balbus, J.L. Gamble, C.B. Beard, J.E. Bell, D. Dodgen, R.J. Eisen, N. Fann, M.D. Hawkins, S.C. Herring, L. Jantarasami, D.M. Mills, S. Saha, M.C. Sarofim, J. Trtanj, and L. Ziska, Eds. U.S. Global Change Research Program, Washington, DC, 312 pp. http://dx.doi.org/10.7930/J0R49NQX

Diepgen TL, Ofenloch RF, Bruze M, Bertuccio P, Cazzaniga S, Coenraads PJ, Elsner P, Goncalo M, Svensson A, Naldi L. Prevalence of contact allergy in the general population in different European regions. Br J Dermatol. 2016;174:319–29.

Donaldson K, Murphy FA, Duffin R, Poland CA. Asbestos, carbon nanotubes and the pleural mesothelium: a review of the hypothesis regarding the role of long fibre retention in the parietal pleura, inflammation and mesothelioma. Part Fibre Toxicol. 2010;7:5.

Fyhrquist N, Lehto E, Lauerma A. New findings in allergic contact dermatitis. Curr Opin Allergy Clin Immunol. 2014;14:430–5.

Galli SJ, Tsai M. IgE and mast cells in allergic disease. Nat Med. 2012;18:693–704.

Gulson B, McCall M, Korsch M, Gomez L, Casey P, Oytam Y, Taylor A, McCulloch M, Trotter J, Kinsley L, Greenoak G. Small Amounts of Zinc from Zinc Oxide Particles in Sunscreens Applied Outdoors Are Absorbed through Human Skin. Toxicol Sci. 2010;118(1):140–149.

Hamilton RF, Buford M, Xiang CC, Wu NQ, Holian A. NLRP3 inflammasome activation in murine alveolar macrophages and related lung pathology is associated with MWCNT nickel contamination. Inhal Toxicol. 2012;24:995–1008.

Horie M, Stowe M, Tabei M, Kuroda E. Pharyngeal aspiration of metal oxide nanoparticles showed potential of allergy aggravation effect to inhaled ovalbumin. Inhal Toxicol. 2015;27:181–90.

Hussain S, Vanoirbeek JA, Hoet PH. Interactions of nanomaterials with the immune system. Wiley Interdiscip Rev Nanomed Nanobiotechnol. 2012;4:169–83.

Ilves M, Palomaki J, Vippola M, Lehto M, Savolainen K, Savinko T, Alenius H. Topically applied ZnO nanoparticles suppress allergen induced skin inflammation but induce vigorous IgE production in the atopic dermatitis mouse model. Part Fibre Toxicol. 2014;11:38.

Ilves M, Vilske S, Aimonen K, Lindberg HK, Pesonen S, Wedin I, Nuopponen M, Vanhala E, Hojgaard C, Winther JR, et al. Nanofibrillated cellulose causes acute pulmonary inflammation that subsides within a month. Nanotoxicology. 2018;12:729–46.

Inoue K. Promoting effects of nanoparticles/materials on sensitive lung inflammatory diseases. Environ Health Prev Med. 2011;16:139–43.

Inoue K, Koike E, Yanagisawa R, Hirano S, Nishikawa M, Takano H. Effects of multi-walled carbon nanotubes on a murine allergic airway inflammation model. Toxicol Appl Pharmacol. 2009;237:306–16.

Inoue K, Yanagisawa R, Koike E, Nishikawa M, Takano H. Repeated pulmonary exposure to single-walled carbon nanotubes exacerbates allergic inflammation of the airway: possible role of oxidative stress. Free Radic Biol Med. 2010;48:924–34.

Izumi R, Azuma K, Izawa H, Morimoto M, Nagashima M, Osaki T, Tsuka T, Imagawa T, Ito N, Okamoto Y, et al. Chitin nanofibrils suppress skin inflammation in atopic dermatitis-like skin lesions in NC/Nga mice. Carbohydr Polym. 2016;146:320–7.

Katwa P, Wang X, Urankar RN, Podila R, Hilderbrand SC, Fick RB, Rao AM, Ke PC, Wingard CJ, Brown JM. A carbon nanotube toxicity paradigm driven by mast cells and the IL-(3)(3)/ST(2) axis. Small (Weinheim an der Bergstrasse, Germany). 2012;8:2904–12.

Kim HY, DeKruyff RH, Umetsu DT. The many paths to asthma: phenotype shaped by innate and adaptive immunity. Nat Immunol. 2010;11:577–84.

Kuempel ED, Jaurand MC, Moller P, Morimoto Y, Kobayashi N, Pinkerton KE, Sargent LM, Vermeulen RC, Fubini B, Kane AB. Evaluating the mechanistic evidence and key data gaps in assessing the potential carcinogenicity of carbon nanotubes and nanofibers in humans. Crit Rev Toxicol. 2017;47:1–58.

Lambrecht BN, Hammad H. The immunology of asthma. Nat Immunol. 2015;16:45–56.

Larsen ST, Roursgaard M, Jensen KA, Nielsen GD. Nano titanium dioxide particles promote allergic sensitization and lung inflammation in mice. Basic Clin Pharmacol Toxicol. 2010;106:114–7.

Lin LL, Grice JE, Butler MK, Zvyagin AV, Becker W, Robertson TA, Soyer HP, Roberts MS, Prow TW. Time-correlated single photon counting for simultaneous monitoring of zinc oxide nanoparticles and NAD(P)H in intact and barrier-disrupted volunteer skin. Pharm Res. 2011;28:2920–30.

Mazurek JM, Syamlal G. Prevalence of asthma, asthma attacks, and emergency department visits for asthma among working adults – National Health Interview Survey, 2011–2016. MMWR Morb Mortal Wkly Rep. 2018;67:377–86.

Menas AL, Yanamala N, Farcas MT, Russo M, Friend S, Fournier PM, Star A, Iavicoli I, Shurin GV, Vogel UB, et al. Fibrillar vs crystalline nanocellulose pulmonary epithelial cell responses: cytotoxicity or inflammation? Chemosphere. 2017;171:671–80.

Meunier E, Coste A, Olagnier D, Authier H, Lefevre L, Dardenne C, Bernad J, Beraud M, Flahaut E, Pipy B. Double-walled carbon nanotubes trigger IL-1 beta release in human monocytes through Nlrp3 inflammasome activation. Nanomed Nanotechnol. 2012;8:987–95.

Mishra V, Baranwal V, Mishra RK, Sharma S, Paul B, Pandey AC. Titanium dioxide nanoparticles augment allergic airway inflammation and Socs3 expression via NF-kappaB pathway in murine model of asthma. Biomaterials. 2016;92:90–102.

Monteiro-Riviere NA, Wiench K, Landsiedel R, Schulte S, Inman AO, Riviere JE. Safety evaluation of sunscreen formulations containing titanium dioxide and zinc oxide nanoparticles in UVB sunburned skin: an in vitro and in vivo study. Toxicol Sci. 2011;123:264–80.
Nadworny PL, Wang J, Tredget EE, Burrell RE. Anti-inflammatory activity of nanocrystalline silver in a porcine contact dermatitis model. Nanomedicine. 2008;4:241–51.
NIOSH. Occupational exposure to carbon nanotubes and nanofibers: current intelligence bulletin 65; 2013.
Palomaki J, Valimaki E, Sund J, Vippola M, Clausen PA, Jensen KA, Savolainen K, Matikainen S, Alenius H. Long, needle-like carbon nanotubes and asbestos activate the NLRP3 inflammasome through a similar mechanism. ACS Nano. 2011;5:6861–70.
Park EJ, Cho WS, Jeong J, Yi J, Choi K, Park K. Pro-inflammatory and potential allergic responses resulting from B cell activation in mice treated with multi-walled carbon nanotubes by intratracheal instillation. Toxicology. 2009;259:113–21.
Park JW, Lee IC, Shin NR, Jeon CM, Kwon OK, Ko JW, Kim JC, Oh SR, Shin IS, Ahn KS. Copper oxide nanoparticles aggravate airway inflammation and mucus production in asthmatic mice via MAPK signaling. Nanotoxicology. 2016;10:445–52.
Porter DW, Hubbs AF, Mercer RR, Wu N, Wolfarth MG, Sriram K, Leonard S, Battelli L, Schwegler-Berry D, Friend S, et al. Mouse pulmonary dose- and time course-responses induced by exposure to multi-walled carbon nanotubes. Toxicology. 2010;269:136–47.
Poulsen SS, Saber AT, Williams A, Andersen O, Kobler C, Atluri R, Pozzebon ME, Mucelli SP, Simion M, Rickerby D, et al. MWCNTs of different physicochemical properties cause similar inflammatory responses, but differences in transcriptional and histological markers of fibrosis in mouse lungs. Toxicol Appl Pharmacol. 2015;284:16–32.
Rossi EM, Pylkkanen L, Koivisto AJ, Nykasenoja H, Wolff H, Savolainen K, Alenius H. Inhalation exposure to nanosized and fine TiO2 particles inhibits features of allergic asthma in a murine model. Part Fibre Toxicol. 2010;7:35.
Rydman EM, Ilves M, Koivisto AJ, Kinaret PA, Fortino V, Savinko TS, Lehto MT, Pulkkinen V, Vippola M, Hameri KJ, et al. Inhalation of rod-like carbon nanotubes causes unconventional allergic airway inflammation. Part Fibre Toxicol. 2014;11:48.
Ryman-Rasmussen JP, Tewksbury EW, Moss OR, Cesta MF, Wong BA, Bonner JC. Inhaled multiwalled carbon nanotubes potentiate airway fibrosis in murine allergic asthma. Am J Respir Cell Mol Biol. 2009;40:349–58.
Sandilands A, Sutherland C, Irvine AD, McLean WH. Filaggrin in the frontline: role in skin barrier function and disease. J Cell Sci. 2009;122:1285–94.
Savolainen K, Alenius H, Norppa H, Pylkkanen L, Tuomi T, Kasper G. Risk assessment of engineered nanomaterials and nanotechnologies–a review. Toxicology. 2010;269:92–104.
Sayers BC, Taylor AJ, Glista-Baker EE, Shipley-Phillips JK, Dackor RT, Edin ML, Lih FB, Tomer KB, Zeldin DC, Langenbach R, et al. Role of cyclooxygenase-2 in exacerbation of allergen-induced airway remodeling by multiwalled carbon nanotubes. Am J Respir Cell Mol Biol. 2013;49:525–35.
Scotton CJ, Martinez FO, Smelt MJ, Sironi M, Locati M, Mantovani A, Sozzani S. Transcriptional profiling reveals complex regulation of the monocyte IL-1 beta system by IL-13. J Immunol (Baltimore, MD: 1950). 2005;174:834–45.
Seltmann J, Roesner LM, von Hesler FW, Wittmann M, Werfel T. IL-33 impacts on the skin barrier by downregulating the expression of filaggrin. J Allergy Clin Immunol. 2015;135:1659–1661 e1654.
Senzui M, Tamura T, Miura K, Ikarashi Y, Watanabe Y, Fujii M. Study on penetration of titanium dioxide (TiO(2)) nanoparticles into intact and damaged skin in vitro. J Toxicol Sci. 2010;35:107–13.
Shershakova N, Baraboshkina E, Andreev S, Purgina D, Struchkova I, Kamyshnikov O, Nikonova A, Khaitov M. Anti-inflammatory effect of fullerene C60 in a mice model of atopic dermatitis. J Nanobiotechnol. 2016;14:8.
Shipkowski KA, Taylor AJ, Thompson EA, Glista-Baker EE, Sayers BC, Messenger ZJ, Bauer RN, Jaspers I, Bonner JC. An allergic lung microenvironment suppresses carbon nanotube-

induced inflammasome activation via STAT6-dependent inhibition of caspase-1. PLoS One. 2015;10(6):e0128888.

Sun BB, Wang X, Ji ZX, Li RB, Xia T. NLRP3 inflammasome activation induced by engineered nanomaterials. Small (Weinheim an der Bergstrasse, Germany). 2013;9:1595–607.

Thompson EA, Sayers BC, Glista-Baker EE, Shipkowski KA, Ihrie MD, Duke KS, Taylor AJ, Bonner JC. Role of signal transducer and activator of transcription 1 in murine allergen-induced airway remodeling and exacerbation by carbon nanotubes. Am J Respir Cell Mol Biol. 2015;53:625–36.

Thomsen SF, Ulrik CS, Kyvik KO, Hjelmborg J, Skadhauge LR, Steffensen I, Backer V. Importance of genetic factors in the etiology of atopic dermatitis: a twin study. Allergy Asthma Proc. 2007;28:535–9.

Weidinger S, Beck LA, Bieber T, Kabashima K, Irvine AD. Atopic dermatitis. Nat Rev Dis Primers. 2018;4:1.

Yanamala N, Farcas MT, Hatfield MK, Kisin ER, Kagan VE, Geraci CL, Shvedova AA. In vivo evaluation of the pulmonary toxicity of cellulose nanocrystals: a renewable and sustainable nanomaterial of the future. ACS Sustain Chem Eng. 2014;2:1691–8.

Chapter 10
Nanomaterial Effects on Viral Infection

Hao Chen, Sara T. Humes, Navid B. Saleh, John A. Lednicky, and Tara Sabo-Attwood

Abstract The potential for environmental and occupational exposures of populations to nanomaterials (NMs) has fostered concerns of associated adverse health effects, with a particular emphasis on pulmonary injury and disease. Many studies have revealed that several types of NMs can evoke a variety of biological responses, such as pulmonary inflammation and oxidative stress, which contribute to allergy, fibrosis, and granuloma formation. Less attention has been paid to health effects that may result from exposure to NMs and additional stressors such as pathogens, with a particular focus on susceptibility to viral infection. This chapter will summarize the current body of literature related to NMs and viral exposures with a primary focus on immune modulation. A summary of the studies performed and major findings to date will be discussed, highlighting proposed molecular mechanisms behind NM-driven host susceptibility, challenges, limitations, and future research needs. Specific mechanisms discussed include direct interaction between NMs and biological molecules, activation of pattern recognition receptors (PRRs) and related signaling pathways, production of oxidative stress and mitochondrial dysfunction, inflammasome activation, and modulation of lipid signaling networks.

Keywords Nanomaterials · Carbon nanotubes · Pathogens · Viruses · Pulmonary · Inflammation · Infection · Oxidative stress · Immune response

H. Chen · S. T. Humes · J. A. Lednicky · T. Sabo-Attwood (✉)
Department of Environmental and Global Health, Center for Environmental and Human Toxicology and Emerging Pathogens Institute, University of Florida, Gainesville, FL, USA
e-mail: sabo@phhp.ufl.edu

N. B. Saleh
Department of Civil, Architectural, and Environmental Engineering, University of Texas, Austin, TX, USA

J. C. Bonner, J. M. Brown (eds.), *Interaction of Nanomaterials with the Immune System*, Molecular and Integrative Toxicology,
https://doi.org/10.1007/978-3-030-33962-3_10

10.1 Introduction

It has been demonstrated that exposure to small particles and fibers (i.e., particulate matter and asbestos), most notably via inhalation, can enhance host susceptibility to microbial infections. The ability of nanomaterials (NMs) to alter the health outcomes associated with infections, specifically those associated with viruses, has not been well characterized; however, several recent reports suggest that select types of NMs have the capacity to modulate and, in some cases, worsen the outcomes of viral infections. In this context, a few reports have begun to address the molecular mechanisms involved that contribute to the impacts that these NMs have on viral infectivity, primarily through modulation of the host immune system. In this chapter, the current state of knowledge regarding these mechanisms in response to varied NMs and viruses will be discussed, highlighting gaps in the science and suggesting where future research is needed.

10.2 Multiple Environmental Stressors

The notion of "multiple environmental stressors" is not new, and several lines of evidence support that exposure to environmental chemicals/particulates and microbes can induce adverse health effects that are distinct from exposure to single agents alone (Feingold et al. 2010). For example, results of epidemiological studies demonstrate that exposure to air pollution, which contains a mixture of chemical and biological agents, is a risk factor for increased incidence of pulmonary infections (Ciencewicki and Jaspers 2007), co-exposure to pesticides and heavy metals is suspected to have a direct impact on the gastrointestinal tract and disrupt the normal microbiome, leading to persistent inflammation and infection (Mapesa et al. 2016), and exposure to both aflatoxin and hepatitis B virus (HBV) (Groopman et al. 2005) can lead to 59.4 times higher risk of developing liver cancer compared to only 7.3 and 3.4 times the risk for HBV or aflatoxin exposure alone (Qian et al. 1994). Small particulates, especially those that are inhaled, have additionally been shown to increase susceptibility to respiratory viruses. Epidemiological studies have demonstrated that human exposure to inhaled diesel exhaust, cigarette and wood smoke, coal dust, silica, particulate matter 2.5 ($PM_{2.5}$), and air pollution of varied composition increases the severity and chance of acquiring respiratory infections (Ciencewicki and Jaspers 2007; Speizer et al. 1980; Clifford et al. 2015; Huang et al. 2016; Jaspers et al. 2009; Chen et al. 2012; Becker and Soukup 1999; Ma et al. 2017; Rabovsky et al. 1986; Rebuli et al. 2018). There is now consistent evidence to support that exposure to air pollution leads to severe lower respiratory infections, especially in children under 5 years of age. Furthermore, this association has been observed after children are exposed to short-term spikes of $PM_{2.5}$ (Horne et al. 2018).

Due to their small size and resemblance to previously mentioned particulates and fibers (i.e., asbestos) that are associated with adverse pulmonary effects, there is remarkable interest in inhalation as a route of exposure to NMs and subsequent toxicity. Airborne exposure to NMs is likely to occur in occupational settings, by direct consumer use of products and indirectly through environmental contamination, and several studies have reported that certain types of NMs, such as carbon nanotubes (CNTs), are present above the National Institute for Occupational Safety and Health (NIOSH) recommended exposure limit in workplace environments (Boonruksa et al. 2016; Thompson et al. 2015; Hedmer et al. 2015). Other potential exposures could arise through release from consumer products (i.e., tires, electronics, cosmetics), product end-of-life, and avenues of disposal (Nowack et al. 2013).

Most of the pulmonary toxicity studies performed to date investigate well-described adverse effects of particulate exposure such as sub-chronic tissue damage, fibrogenesis, granulomatous changes, impaired clearance, robust inflammation, airway hyperreactivity, and airflow obstruction (Guzman et al. 2006; Li et al. 2010; Teeguarden et al. 2010; Saber et al. 2013). These data originate primarily from studies that assess the toxicity of exposure to single-component NMs in isolation of other imposed stressors. Because the scenario of co-exposure to NMs and infectious agents in the ambient environment is likely, understanding the combined health effects caused by exposure to NMs and viruses is highly significant to the field of public health.

10.3 Viral Infections and Public Health Impacts

Viral infections account for a considerable global health burden. Several viruses contribute to this burden and include human immunodeficiency virus (HIV), Ebola, corona, Zika, West Nile, dengue, and hepatitis A–C, among others. Several common viruses, such as influenza viruses, are associated with significant infection rates annually, which include seasonal epidemics and pandemics. Emerging and reemerging viruses, such as those that cause severe acute respiratory syndrome (SARS), can also contribute significantly to health burdens. Acute lower respiratory tract (LRT) infections associated with viruses are a leading cause of death and disability among children and adults globally and are consistently considered one of the top ten global health challenges. Epidemiological evidence shows that lower respiratory tract infections caused 2.74 million deaths and 103 million disability-adjusted life years (DALYs) based on the mortality and morbidity data from 195 countries in 2015 (GBD 2015 LRI Collaborators 2017). Other common respiratory viruses, including respiratory syncytial virus and rhino virus, also contribute considerably to the global health burden (Forum of International Respiratory Societies 2017). In the USA, upper and lower respiratory tract infections are estimated to be responsible for about \$15 billion in direct treatment and \$9 billion due to missed work (Dixon 1985). It is evident that viral infections remain a top priority in the field of public health, and thus understanding how exposure to contaminants, such as nanoparticles (NPs), may impact the morbidity and mortality associated with these infections is vitally important.

10.4 Nanoparticle Exposure and Viral Infection

A literature search on PubMed using terminologies "nanomaterials" and "viruses" revealed 3639 papers. Addition of the term "toxicity" produced 265 hits and the subsequent addition of "infection" resulted in 60 papers. More careful assessment of these latter publications led to three papers that were excluded because they did not actually study viruses. Of the remaining 57, only 15 published from 2010 to 2018 focused specifically on the toxicity of NMs and viral infections (9%) where live viruses were employed. More details for these 15 studies are shown in Table 10.1. The remaining studies investigated nonpathogen-related toxicity (26%) or the application of NMs to therapeutic delivery, vaccine development, and viral detection (79%) which will not be discussed in detail in this chapter (Fig. 10.1).

The studies that have investigated the toxicity of NMs and their ability to modulate viral infections focused on a few select types of NMs that included carbon nanotubes, spherical carbon and carbon black, fullerenes, titanium dioxide (TiO_2), silver, and gold. Specific viruses employed were herpes simplex virus type 2 (HSV-2), influenza A virus (IAV), respiratory syncytial virus (RSV), HIV-1, dengue virus type 2, murine gamma herpesvirus 68 (MHV-68), Epstein–Barr virus, and foot-and-mouth disease virus (FMDV). Interestingly, only six of the published articles addressed NM exposure and increased viral susceptibility, whereas results of the remaining nine papers showcased the ability of NMs to inhibit virus titers either through blocking viral entry or through replication. For example, cells exposed to silver nanoparticles were shown to inhibit dengue virus serotype 2 (DENV-2) viral replication and production of progeny viruses (Murugan et al. 2015), and inhibit replication of HSV-2 (Hu et al. 2014), HIV-1 (Trefry and Wooley 2012), and HSV-2 (Sopova et al. 2010), whereas foot-and-mouth disease virus (FMDV) titers were reduced by gold NP (Rafiei et al. 2016). Another interesting aspect of these studies showed that the effects of carbon NMs on viral infection varied by the type of nanoparticles tested. Exposure to both single-walled carbon nanotubes (SWCNTs) was able to increase virus titers of IAV (Sanpui et al. 2014; Chen et al. 2017). Interestingly, in another study, exposure of mice to multiwalled carbon nanotubes (MWCNTs) exacerbated pneumonia and immune cell infiltration in lungs of RSV-infected mice but did not influence the virus titers compared with RSV treatment only (Hashiguchi et al. 2015a). Our group has also observed a similar effect with MWCNTs in a mouse model of IAV infection where titers of IAV were not increased (in review), despite the significant increase in titers observed for SWCNTs in the same mouse model. Spherical carbon nanoparticles (Barras et al. 2016) and fullerene (C60) derivatives (Shoji et al. 2013) were shown to interfere with HSV-1 and IAV, respectively, whereas exposure to a single dose of TiO_2 nanoparticles exacerbated pneumonia in RSV-infected mice (Hashiguchi et al. 2015a). Lastly, a more recent report showed for the first time the ability of double-walled carbon nanotubes (DWCNTs) to reactivate latent herpesvirus (MHV-68) in cells and mice (Sattler et al. 2017). They also discovered that these DWCNTs could induce more pronounced effects in terms of reactivating latent virus than carbonaceous spherical nanoparticles.

Table 10.1 Literature summary of the impact of NMs on viral susceptibility

NP type	Prepared with	Avg. diameter (nm)	NM exposure	Dose	Experimental model	Virus type	Virus exposure	Dose	Exposure time	Overall findings	Reference
Carbon											
Single-walled CNT	1% Pluronic F-127	0.805	Cell culture media	50 μg/mL	Human small airway epithelial cells	IAV H1N1 strain A/Mexico/4108/2009	Cell media	MOI 0.1–0.5	CNT 24 h + IAV 24 h	CNT increased IAV titers; repressed antiviral genes	Sanpui et al. (2014)
Single-walled CNT	1% Pluronic F-127	0.805	IT	20 μg/mouse	C57BL/6 mouse	IAV H1N1 strain A/Mexico/4108/2009	IN	3.12×10^4 $TCID_{50}$/mouse	CNT 4d + IAV 3d	CNT increased IAV titers and injury in lungs; altered immune markers and repressed antiviral genes	Chen et al. (2017)
Multiwalled CNT	PBS	40–50	IN	0–0.25 mg/kg	BALB/c mouse	Human respiratory syncytial virus (RSV)	IN	3.5×10^5 PFU /mouse	CNT 5d + RSV 5d	CNT exacerbated pneumonia and immune cell infiltration in lungs; no change in virus titers compared to RSV only	Hashiguchi et al. (2015a)
Double-walled CNT	Unknown	1000, 7000, and 15,000 nm length	IN	0–0.25 mg/kg	BALB/c mouse	Human respiratory syncytial virus (RSV)	IN	3.5×10^5 PFU /mouse	CNT 1, 3 5 d + RSV 5d	CNT increased CCL5 in BALF of RSV-infected mice; increased monocyte infiltration in lung tissue	Watanabe et al. (2018)
Double-walled CNT; spherical carbon	1% Pluronic F-127	10	Cell culture media in vitro; IT in vivo	50 μg/mL for in vitro; 50 μg per mouse	BHK-21 cells, MHV-68-infected B cell line S11, ANA-1, MH-S; C57BL/6 mouse	Murine gamma herpesvirus 68	Cell media; IN	MOI 1.0 for in vitro; 5.0×10^4 PFU/mouse	MHV-68 2 h + CNT 2 h in vitro; MHV-68 28d + CNT 24 h in vivo	CNT and spherical carbon restored acute herpesvirus infection; boosted production of lytic viral proteins and inflammatory responses in lungs	Sattler et al. (2017)

(continued)

Table 10.1 (continued)

NP type	Prepared with	Avg. diameter (nm)	NM exposure	Dose	Experimental model	Virus type	Virus exposure	Dose	Exposure time	Overall findings	Reference
Fullerene (C60) derivatives	DMSO	N/A	Cell culture media	25–100 uM	MDCK cells	IAV H1N1 strain A/PR/8/34 or A/Aichi/2/68	Cell media	MOI 1.0	Mix fullerenes + IAV; expose cells 24 h	Fullerenes inhibited endonuclease activity of PA protein (subunit of influenza RNA polymerase); inhibited IAV infection and the expression of nucleoprotein and nonstructural protein 1	Shoji et al. (2013)
Carbon nanodots	Water	10–30	Cell media	100, 200, 300 μg/mL	Vero cells, human lung cancer cells (A549)	Herpes simplex virus type 1	Cell media	10^3 $TCID_{50}$/mL	C-dots and HSV-1 cocultured 2 h	Carbon dots interfered with the virus entry; inhibited infection	Barras et al. (2016)
Carbon black	Ultrapure water	14	Incubate with human cathelicidin (LL-37); cell media	100 μg/mL	WI-38 cells	Rhinovirus (RV1B)	Cell media	Dose determined during assay	Carbon black 1 h + RV1B 1 h	Carbon black altered host defense peptide structure and function	Findlay et al. (2017)
Metal oxides											
TiO_2 NP	PBS	35	IN	0.25 or 2.5 mg/kg	BALB/c mice	Respiratory syncytial virus (RSV)	IN	3.5 × 10^5 PFU/mouse	TiO_2 5d + RSV 5d	TiO_2 NP exacerbated pneumonia in RSV-infected mice; increased IFN-γ, CCL5; increased infiltration of lymphocytes into the alveolar septa	Hashiguchi et al. (2015b)
Silicon dioxide NP; silver NP	N/A	N/A	Cell media	0.1 and 1.0 mg/mL	Vero cells	Herpes simplex virus type 2 (HSV-2)	Cell media	N/A	N/A	SiNP exposure had no influence on HSV-2; AgNP reduced virus titers	Sopova et al. (2010)

Metals											
Silver NP	Ultrapure water	25	Cell media	10, 100, 1000 μg/mL	293 T/17 human embryonic kidney cells; Vero cells	HIV-1 vector (pNL4-3Δ + cass)	NP mixed with HIV-1	$1 - 11 \times 10^2$ $TCID_{50}$ /ml	Mix NP and HIV-1 vector and add to cells 1 h	AgNP inhibited early stage infection	Trefry and Wooley (2012)
Silver NP	Cell media	30–40	Cell media	25 μg/mL	Vero cells	Herpes simplex virus 2 (HSV-2)	NP mixed with HSV-2	5×10^4 $TCID_{50}$/ mL	Cell media 1 h	AgNP inhibited production of progeny viruses and virus replication	Hu et al. (2014)
Silver NP	Deionized water	30–70	Cell media	10, 20, 30 μg/ mL	Vero cells	Dengue virus (DEN-2)	Cell media	5×10^4 $TCID_{50}$ per mL	NP 1 h + DEN-2 1 h and culture for 5d	AgNP inhibited virus replication	Murugan et al. (2015)
Gold NP	Cell media	13	Cell media	1.56 μg/mL	BHK-21 cells	Foot-and mouth disease virus (FMDV)	Cell media	100 PFU	Mix NP and FMDV 1 h and add to cells for 24 h	NP reduced virus titer; inhibition of viral growth at the early stages of infection; inhibited postentry stages of virus replication concomitant with the onset of intracellular viral RNA synthesis	Rafiei et al. (2016)
Gold Nanorods	PBS	10	Cell media; IN	1.25 and 2.5 μg/mL; 20 μg/mouse	Hep-2 cells; BALB/c mice	RSV	Gold nanorods mixed with RSV	100 PFU/mL; 6 $\times 10^5$ PFU / mouse	24 h in vitro; 5d in vivo	Gold nanorods inhibited RSV infection via upregulation of TLR, NOD-like receptor, and RIG-I-like receptors	Bawage et al. (2016)

IT intratracheal aspiration, *IN* intranasal, *IAV* influenza virus, *CNT* carbon nanotubes, *MOI* multiplicity of infection, *PFU* plaque forming units

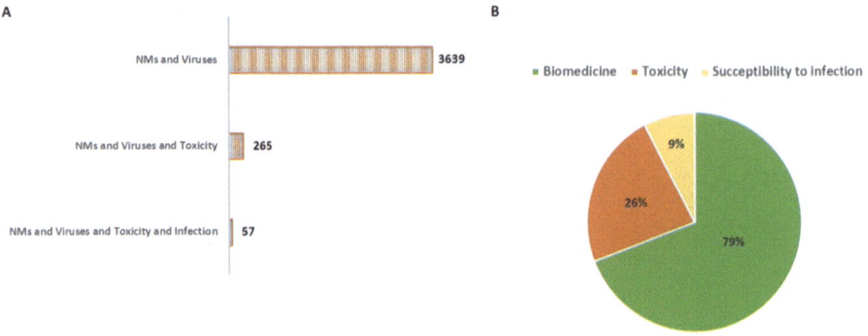

Fig. 10.1 (**a**) Number of articles based on select search terms in the NCBI PubMed database. (**b**) Percentage of scientific articles (out of 57) that investigated NM exposures and susceptibility to infection (9%) compared to those that focused on other toxic endpoints (26%) and biomedical applications (79%)

10.5 Nanoparticle and Viral Exposure Routes and Scenarios

There are several ways whereby individuals can be exposed to NMs and viruses which likely influence resultant health effects: (1) nanoparticles and viruses can impact an organism through separate exposures sequentially (in either order) or (2) they can interact in the environment to form an "exposure mixture," leading to a simultaneous dual exposure scenario. While epidemiological data support the association between particulate exposures (via inhalation) and respiratory infections, it is difficult to define exactly how and when the exposures occurred. There is some evidence however to suggest that the order of exposure is important. For example, it has been demonstrated that small airway epithelial cells infected with IAV produce increased titers only when exposed to pristine SWCNTs prior to infection. The noted increase in IAV titers could not be duplicated with a simultaneous dual exposure or by exposing the cells to the SWCNTs after infection (Sanpui et al. 2014). However, in another study, DWCNTs were able to reactivate latent herpesvirus, suggesting that the infections occurred prior to NP exposure (Sattler et al. 2017). There is not an abundance of evidence to suggest that classic additive, synergistic, or antagonistic toxicity occurs under any of these exposure scenarios, but this observation likely depends on the endpoints that are quantified, which have been limited. However, mixed injury profiles may be much more detrimental compared with effects caused by exposure to single agents (Sanpui et al. 2014; Chen et al. 2017; Walling and Lau 2014; Shvedova et al. 2008). Noteworthy, the exposure scenarios described all likely occur in occupational and environmental settings on some level.

10.6 Immune Response to Viruses

With the knowledge that select NMs can alter host susceptibility to viral infections, it is necessary to understand the molecular mechanisms by which these particles elicit such responses. There is a large focus on modulation of the immune system as a primary target of NMs which requires an understanding of how the host immune system typically responds to viral infections. Viruses depend on host cells to reproduce and the type of targeted host cell depends on the specific strain of virus. For example, cold and flu viruses infect respiratory epithelial cells, whereas HIV attacks T cells of the immune system (Owen et al. 2013). Despite diverse cell types, most viruses go through a lytic cycle that encompasses viral entry into the host cell (a few viruses do not enter the cell but inject their contents), replication using the host cell machinery, and packaging and release of new viral progeny (Owen et al. 2013). As an example, influenza A viruses primarily enter lung cells by binding directly to cell surface receptors (e.g., sialic acid receptors) through which they then enter the host cell through various uptake means such as receptor-mediated endocytosis, clathrin-coated vesicles, and/or macropinocytosis. The virus then releases its contents, including genetic material, into the cell and utilizes host machinery to replicate. Progeny viral particles are assembled beneath the host cell's outer membrane and then released from the cell to initiate a new infection cycle (Edinger et al. 2014).

Once infected, the host immune response to viruses occurs in two phases. The first is a rapid-onset "innate" response that involves the synthesis of proteins called interferons (IFNs) and the stimulation of natural killer lymphocytes. When viruses enter epithelial cells or are phagocytized by immune cells such as macrophages or neutrophils, the cells produce multiple cytokines and chemokines to activate additional innate immune cells. Type I interferons (IFNs) are the principal cytokines that are involved in these responses (Koyama et al. 2008). Once IFNs and downstream cytokines/chemokines are produced, they act as signaling molecules that allow infected cells to warn nearby cells of a viral presence – this signal makes neighboring cells increase the numbers of major histocompatibility complex (MHC) class I molecules upon their surfaces, so that T cells surveying the area can identify and eliminate the viral infection. As infection proceeds, the "adaptive immune response" is initiated, which ultimately results in the synthesis of virus-specific antibodies by B lymphocytes and the cell-mediated response (the synthesis of specific cytotoxic T lymphocytes that kill infected cells).

Upon activation of T cells and B cells specific to the certain types of viruses, an adapted and long-term immune response to the virus will be developed. Antibodies produced by B cells are specific to viral surface antigens and are very important in neutralizing viruses and preventing spread during the acute infection phase (Owen et al. 2013). For example, antibodies can bind to and neutralize viral antigens that have important functions in cellular membrane penetration. $CD8^+$ cytotoxic T (Tc) cells and $CD4^+$ helper T (Th) cells are required for cell-mediated antiviral defense. Activated Th cells produce IFNs, interleukin 2 (IL-2), and other cytokines while activated Tc cells can directly kill infected cells.

10.7 Viral-Related Immune Mechanisms Perturbed by NMs

With the knowledge that select NMs can increase host susceptibility to viral infections, it is necessary to understand the molecular mechanisms by which these particles elicit such responses. While the response can vary somewhat between virus types, we will focus on several distinct immune mechanisms perturbed by NMs that have been reported and may contribute to altered viral infection status. These include: (1) direct interaction between NMs and biological molecules, (2) activation of pattern recognition receptors (PRRs) and related signaling pathways, (3) production of oxidative stress and mitochondrial dysfunction, (4) inflammasome activation, and (5) modulation of lipid signaling networks.

10.7.1 Direct Interaction Between NMs and Biological Molecules

There is limited information regarding the direct interactions between NMs and viruses and how such actions contribute to altered host immunity and/or disrupted viral life cycles, and hence, infection status. The physicochemical properties of NMs, that is, hydrophobicity, surface moieties generated through surface reactions, and the inherent electronic structure that promotes pi–pi interaction with aromatic organics, can influence favorable interactions with biomaterials (Otyepkova et al. 2016; Li et al. 2013a; Wepasnick et al. 2011; Smith et al. 2009; Guo et al. 2014). Certain NMs, such as those that are carbonaceous in composition (i.e., CNT, graphene), have a propensity to interact with biomacromolecules due to these inherent properties, and much work has been focused on the ability of such NPs to form a "corona" in biological systems (Saleh et al. 2010). It is therefore reasonable to assume that NPs can interact with natural bio-particulates, that is, viruses. Due to their size and surface chemistry, these ultrafine bio-particulates may have elongated residence time in environmental media, such as air, water, or soil, which can greatly impact environmental fate and transport and thus human exposure.

10.7.1.1 Interaction of NPs with Protein Corona

"Protein coronas" are a complex of proteins in cell culture media or biological fluids that adsorb to the surface of NMs. Such proteins and other molecules can hitch a ride on NMs, using them as vehicles for transport either to the cell surface or to aid in cell entry. This may also be the case for viruses. For example, sorption of pathogen moieties may be transported by NMs to receptors, such as pathogen recognition receptors (PRRs), to initiate antiviral responses (Monopoli et al. 2011; Yoneyama et al. 2016); however, the opposite could also occur, whereby direct interaction of

NMs and pathogen moieties could limit interaction with immune surveillance systems, impairing normal host defenses.

Protein corona-coated NMs have been recently termed nanomaterial-associated molecular patterns (NAMPs). NAMPs have been shown to be directly recognized by PRRs based on their protein composition (Silva et al. 2017). Proteins undergoing adsorption to NMs can experience possible conformational and functional changes, leading to altered immune responses to either the NMs or pathogens. For example, the epitope of a protein in a corona mixture may be more exposed and thus more easily recognized by the PRR (Saptarshi et al. 2013). Another study discovered very distinctive adsorption and binding patterns of protein corona based on profiles of amino acids (serine, tyrosine, etc.) that may have distinct effects on the activation of immune pathways (Cai et al. 2013).

10.7.1.2 Interaction of NMs with Viruses

Numerous studies have shown that NMs are excellent vehicles for targeting viruses as part of their design (i.e., adding antibodies to select viruses), and as a result are widely employed in diagnostic or therapeutic platforms (Lehner et al. 2013). However, very few studies have addressed the ability of NMs in their native form to interact directly with viruses or viral particles, either in the ambient environment or within a host. A recent study utilized MS2 bacteriophages as a model surrogate for norovirus, which is known to cause 18% of all diarrheal disease incidences worldwide (Ahmed et al. 2014), to investigate their ability to interact with carbon NMs (Hoelzer et al. 2013). MS2 capsids consist of small icosahedral shells ($T = 3$), composed of 90 dimers of the coat protein, which protects the genomic RNA inside (Valegard et al. 1990). The viral coat proteins include various amino acid residues, both on the interior (20–90 residues) and the exterior (1–20 and 90–129 residues) of the capsids (Penrod et al. 1996). These amino acids present positively (e.g., protonated amine and diaminoiminium groups) and negatively charged moieties (e.g., carboxylates), which display a pH-dependent net surface charge on the MS2 (Penrod et al. 1995). Furthermore, a number of residues on the capsid exterior exhibit positive hydropathy index values, indicating strong hydrophobicity (Penrod et al. 1995, 1996; Kyte and Doolittle 1982). Strong hydrophobicity and presence of surface moieties on these colloidally stable MS2 capsids can facilitate attachment to carbon NMs. A recent report from our groups showed that interaction between MS2 capsids and multiwalled carbon nanotubes occurs in water and that this interaction was stable, even in high salinity conditions (Merryman et al. 2019). The implication of this work may extend beyond MS2 capsids and into other similar waterborne viruses (e.g., rotavirus, adenovirus, etc.) since many such virus capsids are decorated with similar amino acid residues, containing similar functional moieties (Dawson et al. 2005; Armanious et al. 2016). Similar studies to address NM–viral interactions in air or in hosts have not been documented. Such interactions in the environmental space will likely have a profound impact in controlling the fate and transport of viruses.

10.7.1.3 Interaction of NMs with Host Cells

NMs have been shown to interact with many cell types through a variety of different mechanisms. In some cases, NMs interact with the cell surface directly through spontaneous membrane penetration, essentially piercing the cell membrane, which can lead to complete internalization (Li et al. 2013b; Shi et al. 2011). This has been primarily studied for carbon-based NMs that are long and fiber-like. In contrast, there are numerous studies that support the cellular uptake of NMs through receptor-mediated mechanisms that lead to endocytosis.

Several cell types that are known targets for viruses and/or provide defense against viral infections have been shown to engulf NMs where the cell fate depends on the type of particle, dose, and cell type. For example, NMs can be phagocytosed and degraded by pulmonary macrophages (Lahiani et al. 2017; Meng et al. 2015), epithelial cells (Foldbjerg et al. 2014), brain microglia cells (Shigemoto-Mogami et al. 2016), and lymphocytes (Brown et al. 2007), among others. Macrophages and epithelial cells usually serve as first-line defense barriers for viral infections of the respiratory tract, and elsewhere, where they engulf NMs and activate cellular signaling pathways to induce inflammatory responses (Pondman et al. 2015). Once engulfed, both macrophages (and eosinophils) can degrade and clear NMs through peroxidase-driven mechanisms (Kotchey et al. 2013; Andon et al. 2013). However, the degradation is largely dependent on the specific NM-type and is highly influenced by size, surface chemistry, and surface area. For example, SWCNTs with oxygen moieties, but not the pristine form of SWCNTs, can undergo degradation more readily (Allen et al. 2009). Thus, CNTs with a lower degradability may stay longer and induce prolonged impacts on disease susceptibility within a host. It can be hypothesized that the ability of NMs to alter the abundance and response of cell populations to viral infections could lead to more (or less) severe infections. For example, limiting the number of macrophages available to clear virus particles or impairing their ability to recruit additional immune cells would essentially diminish the normal host antiviral defenses.

Specific to viruses, the ability of NMs to influence specific viral entry receptors has only recently been studied. For example, in order to enter cells, IAV binds primarily to sialic acid receptors which are small sugars attached to proteins within cell membranes. A number of chemically distinct forms of sialic acids exist which differ in the linkage of sialic acid to the sugar galactose. Joining the carbon atom at position 2 of the sialic acid hexose via an oxygen atom to the carbon at position 3 or position 6 of the hexose of galactose creates alpha 2,3 or alpha 2,6 linkages, respectively. These linkages differ in both their distribution in tissues (i.e., lung) and affinity for human IAV strains with preferential affinity for alpha 2,6 linkages (Ramos and Fernandez-Sesma 2012). Once viruses enter cells through these receptors, they form a replication complex and using the host cell machinery generate copies of their genomes that are subsequently packaged into viral progeny particles that are released from the cell. Recent observations by our group (manuscript in preparation) have shown that while CNTs do not alter the expression of the enzymes that add the receptor-specific moieties in lung epithelial cells, they can alter the number of receptors on the cell surface. This is the first observation that NPs can influence select viral-specific receptor localization which may help to explain why these same NPs alter IAV infectivity. In other cases, NPs have been shown to significantly interfere with viral entry, although precise mechanisms have not been identified. For example, carbon nanodots formed with 4-aminophenylboronic acid hydrochloride impaired entry of herpesvirus into both Vero and A549 cells that was most prominent during the early stages of interaction between the virus and cell receptors (Barras et al. 2016).

10.7.2 Activation of Pattern Recognition Receptors (PRRs) and Related Signaling Pathways

Initial recognition of viruses by host cells is mediated by intracellular pathogen recognition receptors (PRRs) that interact directly with conserved molecular patterns present in lipid, lipoprotein, protein, carbohydrates, and nucleic acid moieties called pathogen-associated molecular patterns (PAMPs Fig. 10.2) (Kawai and Akira 2009). Activation of these receptors, including toll-like receptors (TLRs), C-type lectin receptors (CLRs), retinoic acid-inducible gene-I (RIG-I)-like receptors

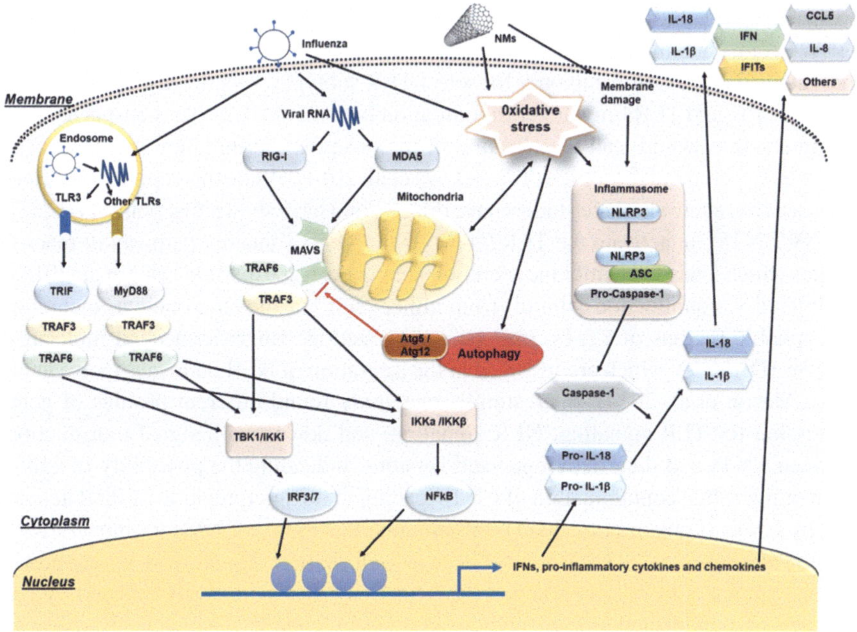

Fig. 10.2 Diagram showing the interplay of several signaling mechanisms that may be perturbed by NMs and viruses including endosomal PRRs (TLRs) and PRRs that work through the mitochondria (RIG-I–MAVS), inflammasomes, and how ROS may serve as a trigger

(RLRs), nucleotide-binding oligomerization domain (NOD)-like receptors (NLRs), and absent in melanoma 2-like receptors (AIM2) (Silva et al. 2017; Kawai and Akira 2009), leads to the production and release of IFNs that interrupt viral life cycles by directly interfering with their ability to replicate within an infected host cell. For example, both viral DNA and RNA can be recognized by several PRRs. Endosomal TLRs, such as TLR3, can recognize double-stranded RNA of viruses, while TLR7 and TLR8 can recognize single-stranded viral RNA (Owen et al. 2013). Surface TLRs such as TLR2 and TLR4 have also been shown to detect components of viruses that cause measles, hepatitis C, herpes, and respiratory syncytial virus which initiate pro-inflammatory immune responses (Koyama et al. 2008). While TLRs detect extracellular DNA and RNA endocytosed by cells, RLRs, including RIG-I and Melanoma Differentiation-Associated protein 5 (MDA5), are cytosolic receptors required for detection of intracellular viral RNA (and sometimes viral DNA) from actively replicating cells (Dixit and Kagan 2013). RIG-I and MDA5 can form a complex that activates mitochondrial antiviral signaling (MAVS) to induce type I IFNs and pro-inflammatory cytokines.

An interesting hypothesis has emerged that suggests NMs can modulate PAMP signaling networks either through direct interaction with receptors or indirectly through production of endogenous ligands. With respect to the former, one study proposed that CNTs interact directly with TLR homo- and heterodimer extracellular domains based on a computational model (Turabekova et al. 2014). By using X-ray structuring and mathematical modeling, they discovered that the internal hydrophobic pockets of some TLRs may be capable of binding select carbon nanostructures, such as SWCNTs, containing 11 carbon atom layers (Turabekova et al. 2014). Furthermore, they showed this relationship only for select TLR subtypes, specifically TLR1/TLR2, TLR2/TLR6, and TLR4/myeloid differentiation factor-2 (MD-2). They suspected that this interaction would lead to activation of the receptor and initiation of subsequent downstream immune responses (Turabekova et al. 2014). While this report did not test TLR activity, a few alternate studies have done so and have shown that both MWCNTs and SWCNTs can activate the TLR2/4–MyD88–NFκB signaling pathway in macrophages which leads to chemokine secretion (Meng et al. 2015; Mukherjee et al. 2018). Additionally, noninfectious stimuli from either "self" or the environment, including hydrophobic regions of CNTs, may be categorized as danger-associated molecular patterns (DAMPs), which are involved in the activation of NLR pathways (Silva et al. 2017; Mason et al. 2012). Interestingly, one study found different profiles of gene expression for TLR signaling, NLR signaling, and downstream signal transduction between CNTs and their depyrogenized versions, indicating the possibility of endotoxin or microbial contamination of CNTs leading to misinterpretation of PRR activation by CNTs (Lahiani et al. 2017). Importantly, the ability of NMs to impact PRRs specific to viruses (i.e., TLR3) has not been as well studied. Recent work by our group has shown that SWCNTs can impair TLR3 activity that is normally induced by the viral analog polyinosinic:polycytidylic acid (poly I:C) and IAV which leads to downstream expression of genes that combat viral replication. Furthermore, these same NMs also impact RIG-I–MDA5 signaling by impairing the ability of the RIG-I–MDA5 complex to "dock" with mitochondrial antiviral signaling (MAVS) protein on the mitochondrial membrane (Chen et al. 2019). Interruption of this docking mechanism leads to the reduced expression of antiviral genes, leaving the host cells more susceptible to increased virus titers.

10.7.3 Production of Oxidative Stress and Mitochondrial Dysfunction

Oxidative stress occurs when antioxidant mechanisms in a cell or body fail to scavenge overproduced reactive oxygen species (ROS), causing further organelle damage and cell apoptosis (Dong and Ma 2015; Shvedova et al. 2012). Such cellular stress can cause oxidation of biomolecules, such as lipids, proteins, and DNA, and can either activate or impair signal transduction of immune responses (Weinberg et al. 2015).

As mentioned above, RIG-I–MDA5 controls antiviral signaling pathways through docking with MAVS which is located on the mitochondrial membrane, and its function is highly dependent on mitochondrial structure and function (Weinberg et al. 2015). Because the mitochondria are known to serve as a source of ROS production, they can also be damaged by ROS. As many NMs are known to generate ROS production, it is possible that they can induce mitochondrial dysfunction and as a result, affect downstream innate immune responses (Kim et al. 2015). In a recent study, ROS produced by SWCNTs was shown to alter mitochondrial membrane potential and the expression of genes involved in mitochondrial function (fission, fusion). Furthermore, these same SWCNTs impaired mitochondrial-dependent antiviral pathways (RIG-I–MAVS) which led to high IAV titers (Chen et al. 2019).

The literature is not consistent on the role of ROS in response to viral infections as some strains cause ROS and others do not. However, it is known that oxidative stress-related protein complexes called stress granules (SGs) commonly assemble in infected cells. These SGs contain several ribonucleoproteins (mRNPs) and also the same RIG-I–MDA5 proteins found in the mitochondria which can induce the activation of protein kinase receptor (PKR)/eukaryotic translation initiation factor 2a (eIF2a) pathways (Yoneyama et al. 2016). Thus, oxidative stress induced by NMs may also affect the formation and/or function of SGs, formed in response to viruses which also may interfere with the normal defense response to an infection.

10.7.3.1 Stimulation of Autophagy

Autophagy is a cellular process that promotes energy efficiency and mediates damage control of cells that experience various kinds of stimuli or stresses, such as oxidative stress (Glick et al. 2010). Autophagy is also associated with disease susceptibility and related outcomes to pathogenic infections; for example, the hemagglutinin protein of influenza virus can lead to autophagy-related cell death and exacerbate viral infections in mice (Sun et al. 2012). Autophagy is proven to be a mechanism for negative-stranded RNA viruses to evade host's immune responses as the viruses can regulate and exploit the cellular state of autophagy (Wang et al. 2018).

Studies have shown that NMs can induce autophagy. Park and colleagues found that exposure of BEAS-2B cells to SWCNTs led to higher expression of autophagic genes and proteins with increased formation of autophagosome-like vacuoles (Park et al. 2014a). Similarly, another study discovered that high-pressure carbon monoxide conversion (HiPco SWCNTs) can induce autophagosome formation in a macrophage cell line, RAW264.7 (Park et al. 2014b). As discussed earlier, functionalization of CNTs is a major factor in determining their toxicity. SWCNTs functionalized with carboxylic acid induced formation of autophagosomes through the protein kinase B (AKT) - tuberous sclerosis complex 2 (TSC2) - mammalian target of rapamycin (mTOR) pathway in A549 cells (Liu et al. 2011). An autophagy study on 81 CNTs with different functional groups showed that the modified surface chemistry on CNTs determined the capacity for and specific pathways of inducing autophagy (Wu et al. 2014). Thus, autophagy as a result of NM exposure may push cells to a state where pathogens can benefit, leading to exacerbation of infection. Interestingly, some key proteins that regulate autophagy are also important in controlling pathogen-induced innate immune pathways. For example, the higher expression of *Atg5* and *Atg12* is negatively associated with interferon production through the RLR/MAVS signaling pathway (Takeshita et al. 2008).

10.7.4 Inflammasome Activation

Several types of NMs, such as CNTs, can cause pulmonary inflammation through activation of several innate immune sensors associated with oxidative stress, such as nucleotide-binding oligomerization domain (NOD)-like receptor pyrin domain-containing 3 (NLRP3) inflammasome (Girtsman et al. 2014; Dong and Ma 2016; Palomaki et al. 2011). The NLRP3 inflammasome is the most extensively studied inflammasome complex, and is activated by a wide range of environmental stimuli, including bacteria, viruses, and ROS (Dong and Ma 2015; Palomaki et al. 2011; Chen et al. 2015; Sarvestani and McAuley 2017). The main function of NLRP3 is cleaving protease-1 to activate caspase-1 and trigger the secretion of downstream pro-inflammatory cytokines, interleukin (IL)-1β and IL-18 (Sarvestani and McAuley 2017).

Studies have shown that *Chlamydia pneumoniae* and SWCNT co-treatment synergistically enhanced IL-1β secretion in macrophages in a time- and dose-dependent manner via the NLRP3 inflammasome and caspase-1 activation pathways (Matsuo et al. 2015). This IL-1β secretion can be reduced by either inhibition of NLRP3 expression or blocking ROS production by N-acetyl-L-cysteine (Matsuo et al. 2015; Geiler et al. 2010). Another study indicated that CNTs and TLR agonists synergistically promoted activation of NLRP3 inflammasome and production of IL-1β, an essential process to enhance inflammatory activity (Yang et al. 2013). Interestingly, the secretion of downstream IL-1β was not affected by SWCNTs, either alone or in combination with IAV in mouse bronchoalveolar lavage fluid (BALF) (Chen et al. 2017). In this study, the length of exposure to CNTs might be a factor (Chen et al. 2017; Shvedova et al. 2008), since other studies indicated a much earlier peak (day 1) in secreted IL-1β in the BALF of mice treated with SWCNTs (Shvedova et al. 2005). SWCNTs and *Listeria monocytogenes* or *Toxoplasma gondii* have also been shown to synergistically induce IL-1β (Shvedova et al. 2008; Swedin et al. 2012), but with lowered ROS production compared to those treated with CNTs only. SWCNTs can also serve as a negative regulator of NLRP3 inflammasome activation via ROS, leading to reduced inflammatory responses in Caco-2 cells infected by *Escherichia coli* and *Staphylococcus aureus* (Chen et al. 2015). The synergism of NLRP3 activation in the co-exposure scenario might result from mechanisms that include PRR pathways with ROS production (Sarvestani and McAuley 2017). This also indicates that the mechanism of NM impacts on pathogen infection likely involves multiple signaling pathways. It is noteworthy to mention that most of the studies performed to date related to inflammasome regulation by NMs and pathogens involve bacteria and fungi, with little focus on viruses.

10.7.5 Modulation of Lipid Signaling Networks

Lipids are naturally occurring hydrophobic biomolecules that play an essential role in cellular life from membrane formation to energy storage (Voelker 1991). In addition, lipids also serve as effectors and second messengers in a variety of biological processes that are associated with the immune system, both innate and adaptive (Cabral 2005; Getz and Reardon 2014). For example, the Class I P13-kinase (P13K) signaling pathway can be activated by a variety of viruses, such as influenza virus, to dock proteins carrying lipid domains and is also essential for viral entry and fusion (Mazzon and Mercer 2014). The activation, differentiation, and function of immune cells, such as macrophages and T lymphocytes, is also highly regulated by lipids such as cholesterol and fatty acids (Hubler and Kennedy 2016). In response to respiratory viruses, specifically, lipids are known to modulate host–pathogen interactions in several ways, acting as major players in viral-induced membrane fluidity, lipid rafts and viral entry receptors, formation of viral replication complexes, and modulation of host cell metabolism – which in turn afford advantages to the virus in completing its life cycle (Garcia-Sastre 2013; Heaton and Randall 2011; Tanner et al. 2014).

The importance of the lipid landscape in lung disease states has only recently been underscored. Several contemporary studies have begun to decipher the major classes of lipids, and in some cases, quantify select lipids that are present in normal lung cells and tissues. In the lung, significant changes of the lipidome are associated with asthma, chronic obstructive pulmonary disease (COPD), cystic fibrosis, and pathogenic infection (Zehethofer et al. 2015; Zhao et al. 2014; Köberlin et al. 2015). In fact, lipids are now recognized as major regulators of immune responses, and lipid networks are being considered as predictive markers of pathogen responses in patients, as well as in rodents (Köberlin et al. 2015). Recent advancements in liquid chromatography-mass spectrometry (LC-MS) technologies have allowed for global "omic" analysis of lipids. Studies performed to date utilizing this technology have shown that lipid biosynthesis, metabolism, and the profile of the lipids present can all be perturbed by disease state. Sphingolipid metabolism was strongly regulated when pathogen sensing pathways were perturbed by commercial agonists (to Toll-like receptors) in human lung cells or in a mouse model of experimental asthma (Zehethofer et al. 2015; Köberlin et al. 2015). Furthermore, changes in biosynthesis of these perturbed lipid profiles correlated with the expression of related lipid biosynthesis genes. Specific to influenza, exposure of ferrets to both historic and contemporary strains of pandemic influenza A virus revealed relationships between altered lipid metabolism and inflammation (Tisoncik-Go et al. 2016). For example, the sphingomyelin module (group of species with similar co-expression relationships) was positively correlated with innate immune and lipid metabolism gene modules, as well as virus titer and levels of viral mRNA (Tisoncik-Go et al. 2016). In mouse models, distinct patterns of lipid metabolites were associated with the pathogenic (5-lipoxygenase metabolites) and resolution phases (12/15-lipoxygenase metabolites) of influenza A infection (Tam 2013). Several reports now show that cells infected specifically with IAV increase de novo fatty acid and sphingolipid biosynthesis and unique cholesterol compositions while lowering levels of phosphatidylethanolamine (PE) and phosphatidylcholine (PC) species and omega-3 polyunsaturated fatty acid (PUFA)-derived lipid mediator, protectin D1 (Tanner et al. 2014; Murray et al. 2009). These changes have been linked with regulation of viral budding, viral replication and pathogenicity.

Additionally, there is substantial interest in the use of lipids as diagnostic and prognostic biomarkers (t'Kindt et al. 2015). Several studies have provided evidence that lipids (and/or metabolites) can be used to differentiate between disease states, such as community acquired and noncommunity-acquired pneumonia cases (To et al. 2016), and between healthy individuals and asthmatic patients (Kang et al. 2014). Previous work by t'Kindt and colleagues was able to identify 1584 unique lipids in induced sputum and demonstrated that sphingolipid profiles in sputum samples were similar between smokers without COPD and never-smokers but were very different from profiles of COPD patients (t'Kindt et al. 2015). Taken together, these studies demonstrate a substantial role for alterations in lipid metabolism during the course of viral infection and disease progression, and suggest that these patterns of alterations are distinct enough to have clinical relevance in diagnosis, prognosis, and treatment.

Though little is known, a few studies provide evidence to support that NMs are capable of altering lipid profiles in cells with impacts on the immune system. For example, long and stiff CNTs can damage membranes of lysosomes, which are responsible for breaking down diverse biomolecules and debris, including phagocytosed pathogens (Zhu et al. 2016). The interaction between pristine or functionalized CNTs and cellular membranes was investigated using computational models, and the results indicated that the distribution and orientation of CNTs within the lipid bilayer membranes were largely dependent on the functionalization of the CNTs (Baoukina et al. 2013). A lipidomic study on all major phospholipid classes revealed that lipid peroxidation in mouse lung tissues exposed to SWCNTs through inhalation was the major oxidative damage mechanism (Tyurina et al. 2011). The characterized specific phospholipid peroxidation pathway pointed to the mitochondria as the target of CNT toxicity (Tyurina et al. 2011). Another study in which mice were exposed to SWCNTs via a single intratracheal instillation revealed an upregulation of fatty acid β-oxidation 13 weeks postexposure, a change that is likely due to stress in the endoplasmic reticulum and mitochondria (Park et al. 2016). These studies support and circle back to the notion of how oxidative stress induced by CNTs may play an important role in immune responses to pathogenic infections, which was previously discussed. Another report investigated the ability of CNTs to change global phospholipids in SWCNT-exposed cells and demonstrated the modification of several lipid classes (Wu et al. 2011) that are highly relevant to IAV infections. No studies to date have studied the changing lipid landscape in response to varied carbonaceous NMs and how such changes modulate viral infections.

Nano-TiO_2 exposure has also been shown to impact lipid metabolism and profiles in ways that could have downstream impacts on immune function. For example, mice exposed intranasally to nano-TiO_2 exhibited increased peroxidation of lipids as a result of ROS production, as well as increased expression of arachidonate 5-lipoxygenase-activating proteins (ALOX5AP) that regulate biosynthesis of leukotrienes, pro-inflammatory lipid mediators secreted by inflammatory cells (Li et al. 2013c). A study on liver cells (HepG2) of several nano-TiO_2 particles of differing sizes and surface areas demonstrated impacts on cholesterol biosynthesis, bile acid synthesis, zymosterol (cholesterol precursor) biosynthesis, free fatty acid availability, and fatty acid activation after exposure to high doses of nano-TiO_2 (Thai et al. 2016).

10.8 Future Perspectives

The advent of emergent technologies, such as the development of NMs, has and will continue to greatly enhance technological advancements in industrial, consumer, environmental, and biomedical sectors, among others. However, it is important that we understand the potential health and safety impacts that are relevant to real work settings. These settings include both single exposures to NMs and co-exposures with relevant viruses that carry tremendous health burden. As discussed, NMs have the ability to modulate normal immune responses to viruses and some reports suggest that this leads to worsened infections. However, the number of studies that have analyzed such outcomes are still relatively small, despite the potential consequences to exposed populations. Much of the work to date has focused heavily on CNTs and a few select viruses (i.e., IAV, RSV) but there is much work to be done regarding the commonalities in immune modulation between different types of NMs and viruses. Defining which stages of the viral life cycle are impacted by NMs would offer tremendous knowledge that is not only useful to the health and safety field, but would uncover mechanisms that could be essential in developing vaccines or other viral therapeutics. Very little data exist regarding the ability of NMs to interact directly with viruses or viral moieties and how such interactions alter environmental fate and transport (and ultimately exposure) and manipulation of normal immune responses to NMs or viruses singly which are likely distinct to some extent. Furthermore, the ability of NMs to alter interactions of viruses with their cell surface receptors and PRRs is an intriguing mechanism that warrants further investigation. The development of technology for generation of lipid profiles has highlighted the importance of lipids as both structural and signaling molecules that play important roles in viral life cycles. This suggests yet another mechanism that can be considered in NM–viral studies. Finally, model systems for testing viral infections with nanoparticulates need further development as standard models used to assess toxicity of one stressor may not be suitable for the other (i.e., some viruses cannot be adapted to mice). In addition, testing of dual stressors has not yet made use of more complex (i.e., 3D) tissue culture systems or organoids which may provide a valuable platform for testing the wide array of NMs and viruses.

References

Ahmed SM, Hall AJ, Robinson AE, Verhoef L, Premkumar P, Parashar UD, Koopmans M, Lopman BA. Global prevalence of norovirus in cases of gastroenteritis: a systematic review and meta-analysis. Lancet Infect Dis. 2014;14(8):725–30. https://doi.org/10.1016/S1473-3099(14)70767-4. PubMed PMID: WOS:000339461700031.

Allen BL, Kotchey GP, Chen Y, Yanamala NV, Klein-Seetharaman J, Kagan VE, Star A. Mechanistic investigations of horseradish peroxidase-catalyzed degradation of single-walled carbon nanotubes. J Am Chem Soc. 2009;131(47):17194–205. https://doi.org/10.1021/ja9083623. Epub 2009/11/07. PubMed PMID: 19891488.

Andon FT, Kapralov AA, Yanamala N, Feng W, Baygan A, Chambers BJ, Hultenby K, Ye F, Toprak MS, Brandner BD, Fornara A, Klein-Seetharaman J, Kotchey GP, Star A, Shvedova AA, Fadeel B, Kagan VE. Biodegradation of single-walled carbon nanotubes by eosinophil peroxidase. Small. 2013;9(16):2721–9. https://doi.org/10.1002/smll.201202508. Epub 2013/03/01. PubMed PMID: 23447468; PMCID: 4039041.

Armanious A, Aeppli M, Jacak R, Refardt D, Sigstam T, Kohn T, Sander M. Viruses at solid-water interfaces: a systematic assessment of interactions driving adsorption. Environ Sci Technol. 2016;50(2):732–43. https://doi.org/10.1021/acs.est.5b04644. PubMed PMID: 26636722.

Baoukina S, Monticelli L, Tieleman DP. Interaction of pristine and functionalized carbon nanotubes with lipid membranes. J Phys Chem B. 2013;117(40):12113–23. https://doi.org/10.1021/jp405732k. Epub 2013/09/13. PubMed PMID: 24024494.

Barras A, Pagneux Q, Sane F, Wang Q, Boukherroub R, Hober D, Szunerits S. High efficiency of functional carbon nanodots as entry inhibitors of herpes simplex virus type 1. ACS Appl Mater Interfaces. 2016;8(14):9004–13. https://doi.org/10.1021/acsami.6b01681. Epub 2016/03/26/. PubMed PMID: 27015417.

Bawage SS, Tiwari PM, Singh A, Dixit S, Pillai SR, Dennis VA, Singh SR. Gold nanorods inhibit respiratory syncytial virus by stimulating the innate immune response. Nanomedicine. 2016;12(8):2299–310. https://doi.org/10.1016/j.nano.2016.06.006. PubMed PMID: 27381068; PMCID: PMC5116403.

Becker S, Soukup JM. Exposure to urban air particulates alters the macrophage-mediated inflammatory response to respiratory viral infection. J Toxicol Environ Health A. 1999;57(7):445–57. Epub 1999/09/24. PubMed PMID: 10494914.

Boonruksa P, Bello D, Zhang J, Isaacs JA, Mead JL, Woskie SR. Characterization of potential exposures to nanoparticles and fibers during manufacturing and recycling of carbon nanotube reinforced polypropylene composites. Ann Occup Hyg. 2016;60(1):40–55. https://doi.org/10.1093/annhyg/mev073. Epub 2015/10/09. PubMed PMID: 26447230.

Brown DM, Kinloch IA, Bangert U, Windle AH, Walter DM, Walker GS, Scotchford CA, Donaldson K, Stone V. An in vitro study of the potential of carbon nanotubes and nanofibres to induce inflammatory mediators and frustrated phagocytosis. Carbon. 2007;45(9):1743–56. https://doi.org/10.1016/j.carbon.2007.05.011.

Cabral GA. Lipids as bioeffectors in the immune system. Life Sci. 2005;77(14):1699–710. https://doi.org/10.1016/j.lfs.2005.05.013. Epub 2005/06/14. PubMed PMID: 15950989.

Cai XN, Ramalingam R, Wong HS, Cheng JP, Ajuh P, Cheng SH, Lam YW. Characterization of carbon nanotube protein corona by using quantitative proteomics. Nanomed-Nanotechnol. 2013;9(5):583–93. https://doi.org/10.1016/j.nano.2012.09.004. PubMed PMID: ISI:000320593600001.

Chen W, Liu Y, Wang H, Hnizdo E, Sun Y, Su L, Zhang X, Weng S, Bochmann F, Hearl FJ, Chen J, Wu T. Long-term exposure to silica dust and risk of total and cause-specific mortality in Chinese workers: a cohort study. PLoS Med. 2012;9(4):e1001206. https://doi.org/10.1371/journal.pmed.1001206. Epub 2012/04/25. PubMed PMID: 22529751; PMCID: 3328438.

Chen H, Wang B, Zhao R, Gao D, Guan M, Zheng L, Zhou X, Chai Z, Zhao Y, Feng W. Coculture with low-dose SWCNT attenuates bacterial invasion and inflammation in human enterocyte-like Caco-2 cells. Small. 2015;11(34):4366–78. https://doi.org/10.1002/smll.201500136. Epub 2015/06/23. PubMed PMID: 26097125.

Chen H, Zheng X, Nicholas J, Humes ST, Loeb JC, Robinson SE, Bisesi JH Jr, Das D, Saleh NB, Castleman WL, Lednicky JA, Sabo-Attwood T. Single-walled carbon nanotubes modulate pulmonary immune responses and increase pandemic influenza a virus titers in mice. Virol J. 2017;14(1):242. https://doi.org/10.1186/s12985-017-0909-z. Epub 2017/12/24. PubMed PMID: 29273069; PMCID: 5741862.

Chen H, Humes ST, Robinson SE, Loeb JC, Sabaraya IV, Saleh NB, Khattri RB, Merritt ME, Martyniuk CJ, Lednicky JA, Sabo-Attwood T. Single-walled carbon nanotubes repress viral-induced defense pathways through oxidative stress. Nanotoxicology 2019;13(9):1176–1196

Ciencewicki J, Jaspers I. Air pollution and respiratory viral infection. Inhal Toxicol. 2007;19(14):1135–46. https://doi.org/10.1080/08958370701665434. Epub 2007/11/08. PubMed PMID: 17987465.

Clifford HD, Perks KL, Zosky GR. Geogenic PM(1)(0) exposure exacerbates responses to influenza infection. Sci Total Environ. 2015;533:275–82. https://doi.org/10.1016/j.scitotenv.2015.07.001. Epub 2015/07/15. PubMed PMID: 26172594.

Dawson DJ, Paish A, Staffell LM, Seymour IJ, Appleton H. Survival of viruses on fresh produce, using MS2 as a surrogate for norovirus. J Appl Microbiol. 2005;98(1):203–9. https://doi.org/10.1111/j.1365-2672.2004.02439.x. PubMed PMID: WOS:000225791600022.

Dixit E, Kagan JC. Intracellular pathogen detection by RIG-I-like receptors. Adv Immunol. 2013;117:99–125. https://doi.org/10.1016/B978-0-12-410524-9.00004-9. Epub 2013/04/25. PubMed PMID: 23611287; PMCID: 3947775.

Dixon RE. Economic costs of respiratory tract infections in the United States. Am J Med. 1985;78(6B):45–51. PubMed PMID: 4014287.

Dong J, Ma Q. Advances in mechanisms and signaling pathways of carbon nanotube toxicity. Nanotoxicology. 2015;9(5):658–76. https://doi.org/10.3109/17435390.2015.1009187. Epub 2015/02/14. PubMed PMID: 25676622; PMCID: 4707672.

Dong J, Ma Q. Suppression of basal and carbon nanotube-induced oxidative stress, inflammation and fibrosis in mouse lungs by Nrf2. Nanotoxicology. 2016;10(6):699–709. https://doi.org/10.3109/17435390.2015.1110758. Epub 2015/11/26. PubMed PMID: 26592091.

Edinger TO, Pohl MO, Stertz S. Entry of influenza A virus: host factors and antiviral targets. J Gen Virol. 2014;95(Pt 2):263–77. https://doi.org/10.1099/vir.0.059477-0. Epub 2013/11/15. PubMed PMID: 24225499.

Feingold BJ, Vegosen L, Davis M, Leibler J, Peterson A, Silbergeld EK. A niche for infectious disease in environmental health: rethinking the toxicological paradigm. Environ Health Perspect. 2010;118(8):1165–72. https://doi.org/10.1289/ehp.0901866. Epub 2010/04/14. PubMed PMID: 20385515; PMCID: 2920090.

Findlay F, Pohl J, Svoboda P, Shakamuri P, McLean K, Inglis NF, Proudfoot L, Barlow PG. Carbon nanoparticles inhibit the antimicrobial activities of the human cathelicidin LL-37 through structural alteration. J Immunol. 2017;199(7):2483–90. https://doi.org/10.4049/jimmunol.1700706. PubMed PMID: 28814602; PMCID: PMC5602159.

Foldbjerg R, Irving ES, Wang J, Thorsen K, Sutherland DS, Autrup H, Beer C. The toxic effects of single-walled carbon nanotubes are linked to the phagocytic ability of cells. Toxicol Res. 2014;3(4):228–41. https://doi.org/10.1039/c3tx50099c.

Forum of International Respiratory Societies. The global impact of respiratory disease. 2nd ed. Sheffield: European Respiratory Society; 2017.

Garcia-Sastre A. Lessons from lipids in the fight against influenza. Cell. 2013;154(1):22–3. https://doi.org/10.1016/j.cell.2013.06.024. PubMed PMID: 23827671.

GBD 2015 LRI Collaborators. Estimates of the global, regional, and national morbidity, mortality, and aetiologies of lower respiratory tract infections in 195 countries: a systematic analysis for the Global Burden of Disease Study 2015. Lancet Infect Dis. 2017;17:1133–61.

Geiler J, Michaelis M, Naczk P, Leutz A, Langer K, Doerr HW, Cinatl J Jr. N-acetyl-L-cysteine (NAC) inhibits virus replication and expression of pro-inflammatory molecules in A549 cells infected with highly pathogenic H5N1 influenza A virus. Biochem Pharmacol. 2010;79(3):413–20. https://doi.org/10.1016/j.bcp.2009.08.025. Epub 2009/09/08. PubMed PMID: 19732754.

Getz GS, Reardon CA. The mutual interplay of lipid metabolism and the cells of the immune system in relation to atherosclerosis. Clin Lipidol. 2014;9(6):657–71. https://doi.org/10.2217/clp.14.50. Epub 2015/02/24. PubMed PMID: 25705263; PMCID: 4335315.

Girtsman TA, Beamer CA, Wu N, Buford M, Holian A. IL-1R signalling is critical for regulation of multi-walled carbon nanotubes-induced acute lung inflammation in C57Bl/6 mice. Nanotoxicology. 2014;8(1):17–27. https://doi.org/10.3109/17435390.2012.744110. Epub 2012/10/26. PubMed PMID: 23094697; PMCID: 4080682.

Glick D, Barth S, Macleod KF. Autophagy: cellular and molecular mechanisms. J Pathol. 2010;221(1):3–12. https://doi.org/10.1002/path.2697. Epub 2010/03/13. PubMed PMID: 20225336; PMCID: 2990190.

Groopman JD, Johnson D, Kensler TW. Aflatoxin and hepatitis B virus biomarkers: a paradigm for complex environmental exposures and cancer risk. Cancer Biomark. 2005;1(1):5–14. Epub 2006/12/29. PubMed PMID: 17192028.

Guo JJ, Yao XJ, Ning LL, Wang QQ, Liu HX. The adsorption mechanism and induced conformational changes of three typical proteins with different secondary structural features on graphene. Rsc Adv. 2014;4(20):9953–62. https://doi.org/10.1039/c3ra45876h. PubMed PMID: WOS:000332061300009.

Guzman KAD, Taylor MR, Banfield JF. Environmental risks of nanotechnology: National nanotechnology initiative funding, 2000–2004. Environ Sci Technol. 2006;40(5):1401–7. PubMed PMID: ISI:000235796100011.

Hashiguchi S, Yoshida H, Akashi T, Hirose A, Sugita C, Kurokawa M, Watanabe W. Effects of multi-walled carbon nanotubes on the pneumonia in respiratory syncytial virus-infected mice. Toxicol Lett. 2015a;238(2, Supplement):S201. https://doi.org/10.1016/j.toxlet.2015.08.603.

Hashiguchi S, Yoshida H, Akashi T, Komemoto K, Ueda T, Ikarashi Y, Miyauchi A, Konno K, Yamanaka S, Hirose A, Kurokawa M, Watanabe W. Titanium dioxide nanoparticles exacerbate pneumonia in respiratory syncytial virus (RSV)-infected mice. Environ Toxicol Pharmacol. 2015b;39(2):879–86. https://doi.org/10.1016/j.etap.2015.02.017. PubMed PMID: 25795424.

Heaton NS, Randall G. Multifaceted roles for lipids in viral infection. Trends Microbiol. 2011;19(7):368–75. https://doi.org/10.1016/j.tim.2011.03.007. PubMed PMID: 21530270; PMCID: PMC3130080.

Hedmer M, Ludvigsson L, Isaxon C, Nilsson PT, Skaug V, Bohgard M, Pagels JH, Messing ME, Tinnerberg H. Detection of multi-walled carbon nanotubes and carbon nanodiscs on workplace surfaces at a small-scale producer. Ann Occup Hyg. 2015;59(7):836–52. https://doi.org/10.1093/annhyg/mev036. PubMed PMID: ISI:000361490900002.

Hoelzer K, Fanaselle W, Pouillot R, Van Doren JM, Dennis S. Virus inactivation on hard surfaces or in suspension by chemical disinfectants: systematic review and meta-analysis of norovirus surrogates. J Food Protect. 2013;76(6):1006–16. https://doi.org/10.4315/0362-028x.Jfp-12-438. PubMed PMID: WOS:000326080900013.

Horne BD, Joy EA, Hofmann MG, Gesteland PH, Cannon JB, Lefler JS, Blagev DP, Korgenski EK, Torosyan N, Hansen GI, Kartchner D, Pope CA 3rd. Short-term elevation of fine particulate matter air pollution and acute lower respiratory infection. Am J Respir Crit Care Med. 2018;198(6):759–66. https://doi.org/10.1164/rccm.201709-1883OC. PubMed PMID: 29652174.

Hu RL, Li SR, Kong FJ, Hou RJ, Guan XL, Guo F. Inhibition effect of silver nanoparticles on herpes simplex virus 2. Genet Mol Res. 2014;13(3):7022–8. https://doi.org/10.4238/2014.March.19.2. Epub 2014/04/01. PubMed PMID: 24682984.

Huang L, Zhou L, Chen J, Chen K, Liu Y, Chen X, Tang F. Acute effects of air pollution on influenza-like illness in Nanjing, China: a population-based study. Chemosphere. 2016;147:180–7. https://doi.org/10.1016/j.chemosphere.2015.12.082. Epub 2016/01/15. PubMed PMID: 26766354.

Hubler MJ, Kennedy AJ. Role of lipids in the metabolism and activation of immune cells. J Nutr Biochem. 2016;34:1–7. https://doi.org/10.1016/j.jnutbio.2015.11.002. Epub 2016/07/18. PubMed PMID: 27424223; PMCID: 5694687.

Jaspers I, Sheridan PA, Zhang W, Brighton LE, Chason KD, Hua X, Tilley SL. Exacerbation of allergic inflammation in mice exposed to diesel exhaust particles prior to viral infection. Part Fibre Toxicol. 2009;6:22. https://doi.org/10.1186/1743-8977-6-22. Epub 2009/08/18. PubMed PMID: 19682371; PMCID: 2739151.

Kang YP, Lee WJ, Hong JY, Lee SB, Park JH, Kim D, Park S, Park CS, Park SW, Kwon SW. Novel approach for analysis of bronchoalveolar lavage fluid (BALF) using HPLC-QTOF-MS-based lipidomics: lipid levels in asthmatics and corticosteroid-treated asthmatic patients. J Proteome Res. 2014;13(9):3919–29. https://doi.org/10.1021/pr5002059. PubMed PMID: WOS:000341345000006.

Kawai T, Akira S. The roles of TLRs, RLRs and NLRs in pathogen recognition. Int Immunol. 2009;21(4):317–37. https://doi.org/10.1093/intimm/dxp017. Epub 2009/02/28. PubMed PMID: 19246554; PMCID: 2721684.

Kim S, Kim MJ, Park DY, Chung HJ, Kim CH, Yoon JH, Kim HJ. Mitochondrial reactive oxygen species modulate innate immune response to influenza A virus in human nasal epithelium. Antivir Res. 2015;119:78–83. https://doi.org/10.1016/j.antiviral.2015.04.011. Epub 2015/05/02. PubMed PMID: 25930096.

Köberlin MS, Snijder B, Heinz LX, Baumann CL, Fauster A, Vladimer GI, Gavin AC, Superti-Furga G. A conserved circular network of coregulated lipids modulates innate immune responses. Cell. 2015;162(1):170–83. https://doi.org/10.1016/j.cell.2015.05.051. Epub 2015/06/23. PubMed PMID: 26095250; PMCID: 4523684.

Kotchey GP, Zhao Y, Kagan VE, Star A. Peroxidase-mediated biodegradation of carbon nanotubes in vitro and in vivo. Adv Drug Deliv Rev. 2013;65(15):1921–32. https://doi.org/10.1016/j.addr.2013.07.007. Epub 2013/07/17. PubMed PMID: 23856412; PMCID: 3855904.

Koyama S, Ishii KJ, Coban C, Akira S. Innate immune response to viral infection. Cytokine. 2008;43(3):336–41. https://doi.org/10.1016/j.cyto.2008.07.009. Epub 2008/08/13. PubMed PMID: 18694646.

Kyte J, Doolittle RF. A simple method for displaying the hydropathic character of a protein. J Mol Biol. 1982;157(1):105–32.

Lahiani MH, Gokulan K, Williams K, Khodakovskaya MV, Khare S. Graphene and carbon nanotubes activate different cell surface receptors on macrophages before and after deactivation of endotoxins. J Appl Toxicol: JAT. 2017;37(11):1305–16. https://doi.org/10.1002/jat.3477. Epub 2017/05/10. PubMed PMID: 28485474.

Lehner R, Wang XY, Marsch S, Hunziker P. Intelligent nanomaterials for medicine: carrier platforms and targeting strategies in the context of clinical application. Nanomed-Nanotechnol. 2013;9(6):742–57. https://doi.org/10.1016/j.nano.2013.01.012. PubMed PMID: WOS:000324046100005.

Li JJ, Muralikrishnan S, Ng CT, Yung LY, Bay BH. Nanoparticle-induced pulmonary toxicity. Exp Biol Med (Maywood). 2010;235(9):1025–33. https://doi.org/10.1258/ebm.2010.010021. Epub 2010/08/20. doi: ebm.2010.010021 [pii]. PubMed PMID: 20719818.

Li ZT, Wang YJ, Kozbial A, Shenoy G, Zhou F, McGinley R, Ireland P, Morganstein B, Kunkel A, Surwade SP, Li L, Liu HT. Effect of airborne contaminants on the wettability of supported graphene and graphite. Nat Mater. 2013a;12(10):925–31. https://doi.org/10.1038/Nmat3709. PubMed PMID: WOS:000324736000019.

Li YF, Yuan HY, von dem Bussche A, Creighton M, Hurt RH, Kane AB, Gao HJ. Graphene microsheets enter cells through spontaneous membrane penetration at edge asperities and corner sites. Proc Natl Acad Sci U S A. 2013b;110(30):12295–300. https://doi.org/10.1073/pnas.1222276110. PubMed PMID: WOS:000322112300043.

Li B, Ze Y, Sun Q, Zhang T, Sang X, Cui Y, Wang X, Gui S, Tan D, Zhu M, Zhao X, Sheng L, Wang L, Hong F, Tang M. Molecular mechanisms of nanosized titanium dioxide-induced pulmonary injury in mice. PloS One. 2013c;8(2):e55563. https://doi.org/10.1371/journal.pone.0055563. PubMed PMID: 23409001; PMCID: PMC3567101.

Liu HL, Zhang YL, Yang N, Zhang YX, Liu XQ, Li CG, Zhao Y, Wang YG, Zhang GG, Yang P, Guo F, Sun Y, Jiang CY. A functionalized single-walled carbon nanotube-induced autophagic cell death in human lung cells through Akt-TSC2-mTOR signaling. Cell Death Dis. 2011;2:e159. https://doi.org/10.1038/cddis.2011.27. Epub 2011/05/20. PubMed PMID: 21593791; PMCID: 3122114.

Ma JH, Song SH, Guo M, Zhou J, Liu F, Peng L, Fu ZR. Long-term exposure to PM2.5 lowers influenza virus resistance via down-regulating pulmonary macrophage Kdm6a and mediates histones modification in IL-6 and IFN-beta promoter regions. Biochem Biophys Res Commun. 2017;493(2):1122–8. https://doi.org/10.1016/j.bbrc.2017.09.013. Epub 2017/09/10. PubMed PMID: 28887033.

Mapesa JO, Maxwell AL, Ryan EP. An exposome perspective on environmental enteric dysfunction. Environ Health Perspect. 2016;124(8):1121–6. https://doi.org/10.1289/ehp.1510459. Epub 2015/12/30. PubMed PMID: 26713888; PMCID: 4977058.

Mason DR, Beck PL, Muruve DA. Nucleotide-binding oligomerization domain-like receptors and inflammasomes in the pathogenesis of non-microbial inflammation and diseases. J Innate Immun. 2012;4(1):16–30. https://doi.org/10.1159/000334247. Epub 2011/11/10. PubMed PMID: 22067846.

Matsuo J, Nakamura S, Takeda S, Ishida K, Yamazaki T, Yoshida M, Chiba H, Hui SP, Yamaguchi H. Synergistic costimulatory effect of chlamydia pneumoniae with carbon nanoparticles on NLRP3 inflammasome-mediated interleukin-1beta secretion in macrophages. Infect Immun. 2015;83(7):2917–25. https://doi.org/10.1128/IAI.02968-14. Epub 2015/05/06. PubMed PMID: 25939513; PMCID: 4468553.

Mazzon M, Mercer J. Lipid interactions during virus entry and infection. Cell Microbiol. 2014;16(10):1493–502. https://doi.org/10.1111/cmi.12340. Epub 2014/08/19. PubMed PMID: 25131438; PMCID: 4265854.

Meng J, Li X, Wang C, Guo H, Liu J, Xu H. Carbon nanotubes activate macrophages into a M1/M2 mixed status: recruiting naive macrophages and supporting angiogenesis. ACS Appl Mater Interfaces. 2015;7(5):3180–8. https://doi.org/10.1021/am507649n. Epub 2015/01/17. PubMed PMID: 25591447.

Merryman AE, Sabaraya IV, Rowles LS III, Toteja A, Carrillo SI, Sabo-Attwood T, Saleh NB. Interaction between functionalized multiwalled carbon nanotubes and MS2 bacteriophages in water. Sci Total Environ. 2019;670:1140–5. https://doi.org/10.1016/j.scitotenv.2019.03.311.

Monopoli MP, Walczyk D, Campbell A, Elia G, Lynch I, Baldelli Bombelli F, Dawson KA. Physical–chemical aspects of protein corona: relevance to in vitro and in vivo biological impacts of nanoparticles. J Am Chem Soc. 2011;133(8):2525–34.

Mukherjee SP, Bondarenko O, Kohonen P, Andon FT, Brzicova T, Gessner I, Mathur S, Bottini M, Calligari P, Stella L, Kisin E, Shvedova A, Autio R, Salminen-Mankonen H, Lahesmaa R, Fadeel B. Macrophage sensing of single-walled carbon nanotubes via Toll-like receptors. Sci Rep. 2018;8(1):1115. https://doi.org/10.1038/s41598-018-19521-9. Epub 2018/01/20. PubMed PMID: 29348435; PMCID: 5773626.

Murray AR, Kisin E, Leonard SS, Young SH, Kommineni C, Kagan VE, Castranova V, Shvedova AA. Oxidative stress and inflammatory response in dermal toxicity of single-walled carbon nanotubes. Toxicology. 2009;257(3):161–71. https://doi.org/10.1016/j.tox.2008.12.023. PubMed PMID: WOS:000264516000008.

Murugan K, Dinesh D, Paulpandi M, Althbyani AD, Subramaniam J, Madhiyazhagan P, Wang L, Suresh U, Kumar PM, Mohan J, Rajaganesh R, Wei H, Kalimuthu K, Parajulee MN, Mehlhorn H, Benelli G. Nanoparticles in the fight against mosquito-borne diseases: bioactivity of Bruguiera cylindrica-synthesized nanoparticles against dengue virus DEN-2 (in vitro) and its mosquito vector Aedes aegypti (Diptera: Culicidae). Parasitol Res. 2015;114(12):4349–61. https://doi.org/10.1007/s00436-015-4676-8. Epub 2015/08/21. PubMed PMID: 26290219.

Nowack B, David RM, Fissan H, Morris H, Shatkin JA, Stintz M, Zepp R, Brouwer D. Potential release scenarios for carbon nanotubes used in composites. Environ Int. 2013;59:1–11. https://doi.org/10.1016/j.envint.2013.04.003. PubMed PMID: ISI:000324901000001.

Otyepkova E, Lazar P, Cepe K, Tomanec O, Otyepka M. Organic adsorbates have higher affinities to fluorographene than to graphene. Appl Mater Today. 2016;5:142–9. https://doi.org/10.1016/j.apmt.2016.09.016. PubMed PMID: WOS:000392950300017.

Owen JA, Punt J, Stranford SA, Jones PP, Kuby J. Kuby immunology. 7th ed. New York: W.H. Freeman; 2013. xxvii, 692, 109 pp.

Palomaki J, Valimaki E, Sund J, Vippola M, Clausen PA, Jensen KA, Savolainen K, Matikainen S, Alenius H. Long, needle-like carbon nanotubes and asbestos activate the NLRP3 inflammasome through a similar mechanism. ACS Nano. 2011;5(9):6861–70. https://doi.org/10.1021/nn200595c. Epub 2011/08/02. PubMed PMID: 21800904.

Park EJ, Zahari NE, Lee EW, Song J, Lee JH, Cho MH, Kim JH. SWCNTs induced autophagic cell death in human bronchial epithelial cells. Toxicology In Vitro. 2014a;28(3):442–50. https://doi.org/10.1016/j.tiv.2013.12.012. Epub 2014/01/07. PubMed PMID: 24389112.

Park EJ, Zahari NE, Kang MS, Lee S, Lee K, Lee BS, Yoon C, Cho MH, Kim Y, Kim JH. Toxic response of HIPCO single-walled carbon nanotubes in mice and RAW264.7 macrophage cells. Toxicol Lett. 2014b;229(1):167–77. https://doi.org/10.1016/j.toxlet.2014.06.015. Epub 2014/06/15. PubMed PMID: 24929217.

Park EJ, Hong YS, Lee BS, Yoon C, Jeong U, Kim Y. Single-walled carbon nanotubes disturbed the immune and metabolic regulation function 13-weeks after a single intratracheal instillation. Environ Res. 2016;148:184–95. https://doi.org/10.1016/j.envres.2016.03.027. PubMed PMID: WOS:000376712800022.

Penrod SL, Olson TM, Grant SB. Whole particle microelectrophoresis for small viruses. J Colloid Interf Sci. 1995;173(2):521–3. https://doi.org/10.1006/jcis.1995.1354. PubMed PMID: WOS:A1995RJ65300030.

Penrod SL, Olson TM, Grant SB. Deposition kinetics of two viruses in packed beds of quartz granular media. Langmuir. 1996;12(23):5576–87. https://doi.org/10.1021/la950884d. PubMed PMID: WOS:A1996VT71800016.

Pondman KM, Pednekar L, Paudyal B, Tsolaki AG, Kouser L, Khan HA, Shamji MH, Ten Haken B, Stenbeck G, Sim RB, Kishore U. Innate immune humoral factors, C1q and factor H, with differential pattern recognition properties, alter macrophage response to carbon nanotubes. Nanomedicine. 2015;11(8):2109–18. https://doi.org/10.1016/j.nano.2015.06.009. Epub 2015/07/15. PubMed PMID: 26169151.

Qian GS, Ross RK, Yu MC, Yuan JM, Gao YT, Henderson BE, Wogan GN, Groopman JD. A follow-up study of urinary markers of aflatoxin exposure and liver cancer risk in Shanghai, People's Republic of China. Cancer Epidemiol Biomarkers Prev. 1994;3(1):3–10. Epub 1994/01/01. PubMed PMID: 8118382.

Rabovsky J, Judy DJ, Rodak DJ, Petersen M. Influenza virus-induced alterations of cytochrome P-450 enzyme activities following exposure of mice to coal and diesel particulates. Environ Res. 1986;40(1):136–44. Epub 1986/06/01. PubMed PMID: 2423323.

Rafiei S, Rezatofighi SE, Roayaei Ardakani M, Rastegarzadeh S. Gold nanoparticles impair foot-and-mouth disease virus replication. IEEE Trans Nanobioscience. 2016;15(1):34–40. https://doi.org/10.1109/TNB.2015.2508718. Epub 2015/12/20. PubMed PMID: 26685261.

Ramos I, Fernandez-Sesma A. Cell receptors for influenza a viruses and the innate immune response. Front Microbiol. 2012;3:117. https://doi.org/10.3389/fmicb.2012.00117. Epub 2012/04/27. PubMed PMID: 22536196; PMCID: 3332393.

Rebuli ME, Speen AM, Martin EM, Addo KA, Pawlak EA, Glista-Baker E, Robinette C, Zhou H, Noah TL, Jaspers I. Wood smoke exposure alters human inflammatory responses to viral infection in a sex-specific manner: a randomized, placebo-controlled study. Am J Respir Crit Care Med. 2018. https://doi.org/10.1164/rccm.201807-1287OC. Epub 2018/10/27. PubMed PMID: 30360637.

Saber AT, Lamson JS, Jacobsen NR, Ravn-Haren G, Hougaard KS, Nyendi AN, Wahlberg P, Madsen AM, Jackson P, Wallin H, Vogel U. Particle-induced pulmonary acute phase response correlates with neutrophil influx linking inhaled particles and cardiovascular risk. PloS One. 2013;8(7):e69020. https://doi.org/10.1371/journal.pone.0069020. Epub 2013/07/31. PubMed PMID: 23894396; PMCID: 3722244.

Saleh NB, Pfefferle LD, Elimelech M. Influence of biomacromolecules and humic acid on the aggregation kinetics of single-walled carbon nanotubes. Environ Sci Technol. 2010;44(7):2412–8. https://doi.org/10.1021/es903059t. PubMed PMID: WOS:000275993700031.

Sanpui P, Zheng X, Loeb JC, Bisesi JH Jr, Khan IA, Afrooz AR, Liu K, Badireddy AR, Wiesner MR, Ferguson PL, Saleh NB, Lednicky JA, Sabo-Attwood T. Single-walled carbon nanotubes increase pandemic influenza A H1N1 virus infectivity of lung epithelial cells. Part Fibre Toxicol. 2014;11:66. https://doi.org/10.1186/s12989-014-0066-0. Epub 2014/12/17. PubMed PMID: 25497303; PMCID: PMC4318452.

Saptarshi SR, Duschl A, Lopata AL. Interaction of nanoparticles with proteins: relation to bio-reactivity of the nanoparticle. J Nanobiotechnol. 2013;11:26. https://doi.org/10.1186/1477-3155-11-26. Epub 2013/07/23. PubMed PMID: 23870291; PMCID: 3720198.

Sarvestani ST, McAuley JL. The role of the NLRP3 inflammasome in regulation of antiviral responses to influenza A virus infection. Antivir Res. 2017;148:32–42. https://doi.org/10.1016/j.antiviral.2017.10.020. Epub 2017/11/04. PubMed PMID: 29097227.

Sattler C, Moritz F, Chen S, Steer B, Kutschke D, Irmler M, Beckers J, Eickelberg O, Schmitt-Kopplin P, Adler H, Stoeger T. Nanoparticle exposure reactivates latent herpesvirus and restores a signature of acute infection. Part Fibre Toxicol. 2017;14(1):2. https://doi.org/10.1186/s12989-016-0181-1. Epub 2017/01/11. PubMed PMID: 28069010; PMCID: 5223553.

Shi X, von dem Bussche A, Hurt RH, Kane AB, Gao H. Cell entry of one-dimensional nanomaterials occurs by tip recognition and rotation. Nat Nanotechnol. 2011;6(11):714–9. https://doi.org/10.1038/nnano.2011.151. PubMed PMID: 21926979; PMCID: PMC3215144.

Shigemoto-Mogami Y, Hoshikawa K, Hirose A, Sato K. Phagocytosis-dependent and independent mechanisms underlie the microglial cell damage caused by carbon nanotube agglomerates. J Toxicol Sci. 2016;41(4):501–9. https://doi.org/10.2131/jts.41.501. Epub 2016/07/20. PubMed PMID: 27432236.

Shoji M, Takahashi E, Hatakeyama D, Iwai Y, Morita Y, Shirayama R, Echigo N, Kido H, Nakamura S, Mashino T, Okutani T, Kuzuhara T. Anti-influenza activity of c60 fullerene derivatives. PloS One. 2013;8(6):e66337. https://doi.org/10.1371/journal.pone.0066337. PubMed PMID: 23785493; PMCID: PMC3681905.

Shvedova AA, Kisin ER, Mercer R, Murray AR, Johnson VJ, Potapovich AI, Tyurina YY, Gorelik O, Arepalli S, Schwegler-Berry D, Hubbs AF, Antonini J, Evans DE, Ku BK, Ramsey D, Maynard A, Kagan VE, Castranova V, Baron P. Unusual inflammatory and fibrogenic pulmonary responses to single-walled carbon nanotubes in mice. Am J Physiol Lung Cell Mol Physiol. 2005;289(5):L698–708. https://doi.org/10.1152/ajplung.00084.2005. Epub 2005/06/14. PubMed PMID: 15951334.

Shvedova AA, Fabisiak JP, Kisin ER, Murray AR, Roberts JR, Tyurina YY, Antonini JM, Feng WH, Kommineni C, Reynolds J, Barchowsky A, Castranova V, Kagan VE. Sequential exposure to carbon nanotubes and bacteria enhances pulmonary inflammation and infectivity. Am J Respir Cell Mol Biol. 2008;38(5):579–90. https://doi.org/10.1165/rcmb.2007-0255OC. Epub 2007/12/2. PubMed PMID: 18096873; PMCID: PMC2335338.

Shvedova AA, Pietroiusti A, Fadeel B, Kagan VE. Mechanisms of carbon nanotube-induced toxicity: focus on oxidative stress. Toxicol Appl Pharmacol. 2012;261(2):121–33. https://doi.org/10.1016/j.taap.2012.03.023. Epub 2012/04/20. PubMed PMID: 22513272; PMCID: 4686133.

Silva AL, Peres C, Conniot J, Matos AI, Moura L, Carreira B, Sainz V, Scomparin A, Satchi-Fainaro R, Preat V, Florindo HF. Nanoparticle impact on innate immune cell pattern-recognition receptors and inflammasomes activation. Semin Immunol. 2017;34:3–24. https://doi.org/10.1016/j.smim.2017.09.003. Epub 2017/09/25. PubMed PMID: 28941640.

Smith B, Wepasnick K, Schrote KE, Bertele AR, Ball WP, O'Melia C, Fairbrother DH. Colloidal properties of aqueous suspensions of acid-treated, multi-walled carbon nanotubes. Environ Sci Technol. 2009;43(3):819–25. PubMed PMID: 19245021.

Sopova EA, Baranov VI, Gankovskaia OA, Lavrov VF, Zverev VV. Effects of silver and silicon dioxide nanopowders on the development of herpesvirus infection in vitro. Gig Sanit. 2010;4:89–91. Epub 2010/09/30. PubMed PMID: 20873393.

Speizer FE, Ferris B Jr, Bishop YM, Spengler J. Respiratory disease rates and pulmonary function in children associated with NO2 exposure. Am Rev Respir Dis. 1980;121(1):3–10. https://doi.org/10.1164/arrd.1980.121.1.3. Epub 1980/01/01. PubMed PMID: 7352711.

Sun Y, Li C, Shu Y, Ju X, Zou Z, Wang H, Rao S, Guo F, Liu H, Nan W, Zhao Y, Yan Y, Tang J, Zhao C, Yang P, Liu K, Wang S, Lu H, Li X, Tan L, Gao R, Song J, Gao X, Tian X, Qin Y, Xu KF, Li D, Jin N, Jiang C. Inhibition of autophagy ameliorates acute lung injury caused by avian influenza A H5N1 infection. Sci Signal. 2012;5(212):ra16. https://doi.org/10.1126/scisignal.2001931. Epub 2012/02/23. PubMed PMID: 22355189.

Swedin L, Arrighi R, Andersson-Willman B, Murray A, Chen Y, Karlsson MC, Georen SK, Tkach AV, Shvedova AA, Fadeel B, Barragan A, Scheynius A. Pulmonary exposure to single-walled carbon nanotubes does not affect the early immune response against Toxoplasma gondii. Part Fibre Toxicol. 2012;9:16. https://doi.org/10.1186/1743-8977-9-16. Epub 2012/05/25. PubMed PMID: 22621311; PMCID: 3495637.

t'Kindt R, Telenga ED, Jorge L, Van Oosterhout AJ, Sandra P, Ten Hacken NH, Sandra K. Profiling over 1500 lipids in induced lung sputum and the implications in studying lung diseases. Anal Chem. 2015;87(9):4957–64. https://doi.org/10.1021/acs.analchem.5b00732. PubMed PMID: 25884268.

Takeshita F, Kobiyama K, Miyawaki A, Jounai N, Okuda K. The non-canonical role of Atg family members as suppressors of innate antiviral immune signaling. Autophagy. 2008;4(1):67–9. Epub 2007/10/09. PubMed PMID: 17921696.

Tam VC. Lipidomic profiling of bioactive lipids by mass spectrometry during microbial infections. Semin Immunol. 2013;25(3):240–8. https://doi.org/10.1016/j.smim.2013.08.006. PubMed PMID: 24084369; PMCID: PMC3885417.

Tanner LB, Chng C, Guan XL, Lei Z, Rozen SG, Wenk MR. Lipidomics identifies a requirement for peroxisomal function during influenza virus replication. J Lipid Res. 2014;55(7):1357–65. https://doi.org/10.1194/jlr.M049148. PubMed PMID: 24868094; PMCID: PMC4076098.

Teeguarden JG, Webb-Robertson BJ, Waters KM, Murray AR, Kisin ER, Varnum SM, Jacobs JM, Pounds JG, Zanger RC, Shvedova AA. Comparative proteomics and pulmonary toxicity of instilled single-walled carbon nanotubes, crocidolite asbestos, and ultrafine carbon black in mice. Toxicol Sci. 2010;120(1):123–35. https://doi.org/10.1093/toxsci/kfq363. Epub 2010/12/08. doi: kfq363 [pii]. PubMed PMID: 21135415; PMCID: 3044201.

Thai SF, Wallace KA, Jones CP, Ren H, Grulke E, Castellon BT, Crooks J, Kitchin KT. Differential genomic effects of six different TiO2 nanomaterials on human liver HepG2 cells. J Biochem Mol Toxicol. 2016;30(7):331–41. https://doi.org/10.1002/jbt.21798.

Thompson D, Chen SC, Wang J, Pui DY. Aerosol emission monitoring and assessment of potential exposure to multi-walled carbon nanotubes in the manufacture of polymer nanocomposites.

Ann Occup Hyg. 2015;59:1135–51. https://doi.org/10.1093/annhyg/mev044. Epub 2015/07/26. PubMed PMID: 26209597.

Tisoncik-Go J, Gasper DJ, Kyle JE, Eisfeld AJ, Selinger C, Hatta M, Morrison J, Korth MJ, Zink EM, Kim YM, Schepmoes AA, Nicora CD, Purvine SO, Weitz KK, Peng X, Green RR, Tilton SC, Webb-Robertson BJ, Waters KM, Metz TO, Smith RD, Kawaoka Y, Suresh M, Josset L, Katze MG. Integrated omics analysis of pathogenic host responses during pandemic H1N1 influenza virus infection: the crucial role of lipid metabolism. Cell Host Microbe. 2016;19(2):254–66. https://doi.org/10.1016/j.chom.2016.01.002. PubMed PMID: 26867183; PMCID: PMC5271177.

To KKW, Lee KC, Wong SSY, Sze KH, Ke YH, Lui YM, Tang BSF, Li IWS, Lau SKP, Hung IFN, Law CY, Lam CW, Yuen KY. Lipid metabolites as potential diagnostic and prognostic biomarkers for acute community acquired pneumonia. Diagn Microbiol Infect Dis. 2016;85(2):249–54. https://doi.org/10.1016/j.diagmicrobio.2016.03.012. PubMed PMID: WOS:000376822600027.

Trefry JC, Wooley DP. Rapid assessment of antiviral activity and cytotoxicity of silver nanoparticles using a novel application of the tetrazolium-based colorimetric assay. J Virol Methods. 2012;183(1):19–24. https://doi.org/10.1016/j.jviromet.2012.03.014. Epub 2012/04/03. PubMed PMID: 22465243.

Turabekova M, Rasulev B, Theodore M, Jackman J, Leszczynska D, Leszczynski J. Immunotoxicity of nanoparticles: a computational study suggests that CNTs and C60 fullerenes might be recognized as pathogens by Toll-like receptors. Nanoscale. 2014;6(7):3488–95. https://doi.org/10.1039/c3nr05772k. Epub 2014/02/20. PubMed PMID: 24548972.

Tyurina YY, Kisin ER, Murray A, Tyurin VA, Kapralova VI, Sparvero LJ, Amoscato AA, Samhan-Arias AK, Swedin L, Lahesmaa R, Fadeel B, Shvedova AA, Kagan VE. Global phospholipidomics analysis reveals selective pulmonary peroxidation profiles upon inhalation of single-walled carbon nanotubes. ACS Nano. 2011;5(9):7342–53. https://doi.org/10.1021/nn202201j. Epub 2011/08/02. PubMed PMID: 21800898; PMCID: 3321726.

Valegard K, Liljas L, Fridborg K, Unge T. The three-dimensional structure of the bacterial virus MS2. Nature. 1990;345(6270):36–41. https://doi.org/10.1038/345036a0. PubMed PMID: 2330049.

Voelker DR. Organelle biogenesis and intracellular lipid transport in eukaryotes. Microbiol Rev. 1991;55(4):543–60. Epub 1991/12/01. PubMed PMID: 1779926; PMCID: 372837.

Walling BE, Lau GW. Perturbation of pulmonary immune functions by carbon nanotubes and susceptibility to microbial infection. J Microbiol. 2014;52(3):227–34. https://doi.org/10.1007/s12275-014-3695-y. Epub 2014/03/04. PubMed PMID: 24585053.

Wang Y, Jiang K, Zhang Q, Meng S, Ding C. Autophagy in negative-strand RNA virus infection. Front Microbiol. 2018;9:206. https://doi.org/10.3389/fmicb.2018.00206. Epub 2018/03/01. PubMed PMID: 29487586; PMCID: 5816943.

Watanabe W, Akashi T, Hirose A, Miyauchi A, Yoshida H, Kurokawa M. Effects of double-walled carbon nanotubes on the pneumonia in respiratory syncytial virus-infected mice. Toxicol Lett. 2018;295(supplement 1):S209. https://doi.org/10.1016/j.toxlet.2018.06.912.

Weinberg SE, Sena LA, Chandel NS. Mitochondria in the regulation of innate and adaptive immunity. Immunity. 2015;42(3):406–17. https://doi.org/10.1016/j.immuni.2015.02.002. Epub 2015/03/19. PubMed PMID: 25786173; PMCID: 4365295.

Wepasnick KA, Smith BA, Schrote KE, Wilson HK, Diegelmann SR, Fairbrother DH. Surface and structural characterization of multi-walled carbon nanotubes following different oxidative treatments. Carbon. 2011;49(1):24–36. https://doi.org/10.1016/j.carbon.2010.08.034. PubMed PMID: WOS:000284977500005.

Wu S, Metcalf JP, Wu W. Innate immune response to influenza virus. Curr Opin Infect Dis. 2011;24(3):235–40.

Wu D, Huang W, Wang Y, Guan W, Li R, Yang Z, Zhong N. Gene silencing of beta-galactosamide alpha-2,6-sialyltransferase 1 inhibits human influenza virus infection of airway epithelial cells. BMC Microbiol. 2014;14:78. https://doi.org/10.1186/1471-2180-14-78. Epub 2014/03/29. PubMed PMID: 24670114.

Yang M, Flavin K, Kopf I, Radics G, Hearnden CH, McManus GJ, Moran B, Villalta-Cerdas A, Echegoyen LA, Giordani S, Lavelle EC. Functionalization of carbon nanoparticles modulates inflammatory cell recruitment and NLRP3 inflammasome activation. Small. 2013;9(24):4194–206. https://doi.org/10.1002/smll.201300481. Epub 2013/07/11. PubMed PMID: 23839951.

Yoneyama M, Jogi M, Onomoto K. Regulation of antiviral innate immune signaling by stress-induced RNA granules. J Biochem. 2016;159(3):279–86. https://doi.org/10.1093/jb/mvv122. Epub 2016/01/10. PubMed PMID: 26748340; PMCID: 4763080.

Zehethofer N, Bermbach S, Hagner S, Garn H, Muller J, Goldmann T, Lindner B, Schwudke D, Konig P. Lipid analysis of airway epithelial cells for studying respiratory diseases. Chromatographia. 2015;78(5–6):403–13. https://doi.org/10.1007/s10337-014-2787-5. Epub 2015/03/10. PubMed PMID: 25750457; PMCID: 4346681.

Zhao YY, Cheng XL, Lin RC. Lipidomics applications for discovering biomarkers of diseases in clinical chemistry. Int Rev Cell Mol Biol. 2014;313:1–26. https://doi.org/10.1016/B978-0-12-800177-6.00001-3. Epub 2014/11/08. PubMed PMID: 25376488.

Zhu W, von dem Bussche A, Yi X, Qiu Y, Wang Z, Weston P, Hurt RH, Kane AB, Gao H. Nanomechanical mechanism for lipid bilayer damage induced by carbon nanotubes confined in intracellular vesicles. Proc Natl Acad Sci U S A. 2016;113(44):12374–9. https://doi.org/10.1073/pnas.1605030113. Epub 2016/11/03. PubMed PMID: 27791073; PMCID: 5098676.

Chapter 11
Immunotoxicity Testing – In Vitro Cell Culture Models

Barbara Rothen-Rutishauser, Barbara Drasler, and Alke Petri-Fink

Abstract The understanding of how nanomaterials interact with cells, tissues, and organs is important to realize their potential for biomedical applications and for hazard assessment of occupational exposure. Depending on the nanomaterial characteristics, they can trigger various components of the immune system, that is, innate and adaptive immune systems, thereby either activating or suppressing immune system functions. In this chapter, we highlight the currently established immune cell models and readouts to address the possible effects of nanomaterials. In addition to the conventional 2D cultures, the design and use of more complex 3D models is emphasized.

Keywords Immune cell models · Immunotoxicity · Immunoengineering · Nanomaterial–immune cell interactions · Immune response · Human cells · 3D models

11.1 Introduction

The increasing production of nanomaterials provides an advantageous basis for numerous applications such as their use in consumer products (e.g., food additives, cosmetics, and sporting equipment), environmental remediation, and information technology. In addition, the use of nanomaterials in the medical field allows a variety of biomedical applications, for instance, as diagnostic or therapeutic agents, and in novel vaccine formulations (Baetke et al. 2015). However, these new properties and the increasing industrial production have raised concerns about potential adverse effects for human health, and expectations from society, consumer, and regulatory bodies about their safety are increasing. Exposure to nanomaterials might

B. Rothen-Rutishauser (✉) · B. Drasler · A. Petri-Fink
BioNanomaterials Group, Adolphe Merkle Institute, University of Fribourg, Fribourg, Switzerland
e-mail: barbara.rothen@unifr.ch

J. C. Bonner, J. M. Brown (eds.), *Interaction of Nanomaterials with the Immune System*, Molecular and Integrative Toxicology,
https://doi.org/10.1007/978-3-030-33962-3_11

also increase either occupationally during their production and life cycle or by their direct use and application in consumer products (Piccinno et al. 2012). Independent of the exposure, the contact with nanomaterials can occur by means of the lungs via inhalation, the gastrointestinal tract via ingestion, the skin via direct application, or the blood vessels via intravenous administration and then from there being distributed to secondary organs (Bourquin et al. 2018; Kermanizadeh et al. 2015).

Another aspect is that despite all these interesting properties and technology advancements, only very few nanomaterials have been approved by the Food and Drug Administration (FDA) for use in medical applications, whereas the majority have thus far failed to reach the bedside (Venditto and Szoka Jr 2013). Thus, new efforts have to focus on a detailed understanding of the interaction of nanomaterials with cells at the fundamental level as a prerequisite for their safe and successful use in any applications.

Our knowledge is still limited considering the immune responses upon exposure to any nanomaterials, and when nanomaterials come in contact with the immune system, they may trigger various immunomodulatory effects (Smith et al. 2013). The nanomaterial immunogenicity as well as antigenicity (i.e., the ability to induce immune response and the ability to be specifically recognized by the antibodies, respectively) has been overviewed in detail by Ilinskaya and Dobrovolskaia (Ilinskaya and Dobrovolskaia 2016). Current medical trends utilize nanotechnology to engineer nanomaterials with properties suitable for modulating immune responses and, in this respect, the branch of "Immunoengineering" has emerged from engineering approaches of materials with immunomodulatory effects (Swartz and Hubbell 2012; Hubbell and Swartz 2009). Nanomaterials, such as liposomes, polymers, emulsions, fullerenes, and carbon nanotubes, have been utilized for nanomedical applications with a particular focus on immune cell targeting that may modulate specific immune responses (Smith et al. 2013). Nanomaterials can be specifically designed to enhance interaction (i.e., drug release from biodegradable nanoparticles) or avoid interaction through stealth properties (polyethylene (PEG) coating of particles) with immune cells (Smith et al. 2014). On the other side, the undesirable immunogenicity and antigenicity of therapeutics are important for the safe-by-design of the nanomedicine products (Ilinskaya and Dobrovolskaia 2016). Moreover, also the adverse side effects of nanomaterials encountering immune cells need to be considered, such as the immune-mediated destruction or rejection which can result in the elimination of nanomaterials and/or immunotoxicity and thus possible damage to the immune system (Boraschi et al. 2012).

Due to complex immunological homeostasis and its fragile balance that can be interrupted by a contact with nanomaterials, the World Health Organization (WHO) has published the "Guidance for Immunotoxicity Risk Assessment for Chemicals" (WHO, 2008: http://www.who.int/ipcs/methods/harmonization/areas/immunotoxicity/en/). Briefly, this comprehensive report consists of risk assessment methodologies, principles, and experimental methods to assess the potential toxic effect of chemicals to the immune system. This report is currently under revision to expand the guidance to the immunotoxicity testing of nanomaterials. Such a guideline can be a very useful tool for assessing the toxicity and thus the safe use of agents (i.e., nanocarrier-based drugs) that are designed to be used for immunological applications or the potential adverse effects of unintended nanomaterial exposure.

The aim of this chapter is to give an overview about the state-of-the-art of currently available in vitro human immune cell model systems and immunotoxicity endpoints, which can be used for a first evaluation of any nanomaterials to assess the immune hazards or the therapeutic immunomodulatory potential.

11.2 Cellular Elements of the Innate and Adaptive Human Immune Systems

The primary function of the immune system is to protect the organism from potential harmful substances that are recognized as foreign substances, that is, antigens. Antigens can be of microbial origin (i.e., pathogenic bacteria) or nonmicrobial (i.e., toxic compounds and allergens, nanomaterials) that interfere with the immune homeostasis. The immune system utilizes an array of protective mechanisms in order to maintain the immunological stability, and this is achieved by the innate and the adaptive immune systems (Chaplin 2006).

First, the innate immune system is characterized as the organism's "first line of defense," and it, thus, provides rapid sensing and elimination of pathogens. It consists of the physical barriers such as the epithelium lining all body surfaces, cavities, and tubes as well as hematopoietic cells including macrophages, dendritic cells, mast cells, neutrophils, eosinophils, and natural killer (NK) cells, providing mechanisms for a rapid sensing and elimination of pathogens (Chaplin 2006; Turvey and Broide 2010). The cells of the innate immune system express a limited number of receptors on their surface which are able to recognize highly conserved structures expressed by large groups of pathogens, for example, pattern recognition receptors (PRRs), and one type of such receptors is Toll-like receptors (TLRs) (Turvey and Broide 2010; Elsabahy and Wooley 2013).

Antigen-presenting cells (APCs) such as dendritic cells (DCs), macrophages, Langerhans cells (in the skin tissue), or B cells are specialized cells located in the epithelial tissue that capture antigens and transport them to the draining lymphoid tissue. During migration, the cells process the antigens and present them to certain cells such T cells (Abbas and Lichtman 2011). DCs, as the most potent APCs, connect the innate and adaptive immunity (Nicod 1997). DCs capture antigen and present antigen-derived peptides to T cells, a process that is of crucial importance since it triggers the adaptive immune response (vonGarnier and Nicod 2009).

Second, the adaptive immune system provides an antigen-specific defense and is responsible for generation of the immunological memory. Its main feature is a tightly regulated interplay between the APCs, such as macrophages or DCs, and the main actors of the adaptive immunity, T and B cells (or lymphocytes); the latter are responsible for the antigen-specific receptor-mediated antibody production. Generally, the adaptive immune system acts after the innate immune response when B- and T-cell receptors become specific for the antigen after first encounter and start to clonally expand. A subpopulation of these cells may persist as memory B or T cells, and in the case of a re-exposure to the same pathogen, it provides the capacity of a rapid response (based on the principle of the immunological memory) (Chaplin 2006; Bonilla and Oettgen 2010).

11.3 Immunotoxicity Testing

Due to the enormous diversity of nanomaterials being produced and used, the exposure routes to which humans may be potentially subjected to nanomaterials are numerous and primarily include inhalation, injection, ingestion, and permeation through skin (Oberdorster et al. 2007). These biological compartments are designed to act as barriers to the passage of foreign materials into the organism (Bourquin et al. 2018). The epithelium provides a first interface between biological compartments (Rothen-Rutishauser et al. 2012), and after the nanomaterials pass through the epithelial barrier, they may pass through the basement membrane and the subepithelial connective tissue layer and eventually come into contact with immune cells present there or within the epithelial tissue. Biomedical application of nanomaterials via injection directly into the blood stream can result in a direct interaction of the materials with immune cells or the endothelium of the blood vessels (Clift et al. 2015).

Regardless of the route of exposure, nanomaterials can come in contact with cells from the innate as well as adaptive immune system, which might trigger a variety of immune effects (Farrera and Fadeel 2015; Giannakou et al. 2016). Depending on size and properties, nanomaterials also may escape the particle clearing defensive mechanisms, such as phagocytosis, and thus do not trigger a direct inflammatory response (Kunzmann et al. 2011).

Research has shown that nanomaterials can stimulate and/or suppress the immune response (Dobrovolskaia and McNeil 2007). Immunosuppression usually is described as the impairment of any component of the immune system resulting in a suppression of immune cell maturation and/or proliferation, whereas immunostimulation refers to an activation of some components of the immune system (Hussain et al. 2012). Both effects can of course also be used intentionally such as for the development of nanovaccines; however, the immunomodulatory potentials should anyway be routinely assessed for any novel material.

At the moment, there are no specific regulatory documents available for general immunotoxicity assessment of nanomaterials. The adverse effects are not always readily detected using the conventional immunotoxicity methods combining a battery of nanomaterial-specific assays to assess nanomaterial effect on the immune system. Immunotoxicity can be manifested by effects on innate immune response as well as the specific immune response (Elsabahy and Wooley 2013; Giannakou et al. 2016; Dusinska et al. 2017):

- Excessive production of reactive oxygen species (ROS) which might lead to cell death and/or changes in proteins and DNA

- Induction of inflammation through the nuclear factor-kappa B (NF-kB) pathway activation
- Release of pro-inflammatory, inflammatory, and anti-inflammatory mediators
- Hypersensitivity, immunosuppressive, and anti-inflammatory effects
- Direct interaction with the immune system through (TLR)
- Function and activation of immune cells

In addition, cell-specific responses such as interaction of nanomaterials with antigen-presenting maturation (i.e., antigen uptake, processing, and presentation), T-cell activations, and release of antibodies by B cells have been reported (Blank et al. 2011a; Seydoux et al. 2016) and will be covered in the specific subchapters.

It is important to mention that not only the biological systems have to be characterized thoroughly but also the nanomaterials. This includes the description of the material source as well as intrinsic and extrinsic properties, such as primary size, hydrodynamic diameter, shape, surface area zeta potential, and stability in the dispersions used for the experiment (Gao 2018). If immunotoxicity aspects are studied, it is mandatory to include endotoxin tests as described elsewhere (Li et al. 2015; Li and Boraschi 2016) since the presence of, for instance, lipopolysaccharides in the nanomaterial suspensions can result in a wrong interpretation for the nanomaterial hazard (Li et al. 2017). In addition, the use of realistic nanomaterial concentrations as well as the dose delivered into the cell surface, that is, dosimetry, has to be done carefully in order to transform the findings to in vivo findings (Drasler et al. 2017a).

11.4 Alternative Methods to Assess Immunotoxicity

The assessment of potential adverse or off-target effects is currently mainly based on animal testing (Dusinska et al. 2017; Hartung 2009). But recently, the outcomes of animal testing regarding its predictive power for human beings have been questioned because of certain limitations, mainly due to physiological and biochemical species dissimilarities (Shanks et al. 2009). In addition to that, the principle of the 3Rs – Replacement, Reduction, and Refinement – has become an increasing public and legal demand which ethically supports the replacement of animal use with more human-relevant alternatives that do not rely on in vivo testing (Tornqvist et al. 2014). New concepts for efficient, cheaper, and evidence-based testing strategies were proposed, based on the use of computational systems or in vitro systems (Krewski et al. 2010). In addition, the Organisation for Economic Co-operation and Development (OECD) has published test guidelines with endpoints for health effects and in vitro tests of regulatory interest for conventional chemicals (OECD 2013). These in vitro assays cover a broad array of endpoints mimicking the in situ cell response, such as cytotoxicity, oxidative stress, inflammatory responses, or mutagenicity (Drasler et al. 2017b). It is important to add that the predictivity of the assays is still questionable since validation with human data is usually not possible (Frohlich 2018).

Recommendations from the 2008 National Cancer Institute workshop on immunology and nanotechnology have been proposed in terms of the immune cell models, and mono- as well as co-culture systems are well suitable for a first evaluation of nanomaterial immunomodulatory effects (Dobrovolskaia et al. 2009). Independent of the model guidelines, good cell culture practice has to be applied and documented, including the assessment of biological materials used in the experiments, for example, the cultured cells, cell type and species, passage number, the culture medium, and the culture substratum (Gstraunthaler and Hartung 2002). In addition, the duration and level of nanomaterial exposure as well as the way by which the cells are exposed should be noted since the in vitro responses to nanomaterials will heavily depend on all these factors (Park et al. 2009). The choice of cell types should also take into account the possible exposure route, such as inhalation, dermal application, oral exposure, or injection into the blood stream.

It is still a routine practice to perform in vitro testing using a single-cell type (mono-cell culture), which does, however, not represent the complexity of a tissue consisting of multiple cell types relevant for an immune response. An interplay between different cell types in co-cultures is expected to enhance immuno-effects of a tested compound; however, the situation might be different when a nanomaterial is interacting mainly with one specific cell type which can provide protection (Frohlich 2018). Alternative methods such as use of more than one cell type (co-culture) (Rothen-Rutishauser et al. 2012; Fytianos et al. 2016), ex vivo tissue slice cultures (Sauer et al. 2014), or exposure at the air–liquid interface which are relevant for lung cell studies are emerging (Frohlich and Salar-Behzadi 2014; Paur et al. 2011). But only 3D dermal models have been validated for cytotoxicity, genotoxicity, sensitization, and irritation testing (e.g., OECD TG439 (OECD 2015)) and were used to assess the effects of nanomaterials (Choi et al. 2014; Wills et al. 2016). All other cell models are not yet standardized nor validated, and without being validated, it is questionable if such more advanced models are more predictive than 2D models (Frohlich 2018).

11.5 Human Immune Cell Models

A detailed description of individual cell types relevant to the innate and adaptive human immune system has been described above. Herein, specific cell types that are routinely used for assessing in vitro immunotoxicity of nanomaterials are covered; however, the list is not complete since the subset of immune cells depends on the tissue/organ localization as well as on the differentiation and maturation phenotype. It is absolutely relevant that adequate characterization such as cell shape, differentiation status, and verification of the models by positive control assays such as stimulation with lipopolysaccharide (LPS) is necessary before their routine integration into immunological testing of nanomaterials.

One has to keep in mind that mainly research on mice has contributed to our knowledge of the immune system. On the other side, many differences between the mouse and human immune systems (Perlman 2016; Tao and Reese 2017) and interpretation of results found in one species have to be made carefully.

For many studies, human immune cell lines or primary peripheral blood mononuclear cells (PBMCs) are preferred over rodent cells to investigate the effect of nanomaterials. These cells are easily isolated from blood of healthy donors or buffy coats, and companies (e.g., Miltenyi Biotec) offer immunomagnetic beads to enable magnetic isolation of specific cell types directly from buffy coats. One has, however, to keep in mind that these cells have a different composition, phenotype, and activation status than the cells found in the tissue (Kleiveland 2015a). In this chapter, we also mainly focus on the description of human but not rodent immune cell models.

11.5.1 Epithelial Cell Cultures

Epithelial cells express Toll-like receptors on their surface (Wang et al. 2007) and can secrete cytokines (Salazar and Ghaemmaghami 2013) that participate in immune reaction at the primary barrier site; therefore, it is important to verify the interaction of nanomaterials with epithelial cells. The ability of nanomaterials to overcome the epithelial barrier has to be investigated and possible adverse reactions should be studied, such as cytotoxicity, oxidative stress, and (pro-)inflammatory mediator release (Petrarca et al. 2015). Epithelial cell cultures of the primary barrier systems, that is, skin, intestinal tract, and lung, are reliable tools to study possible effects after exposure of those barriers to nanomaterials and have been covered in detail elsewhere (Rothen-Rutishauser et al. 2012; Frohlich 2018; Braakhuis et al. 2015). The choice of cell system also depends on the availability of the cell source, that is, cell line or primary cell source, as well as on the possible exposure route to the nanomaterials. As already mentioned, the effects of nanomaterials at the epithelial barrier should include not only the general toxicity assays but also the (i) translocation of the materials across the barriers, (ii) the barrier structure including epithelial tightness, and (iii) the release of cytokines.

11.5.2 Monocytes/Macrophages

Professional phagocytes such as macrophages can ingest particles up to about several μm's in size by various endocytosis mechanisms (Conner and Schmid 2003) and are key players of the immune response in almost all organs (Abbas and Lichtman 2011). They also express different receptors, for example, scavenger receptors, lectin receptors, Fc receptors, and integrins that allow them to actively recognize a wide array of ligands and molecules. Macrophages have recently been identified as interesting targets for both diagnosis and treatment by designing specific nanocarrier targets (Lameijer et al. 2013).

The use of monocyte/macrophage cell cultures to assess the immune response to nanomaterials has emerged to be a suitable screening tool (Frohlich 2015) and also was recommended as an important assay for predicting short-term inhalation toxicity of nanomaterials (Wiemann et al. 2016).

Besides the human cell line THP-1 (human peripheral blood mononuclear cells (PBMCs), derived from monocytic leukemia patients), J774.1a (mouse macrophages derived from ascites) and RAW264.7 (mouse macrophages derived from ascites) are the frequently used phagocytic cell types for assessing nanomaterial-induced toxicity (Farcal et al. 2015). Monocytes can also be isolated from human peripheral blood, specifically from buffy coats, and differentiated into macrophages (monocyte derived-macrophages, MDMs) with the addition of monocyte colony-stimulating factor (M-CSF) (Fig. 11.1) (Lehmann et al. 2010a). The use of $CD14^+$ MicroBeads has been described, resulting in an increase of monocyte purity (Steiner et al. 2013), and it is also routinely used in many labs nowadays. This primary cell type can be produced in a very reproducible manner and represents a reliable in vitro system (Figs. 11.1 and 11.2).

A panel of in vitro assays of different complexities can assess phagocyte function, such as cytokine secretion, chemotaxis, phagocytosis, as well as respiratory burst which can be measured in all phagocytes (Frohlich 2015). In addition,

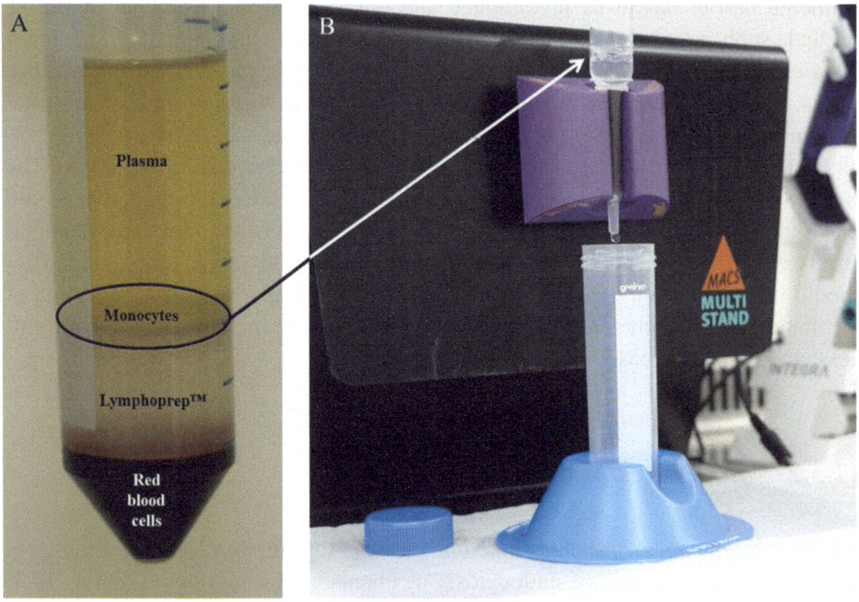

Fig. 11.1 Isolation of primary peripheral blood mononuclear cells. (**a**) Peripheral blood is treated with Lymphoprep™, a density gradient medium, resulting in four layers formed after centrifugation: plasma on the top followed by a thin layer of monocytes, a layer of the gradient containing the remaining white blood cells and red blood cells at the bottom. (**b**) The layer of monocytes is then isolated with a pipette, washed in phosphate buffer saline, and incubated with immunomagnetic beads for magnetic isolation

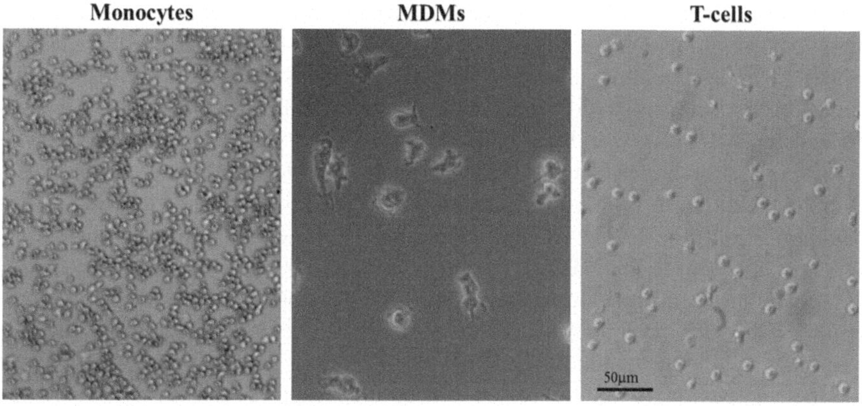

Fig. 11.2 Phase-contrast light microscopy images of various immune cells derived from human peripheral blood. (**a**) Freshly isolated primary monocytes after purification with $CD14^+$ MicroBeads with a round-shaped morphology and small diameter (i.e., about 10 µm). (**b**) Monocytes can be differentiated into monocyte-derived macrophages (MDMs) with the addition of monocyte colony-stimulating factor (M-CSF). Most of the cells are nonadherent, but some attach slightly to the well and start to grow out. (**c**) $CD4^+$ T cells were negatively selected from the $CD14^-$ fraction, and the cells keep the round-shape structure in culture. Scale bar for all images is 50 µm

nanomaterial-induced triggering of the NLRP3 inflammasome in monocytes/macrophages results in secretion of interleukin-1 beta (IL-1β), which is a key pro-inflammatory cytokine (Sun et al. 2013). One of the first studies showing that nanomaterials can activate the inflammasome was done with nano titanium dioxide (Yazdi et al. 2010). Since then, many other publications have revealed the underlying mechanisms of the inflammasome activation upon exposure to nanomaterials. It is again important to mention that the physicochemical characteristics of the nanomaterials, such as size, surface, and shape, are decisive for particle internalization, lysosomal alteration, and subsequent inflammasome/IL-1β processing (Rabolli et al. 2016). The smallest and fiber- or needle-like materials are particularly active to induce inflammasome activation since the interaction of these materials/fibers with macrophages can led to frustrated phagocytosis and ROS production which then activates the cascade (Dostert et al. 2008).

11.5.3 Dendritic Cells

As already mentioned, DCs are the most potent APC in the human body by bridging the innate with the adaptive immune response and are specialized in capturing, processing, and presenting antigens to T cells (Banchereau et al. 2000; Guermonprez et al. 2002). It is important to keep in mind that several phenotypes of DCs exist, and each subset is specialized to respond to a particular antigen and to interact with specific subsets of T cells. A range of surface markers specific for the various subsets have been described, allowing to experimentally identify the specification upon exposure to nanomaterials (Collin and Bigley 2018). Similarly as for macrophages, it is a common practice to isolate monocytes from PMDCs (i.e., from human buffy coats) and to differentiate them to DCs, that is, monocyte-derived dendritic cells (MDDCs), with the addition of growth factors like granulocyte-monocyte colony-stimulating factor (GM-CSF) and interleukin 4 (IL-4) (Lehner et al. 2005). In addition to primary cells, immortalized cell lines such as the histiocytic lymphoma cell type U-937 which exhibits monocytic characteristics (Sundstrom and Nilsson 1976) or the MUTZ-3 (Santegoets et al. 2008) can be used. Since distinct differences in the immunophenotypic characteristics are observed between primary DCs and the immortalized cell lines, these factors should be considered when interpreting the results.

Upon activation of immature DC via antigen capture, cytokines or surface molecules-driven activation, they can differentiate into mature DC (Banchereau et al. 2000). All these parameters can be determined experimentally in vitro by incubating cells with OVA-Alexa Fluor 647 to assess antigen uptake capacity or with BSA-DQ Red to determine antigen processing capacity (Seydoux et al. 2014; Blank et al. 2011b). The phenotypic surface receptors are usually determined by antibody labeling combined with fluorescence flow cytometry (Collin and Bigley 2018).

It has been reported that the nanomaterial shape, size, and surface charge have a significant impact on DCs; particularly positively charged particles induce a higher phagocytic activity (Kumar et al. 2015; Thiele et al. 2001). In addition, our group has recently demonstrated that nanomaterial surface modifications can significantly modulate uptake by human monocyte-derived DCs, though it is difficult to relate this to cellular responses (Fytianos et al. 2015).

11.5.4 T Cells

The targeting of T cells with nanomaterials for immunotherapy to improve antitumor immunity is an interesting approach in comparison to the administration of free drugs (Schmid et al. 2017), and first testing with human-derived T cells can give a fast feedback about possible cellular interactions and adverse effects. However, to the best of our knowledge, there are only few studies addressing nanomaterial interaction with T cells. Similar to PMDC derived macrophages and DCs, T cells can be generated from human buffy coat from the CD14-negative selection ($CD14^-$) with the CD4-positive fraction ($CD4^+$) kept for generation of human T cells in vitro (Fig. 11.2). A possible endpoint of interest in nanomaterial uptake by T cells and efficient uptake of amino-polymeric NPs without an adverse impact of effector functions has recently been reported (Zupke et al. 2015). The most common approach for measuring T-cell activation and proliferation in vitro is via radioactive labeled compounds (such as thymidine incorporation into DNA upon cell proliferation), as described in Blank et al. (2011), who showed that polymer-coated super-paramagnetic iron oxide nanoparticles can modulate the CD4^{+} T-cell activation; however, such experiments can only be done in co-cultures with APCs such as DCs (Blank et al. 2011a).

11.5.5 B Cells

Again, as for the T cells, a very limited number of studies reporting nanomaterial interaction with B cells and possible antibody production are available. Also B cells can also be isolated from human buffy coat in vitro by applying various methods such as negative selection or the positive selection using a CD19^{+} magnetic bead separation method (Heine et al. 2011). Human B cells can then be characterized via surface marker expression: CD19^{+} (Heine et al. 2011) and CD20^{+} (Leandro 2013) surface proteins have been considered as good markers for characterizations of B cells. Regarding engineered nanomaterials, their impact on cell activation and subsequent induction of antibody production by B cells have been reviewed in (Ilinskaya and Dobrovolskaia 2016); for instance, activation of naive B cells was reported to be achieved by, for example, antigen-loaded calcium–phosphate-based nanomaterials (Temchura et al. 2014). Moreover, gold nanoparticles induced the activation of mouse spleen-derived B cells and promoted immunoglobulin G secretion in a size- and time-dependent manner (Lee et al. 2014). It is important to mention that immunizing animals with specific nanomaterials, even in the presence of strong adjuvants, is frequently reported as unsuccessful in terms of generating nanomaterial-specific antibodies; however, it can successfully be induced with conjugation of nanomaterials to a protein carrier (discussed in (Ilinskaya and Dobrovolskaia 2016)). On the other side, different approaches have been described to use nanomaterials for the induction of antigen-specific tolerance, that is, absence of immune response against immunogenic antigen, which is relevant for the treatment of autoimmune diseases and prevention of graft injection (Kishimoto and Maldonado 2018). Therefore, we conclude that the use of isolated B cells to study the nanomaterial–cell interaction might give some first valuable input about possible immunogenic reactions.

11.6 Human Cell Co-cultures

The complexity of the immune system with the interaction of cells, tissues, and soluble factors might not be optimally reflected in simple in vitro experiments in mono-cultures. It has been shown that co-cultures of different cell types have an influence on the outcome of the results, since cells continuously cross talk in vivo through intercellular signaling to maintain homeostasis and to coordinate immune responses (Roggen et al. 2006). In addition, most of the down-stream immune responses are only activated if different cells interact directly via their receptors. For instance, T-cell activation in vivo is tightly controlled and only initiated upon antigen-specific signals presented by APCs (signal 1) and co-stimulation via receptor–receptor interactions (signal 2), together with the stimulation provided by the cytokines in the surrounding environment (signal 3) (Abbas and Lichtman 2011; Monks et al. 1998).

As already mentioned, co-cultures of APCs such as DCs with T cells can be used to assess the activation of the T cells (Blank et al. 2011a). More complex models also considering the spatial microenvironment are, however, needed, and immunoengineering has recently emerged as a new field to address these challenges (Jeanbart and Swartz 2015). Recently, novel approaches using advanced biomaterials, designed to replicate specific properties of the APC surface and thus providing artificial interfaces for APC–T-cell interactions, have been explored (Delcassian et al. 2017).

Since skin, gut, and lung mucosal surfaces present the first line of defense against pathogens, we will cover some examples of immunoengineering approaches for these tissues. The list is, however, not complete, and depending on the research question, other models also can be applied.

11.6.1 Lung Tissue

A triple cell culture in vitro model of the human epithelial lung tissue wall was designed to study the cellular interplay and responses of epithelial cells (i.e., A549 alveolar or 16HBE14o- bronchial cells/primary human alveolar type I cells), MDMs, and MDDCs exposed to nanomaterials (Blank et al. 2007; Lehmann et al. 2010b; Rothen-Rutishauser et al. 2005). Specific targeting of DCs via receptor-mediated mechanisms has already been described as a novel approach for drug delivery purposes in the past, and such co-culture models offer the great opportunity to test novel nanocarriers (Chiodo et al. 2014). Recently, we could show that fluorescently encoded polyvinyl alcohol-coated gold nanoparticles functionalized with a dendritic cell–specific intracellular adhesion molecule 3 (ICAM-3)-grabbing non-integrin (DC-SIGN) antibody on the particle surface enhanced MDDC targeting and activation in such a 3D lung cell model (Fig. 11.3) (Fytianos et al. 2017). The triple cell co-culture model also has been supplemented with human primary autologous $CD4^+$ T cells, and the release of cytokines was assessed after exposure to allergen-loaded subpollen particles. Although the positive control (Concancavalin A) induced a release of most of the mediators, the subpollen particles did not evoke any response (Schleh et al. 2012), and further optimization of the co-cultures will be needed to assess the activation of T cells in such complex models.

In addition, a quadruple culture containing epithelial, endothelial, macrophages, and mast cells has been established (Alfaro-Moreno et al. 2008; Klein et al. 2013) to assess the pro-inflammatory response upon exposure to nanomaterials. As a proof of concept, by coupling this tetraculture model with a native Vitrocell™ aerosol exposure system, aerosolized 50-nm SiO_2-rhodamine nanoparticles were detected in differentiated macrophages sitting on top of the epithelial cells but not in the epithelial cells or mast cells (Klein et al. 2013).

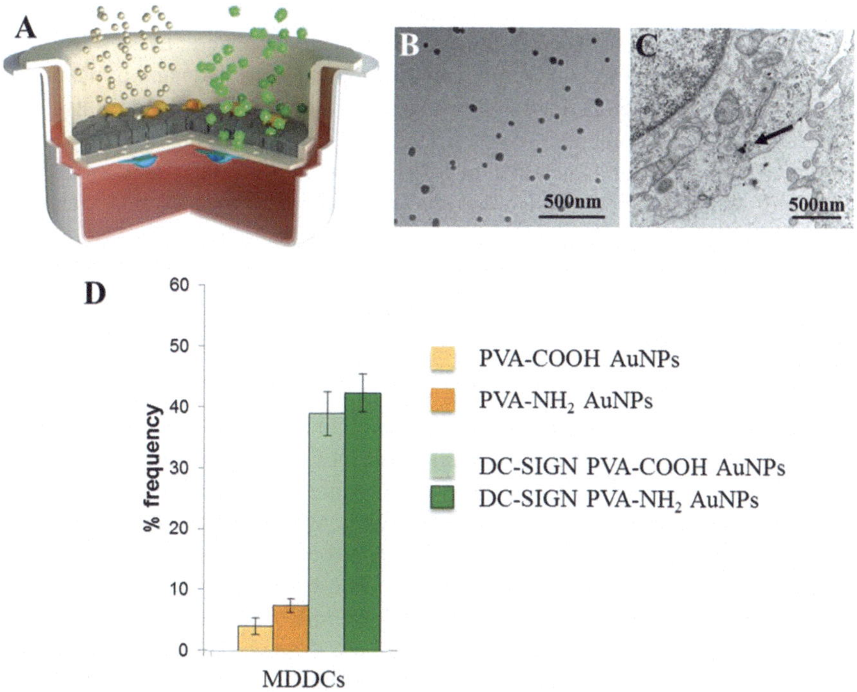

Fig. 11.3 Fluorescently encoded polyvinyl alcohol-coated gold nanoparticles (AuNPs), functionalized with either negative (-COO$^-$) or positive (-NH^{3+}) surface charge functionalized with a DC-SIGN antibody on the particle surface capable of binding to a dendritic cell surface receptor. (**a**) Schematic drawing of the triple cell co-culture model composed of epithelial cells (grey), macrophages (yellow), and dendritic cells (light blue) exposed to different aerosolized gold nanoparticles. (**b**) Visualization of deposited gold nanoparticles by transmission electron microscopy or (**c**) inside the exposed cells (arrow). (**d**) Analysis by flow cytometry revealed that DC-SIGN conjugated particles show an enhanced MDDCs targeting. (Adapted with permission from Fytianos et al. (2017). Copyright 2018 American Chemical Society)

11.6.2 Intestinal Tissue

Also for the intestinal tissue, various co-culture models combining intestinal cells with immune cells have been reported. Usually, the human intestinal cell line Caco-2 is combined with other cells such as mucus-producing cells (HT29) (Walczak et al. 2015; Kleiveland 2015b) and/or with immune cells such as the Raji B cells resembling an M-cell morphology (Kleiveland 2015c). M cells in situ are highly specialized for phagocytosis and transcytosis of gut luminal content across the epithelium. Then, a study by Baleeiro and colleagues (Baleeiro et al. 2016) describes the design of polymeric nanoparticles to deliver the tumor-associated antigen Her2/neu to DCs for induction of T-cell responses by mucosal vaccination. The nanoparticles were efficiently taken up, processed, and presented by DCs inducing CD4$^+$ and CD8$^+$ T-cell proliferation and cytotoxic T cells. In addition, the nanoparticles were applied to the apical side of an intestinal epithelium model and found to be efficiently transported across the epithelial layer to become available to basolateral DCs.

Another approach is to combine human macrophages (THP-1) and human dendritic cells (MUTZ-3) embedded in a collagen scaffold grown on the apical side of transwell permeable membrane inserts with Caco-2 cells seeded on top of this layer, forming a 3D model of the intestinal mucosa. The toxicity of gold nanoparticles was addressed in this system, which was more pronounced in an inflamed intestinal model preactivated with interleukin-1 beta (Susewind et al. 2016).

11.6.3 Skin Tissue

The cytotoxicity testing of chemicals and cosmetics in in vitro skin models goes back to the 1980s (OECD guideline 431). In 2013, the OECD test guideline 431 has been adapted and provides since then an in vitro procedure allowing the identification of noncorrosive and corrosive substances and mixtures. Within this guideline, the use of reconstructed human epidermis, which is composed of several epithelial cell layers mimicking the properties of the upper parts of the human skin, is recommended. Commercially available models such as the EpiDerm™ skin model (www.mattek.com) and the EPISKIN™ (www.loreal.com) are two examples of validated 3D models for this guideline; these models do, however, not contain immune cells. First attempts for co-cultures including immune cells in 3D skin tissue have been described by van den Bogaard who generated 3D skin models to assess the crosstalk between keratinocytes and different T-cell populations to study the pathogenesis of inflammatory skin disease (van den Bogaard et al. 2014). Another 3D organotypic model to understand the pathogenesis of psoriasis has integrated not only keratinocytes and T cells but also cells from innate immune system, for example, macrophages and neutrophils (Lorthois et al. 2017). Although such predictive skin models exist, the majority of experiments with nanomaterials are still done with skin explants from ex vivo human or animal samples or with cell monocultures (Vogt et al. 2014).

11.7 Conclusions

Sophisticated in vitro approaches using relevant immune cells can play a major role in prescreening of nanomaterial safety and/or efficacy since they can provide straightforward and reliable data before animal experimentation is considered. Many well-established monocultures of various immune cells representing the innate as well as the adaptive system exist, which can be used for a first assessment of nanomaterial–immune cell interactions and subsequent immune response. However, considering the highly orchestrated and controlled interaction of immune cells in the human body, the design and application of more complex 3D engineered models allowing the interaction between cells and the extracellular matrix should be emphasized more.

Acknowledgments This work was supported by the Adolphe Merkle Foundation. BRR acknowledges the Swiss National Science Foundation (310030_159847/1), and BD acknowledges the Peter and Traudl Engelhorn Foundation.

Financial Disclosure The authors declare no conflict of interest.

References

Abbas A, Lichtman A. Basic immunology – functions and disorders of the immune system. 3rd ed. Philadelphia: Saunders Elsevier; 2011.

Alfaro-Moreno E, Nawrot TS, Vanaudenaerde BM, Hoylaerts MF, Vanoirbeek JA, Nemery B, et al. Co-cultures of multiple cell types mimic pulmonary cell communication in response to urban PM10. Eur Respir J. 2008;32:1184–94.

Baetke SC, Lammers T, Kiessling F. Applications of nanoparticles for diagnosis and therapy of cancer. Br J Radiol. 2015;88:20150207.

Baleeiro RB, Schweinlin M, Rietscher R, Diedrich A, Czaplewska JA, Metzger M, et al. Nanoparticle-based mucosal vaccines targeting tumor-associated antigens to human dendritic cells. J Biomed Nanotechnol. 2016;12:1527–43.

Banchereau J, Briere F, Caux C, Davoust J, Lebecque S, Liu YJ, et al. Immunobiology of dendritic cells. Annu Rev Immunol. 2000;18:767–811.

Blank F, Rothen-Rutishauser B, Gehr P. Dendritic cells and macrophages form a transepithelial network against foreign particulate antigens. Am J Respir Cell Mol Biol. 2007;36:669–77.

Blank F, Gerber P, Rothen-Rutishauser B, Sakulkhu U, Salaklang J, De PK, et al. Biomedical nanoparticles modulate specific CD4(+) T cell stimulation by inhibition of antigen processing in dendritic cells. Nanotoxicology. 2011a;5:606.

Blank F, Gerber P, Rothen-Rutishauser B, Sakulkhu U, Salaklang J, de Peyer K, et al. Poly(vinylalcohol)-coated super-paramagnetic iron oxide nanoparticles modulate antigen-processing and CD4+ T cell stimulation by human dendritic cells. Nanotoxicology. 2011b;Early Online:1–16.

Bonilla FA, Oettgen HC. Adaptive immunity. J Allergy Clin Immunol. 2010;125:S33–40.

Boraschi D, Costantino L, Italiani P. Interaction of nanoparticles with immunocompetent cells: nanosafety considerations. Nanomedicine (Lond). 2012;7:121–31.

Bourquin J, Milosevic A, Hauser D, Lehner R, Blank F, Petri-Fink A, et al. Biodistribution, clearance, and long-term fate of clinically relevant nanomaterials. Adv Mater. 2018;30:e1704307.

Braakhuis HM, Kloet SK, Kezic S, Kuper F, Park MV, Bellmann S, et al. Progress and future of in vitro models to study translocation of nanoparticles. Arch Toxicol. 2015;89:1469–95.

Chaplin DD. 1. Overview of the human immune response. J Allergy Clin Immunol. 2006;117:S430–5.

Chiodo F, Marradi M, Park J, Ram AF, Penades S, van Die I, et al. Galactofuranose-coated gold nanoparticles elicit a pro-inflammatory response in human monocyte-derived dendritic cells and are recognized by DC-SIGN. ACS Chem Biol. 2014;9:383–9.

Choi J, Kim H, Choi J, Oh SM, Park J, Park K. Skin corrosion and irritation test of sunscreen nanoparticles using reconstructed 3D human skin model. Environ Health Toxicol. 2014;29:e2014004.

Clift MJ, Dechezelles JF, Rothen-Rutishauser B, Petri-Fink A. A biological perspective toward the interaction of theranostic nanoparticles with the bloodstream – what needs to be considered? Front Chem. 2015;3:7.

Collin M, Bigley V. Human dendritic cell subsets: an update. Immunology. 2018;154:3–20.

Conner SD, Schmid SL. Regulated portals of entry into the cell. Nature. 2003;422:37–44.

Delcassian D, Sattler S, Dunlop IE. T cell immunoengineering with advanced biomaterials. Integr Biol (Camb). 2017;9:211–22.

Dobrovolskaia MA, McNeil SE. Immunological properties of engineered nanomaterials. Nat Nanotechnol. 2007;2:469–78.
Dobrovolskaia MA, Germolec DR, Weaver JL. Evaluation of nanoparticle immunotoxicity. Nat Nanotechnol. 2009;4:411–4.
Dostert C, Petrilli V, Van BR, Steele C, Mossman BT, Tschopp J. Innate immune activation through Nalp3 inflammasome sensing of asbestos and silica. Science (New York, NY). 2008;320:674–7.
Drasler B, Sayre P, Steinhauser KG, Petri-Fink A, Rothen-Rutishauser B. In vitro approaches to assess the hazard of nanomaterials. NanoImpact. 2017a;8:99–116.
Drasler B, Sayre P, Steinhäuser KG, Petri-Fink A, Rothen-Rutishauser B. In vitro approaches to assess the hazard of nanomaterials. NanoImpact. 2017b;8:99–116.
Dusinska M, Tulinska J, El YN, Kuricova M, Liskova A, Rollerova E, et al. Immunotoxicity, genotoxicity and epigenetic toxicity of nanomaterials: new strategies for toxicity testing? Food Chem Toxicol. 2017;109:797–811.
Elsabahy M, Wooley KL. Cytokines as biomarkers of nanoparticle immunotoxicity. Chem Soc Rev. 2013;42:5552–76.
Farcal L, Torres AF, Di CL, Rotoli BM, Bussolati O, Bergamaschi E, et al. Comprehensive in vitro toxicity testing of a panel of representative oxide nanomaterials: first steps towards an intelligent testing strategy. PLoS One. 2015;10:e0127174.
Farrera C, Fadeel B. It takes two to tango: understanding the interactions between engineered nanomaterials and the immune system. Eur J Pharm Biopharm. 2015;95:3–12.
Frohlich E. Value of phagocyte function screening for immunotoxicity of nanoparticles in vivo. Int J Nanomedicine. 2015;10:3761–78.
Frohlich E. Comparison of conventional and advanced in vitro models in the toxicity testing of nanoparticles. Artif Cells Nanomed Biotechnol. 2018;46(sup2):1091–107.
Frohlich E, Salar-Behzadi S. Toxicological assessment of inhaled nanoparticles: role of in vivo, ex vivo, in vitro, and in silico studies. Int J Mol Sci. 2014;15:4795–822.
Fytianos K, Rodriguez-Lorenzo L, Clift MJ, Blank F, Vanhecke D, von GC, et al. Uptake efficiency of surface modified gold nanoparticles does not correlate with functional changes and cytokine secretion in human dendritic cells in vitro. Nanomedicine. 2015;11:633–44.
Fytianos K, Drasler B, Blank F, Garnier CV, Seydoux E, Rodriguez-Lorenzo L, et al. Current in vitro approaches to assess nanoparticle interactions with lung cells. Nanomedicine (Lond). 2016;11:2457.
Fytianos K, Chortarea S, Rodriguez-Lorenzo L, Blank F, von GC, Petri-Fink A, et al. Aerosol delivery of functionalized gold nanoparticles target and activate dendritic cells in a 3D lung cellular model. ACS Nano. 2017;11:375–83.
Gao LG. Progress towards standardized and validated characterizations for measuring physicochemical properties of manufactured nanomaterials relevant to nano health and safety risks. NanoImpact. 2018;9:14–30.
Giannakou C, Park MV, De Jong WH, van LH, Vandebriel RJ, Geertsma RE. A comparison of immunotoxic effects of nanomedicinal products with regulatory immunotoxicity testing requirements. Int J Nanomedicine. 2016;11:2935–52.
Gstraunthaler G, Hartung T. Good cell culture practice: good laboratory practice in the cell culture laboratory for the standardization and quality assurance of in vitro studies. In: Lehr C-M, editor. Cell culture models of biological barriers. In-vitro test systems for drug absorption and delivery. London, New York: Taylor and Francis; 2002. p. 112–20.
Guermonprez P, Valladeau J, Zitvogel L, Thery C, Amigorena S. Antigen presentation and T cell stimulation by dendritic cells. Annu Rev Immunol. 2002;20:621–67.
Hartung T. Toxicology for the twenty-first century. Nature. 2009;460:208–12.
Heine G, Sims GP, Worm M, Lipsky PE, Radbruch A. Isolation of human B cell populations. Curr Protoc Immunol. 2011;Chapter 7:Unit7.
Hubbell JA, Swartz MA. Materials engineering for immunomodulation. Nature. 2009;462:449–60.
Hussain S, Vanoirbeek JA, Hoet PH. Interactions of nanomaterials with the immune system. Wiley Interdiscip Rev Nanomed Nanobiotechnol. 2012;4:169–83.

Ilinskaya AN, Dobrovolskaia MA. Understanding the immunogenicity and antigenicity of nanomaterials: past, present and future. Toxicol Appl Pharmacol. 2016;299:70–7.
Jeanbart L, Swartz MA. Engineering opportunities in cancer immunotherapy. Proc Natl Acad Sci U S A. 2015;112:14467–72.
Kermanizadeh A, Balharry D, Wallin H, Loft S, Moller P. Nanomaterial translocation–the biokinetics, tissue accumulation, toxicity and fate of materials in secondary organs–a review. Crit Rev Toxicol. 2015;45:837–72.
Kishimoto TK, Maldonado RA. Nanoparticles for the induction of antigen-specific immunological tolerance. Front Immunol. 2018;9:230.
Klein SG, Serchi T, Hoffmann L, Blomeke B, Gutleb AC. An improved 3D tetraculture system mimicking the cellular organisation at the alveolar barrier to study the potential toxic effects of particles on the lung. Part Fibre Toxicol. 2013;10:31.
Kleiveland CR. Peripheral blood mononuclear. Cell. 2015a:161–7.
Kleiveland CR: Co-cultivation of Caco-2 and HT-29MTX. 2015b:135–140.
Kleiveland CR. Co-culture Caco-2/Immune. Cell. 2015c:197–205.
Krewski D, Acosta D Jr, Andersen M, Anderson H, Bailar JC III, Boekelheide K, et al. Toxicity testing in the 21st century: a vision and a strategy. J Toxicol Environ Health B Crit Rev. 2010;13:51–138.
Kumar S, Anselmo AC, Banerjee A, Zakrewsky M, Mitragotri S. Shape and size-dependent immune response to antigen-carrying nanoparticles. J Control Release. 2015;220:141–8.
Kunzmann A, Andersson B, Vogt C, Feliu N, Ye F, Gabrielsson S, et al. Efficient internalization of silica-coated iron oxide nanoparticles of different sizes by primary human macrophages and dendritic cells. Toxicol Appl Pharmacol. 2011;253:81–93.
Lameijer MA, Tang J, Nahrendorf M, Beelen RH, Mulder WJ. Monocytes and macrophages as nanomedicinal targets for improved diagnosis and treatment of disease. Expert Rev Mol Diagn. 2013;13:567–80.
Leandro MJ. B-cell subpopulations in humans and their differential susceptibility to depletion with anti-CD20 monoclonal antibodies. Arthritis Res Ther. 2013;15 Suppl 1:S3.
Lee CH, Syu SH, Chen YS, Hussain SM, Aleksandrovich OA, Chen WL, et al. Gold nanoparticles regulate the blimp1/pax5 pathway and enhance antibody secretion in B-cells. Nanotechnology. 2014;25:125103.
Lehmann A, Brandenberger C, Blank F, Gehr P, Rothen-Rutishauser B. A 3D model of the human epithelial airway barrier. In: Maguire T, Novik E, editors. Methods in bioengineering. Alternative technologies to animal testing. Boston, London: Artech House; 2010a. p. 239–60.
Lehmann AD, Daum N, Lehr CM, Bur M, Parak W, Gehr P, et al. An *in vitro* triple cell co-culture model of the human alveolar epithelial barrier including primary cells. Submitted to J R Soc Interface. 2010b.
Lehmann AD, Daum N, Lehr CM, Bur M, Parak W, Gehr P, et al., An in vitro triple cell co-culture model with primary cells mimicking the human alveolar epithelial barrier. European Journal of Pharmaceutics and Biopharmaceutics. 2011;77(3):398–406.
Lehner M, Morhart P, Stilper A, Holter W. Functional characterization of monocyte-derived dendritic cells generated under serum free culture conditions. Immunol Lett. 2005;99:209–16.
Li Y, Boraschi D. Endotoxin contamination: a key element in the interpretation of nanosafety studies. Nanomedicine (Lond). 2016;11:269–87.
Li Y, Italiani P, Casals E, Tran N, Puntes VF, Boraschi D. Optimising the use of commercial LAL assays for the analysis of endotoxin contamination in metal colloids and metal oxide nanoparticles. Nanotoxicology. 2015;9:462–73.
Li Y, Fujita M, Boraschi D. Endotoxin contamination in nanomaterials leads to the misinterpretation of immunosafety results. Front Immunol. 2017;8:472.
Lorthois I, Asselineau D, Seyler N, Pouliot R. Contribution of in vivo and organotypic 3D models to understanding the role of macrophages and neutrophils in the pathogenesis of psoriasis. Mediat Inflamm. 2017;2017:7215072.
Monks CR, Freiberg BA, Kupfer H, Sciaky N, Kupfer A. Three-dimensional segregation of supramolecular activation clusters in T cells. Nature. 1998;395:82–6.

Nicod LP. Function of human lung dendritic cells. In: Lung macrophages and dendritic cells in health and disease. New York: Marcel Dekker, Inc.; 1997. p. 311–34.

Oberdorster G, Stone V, Donaldson K. Toxicology of nanoparticles: a historical perspective. Nanotoxicology. 2007;1:2–25.

OECD. OECD guidelines for the testing of chemicals; 2013.

OECD. Test No. 439. In: Vitro skin irritation: reconstructed human epidermis test method. Paris: OECD Publishing; 2015.

Park EJ, Cho WS, Jeong J, Yi J, Choi K, Park K. Pro-inflammatory and potential allergic responses resulting from B cell activation in mice treated with multi-walled carbon nanotubes by intratracheal instillation. Toxicology. 2009;259:113–21.

Paur HR, Cassee FR, Teeguarden JG, Fissan H, Diabate S, Aufderheide M, et al. In-vitro cell exposure studies for the assessment of nanoparticle toxicity in the lung—a dialog between aerosol science and biology. J Aerosol Sci. 2011;42:668–92.

Perlman RL. Mouse models of human disease: an evolutionary perspective. Evol Med Public Health. 2016;2016:170–6.

Petrarca C, Clemente E, Amato V, Pedata P, Sabbioni E, Bernardini G, et al. Engineered metal based nanoparticles and innate immunity. Clin Mol Allergy. 2015;13:13.

Piccinno F, Gottschalk F, Seeger S, Nowack B. Industrial production quantities and uses of ten engineered nanomaterials in Europe and the world. J Nanopart Res. 2012;14:1–11.

Rabolli V, Lison D, Huaux F. The complex cascade of cellular events governing inflammasome activation and IL-1beta processing in response to inhaled particles. Part Fibre Toxicol. 2016;13:40.

Roggen EL, Soni NK, Verheyen GR. Respiratory immunotoxicity: an in vitro assessment. Toxicol In Vitro. 2006;20:1249–64.

Rothen-Rutishauser BM, Kiama SG, Gehr P. A three-dimensional cellular model of the human respiratory tract to study the interaction with particles. Am J Respir Cell Mol Biol. 2005;32:281–9.

Rothen-Rutishauser B, Clift MJD, Jud C, Fink A, Wick P. Human epithelial cells in vitro – are they an advantageous tool to help understand the nanomaterial-biological barrier interaction? ENTL. 2012;1:1–20.

Salazar F, Ghaemmaghami AM. Allergen recognition by innate immune cells: critical role of dendritic and epithelial cells. Front Immunol. 2013;4:356.

Santegoets SJ, van den Eertwegh AJ, van de Loosdrecht AA, Scheper RJ, de Gruijl TD. Human dendritic cell line models for DC differentiation and clinical DC vaccination studies. J Leukoc Biol. 2008;84:1364–73.

Sauer UG, Vogel S, Aumann A, Hess A, Kolle SN, Ma-Hock L, et al. Applicability of rat precision-cut lung slices in evaluating nanomaterial cytotoxicity, apoptosis, oxidative stress, and inflammation. Toxicol Appl Pharmacol. 2014;276:1–20.

Schleh C, Rothen-Rutishauser BM, Blank F, Lauenstein HD, Nassimi M, Krug N, et al. Surfactant Protein D modulates allergen particle uptake and inflammatory response in a human epithelial airway model. Respir Res. 2012;13:8.

Schmid D, Park CG, Hartl CA, Subedi N, Cartwright AN, Puerto RB, et al. T cell-targeting nanoparticles focus delivery of immunotherapy to improve antitumor immunity. Nat Commun. 2017;8:1747.

Seydoux E, Rothen-Rutishauser B, Nita IM, Balog S, Gazdhar A, Stumbles PA, et al. Size-dependent accumulation of particles in lysosomes modulates dendritic cell function through impaired antigen degradation. Int J Nanomedicine. 2014;9:3885–902.

Seydoux E, Rodriguez-Lorenzo L, Blom RA, Stumbles PA, Petri-Fink A, Rothen-Rutishauser BM, et al. Pulmonary delivery of cationic gold nanoparticles boost antigen-specific CD4(+) T Cell Proliferation. Nanomedicine. 2016;12:1815–26.

Shanks N, Greek R, Greek J. Are animal models predictive for humans? Philos Ethics Humanit Med. 2009;4:2.

Smith DM, Simon JK, Baker JR Jr. Applications of nanotechnology for immunology. Nat Rev Immunol. 2013;13:592–605.

Smith MJ, Brown JM, Zamboni WC, Walker NJ. From immunotoxicity to nanotherapy: the effects of nanomaterials on the immune system. Toxicol Sci. 2014;138:249–55.

Steiner S, Czerwinski J, Comte P, Popovicheva O, Kireeva E, Mueller L, et al. Comparison of the toxicity of diesel exhaust produced by bio- and fossil diesel combustion in human lung cells in vitro. Atmos Environ. 2013;81:380–8.

Sun B, Wang X, Ji Z, Li R, Xia T. NLRP3 inflammasome activation induced by engineered nanomaterials. Small. 2013;9:1595–607.

Sundstrom C, Nilsson K. Establishment and characterization of a human histiocytic lymphoma cell line (U-937). Int J Cancer. 1976;17:565–77.

Susewind J, de SC-W RU, Collnot EM, Schneider-Daum N, Griffiths GW, et al. A 3D co-culture of three human cell lines to model the inflamed intestinal mucosa for safety testing of nanomaterials. Nanotoxicology. 2016;10:53–62.

Swartz MA, Hubbell JA. Engineering eapproaches to immunotherapy. Sci Transl Med. 2012;4:148rv9.

Tao L, Reese TA. Making mouse models that reflect human immune responses. Trends Immunol. 2017;38:181–93.

Temchura VV, Kozlova D, Sokolova V, Uberla K, Epple M. Targeting and activation of antigen-specific B-cells by calcium phosphate nanoparticles loaded with protein antigen. Biomaterials. 2014;35:6098–105.

Thiele L, Rothen-Rutishauser B, Jilek S, Wunderli-Allenspach H, Merkle HP, Walter E. Evaluation of particle uptake in human blood monocyte-derived cells in vitro. Does phagocytosis activity of dendritic cells measure up with macrophages? J Control Release. 2001;76:59–71.

Tornqvist E, Annas A, Granath B, Jalkesten E, Cotgreave I, Oberg M. Strategic focus on 3R principles reveals major reductions in the use of animals in pharmaceutical toxicity testing. PLoS One. 2014;9:e101638.

Turvey SE, Broide DH. Innate immunity. J Allergy Clin Immunol. 2010;125:S24–32.

van den Bogaard EH, Tjabringa GS, Joosten I, Vonk-Bergers M, van RE, Tijssen HJ, et al. Crosstalk between keratinocytes and T cells in a 3D microenvironment: a model to study inflammatory skin diseases. J Invest Dermatol. 2014;134:719–27.

Venditto VJ, Szoka FC Jr. Cancer nanomedicines: so many papers and so few drugs! Adv Drug Deliv Rev. 2013;65:80–8.

Vogt A, Rancan F, Ahlberg S, Nazemi B, Choe CS, Darvin ME, et al. Interaction of dermatologically relevant nanoparticles with skin cells and skin. Beilstein J Nanotechnol. 2014;5:2363–73.

vonGarnier C, Nicod LP. Immunology taught by lung dendritic cells. Swiss Med Wkly. 2009;139:186–92.

Walczak AP, Kramer E, Hendriksen PJ, Helsdingen R, van der Zande M, Rietjens IM, et al. In vitro gastrointestinal digestion increases the translocation of polystyrene nanoparticles in an in vitro intestinal co-culture model. Nanotoxicology. 2015;9:886–94.

Wang J, Matsukura S, Watanabe S, Adachi M, Suzaki H. Involvement of Toll-like receptors in the immune response of nasal polyp epithelial cells. Clin Immunol. 2007;124:345–52.

Wiemann M, Vennemann A, Sauer UG, Wiench K, Ma-Hock L, Landsiedel R. An in vitro alveolar macrophage assay for predicting the short-term inhalation toxicity of nanomaterials. J Nanobiotechnol. 2016;14:16.

Wills JW, Hondow N, Thomas AD, Chapman KE, Fish D, Maffeis TG, et al. Genetic toxicity assessment of engineered nanoparticles using a 3D in vitro skin model (EpiDerm). Part Fibre Toxicol. 2016;13:50.

Yazdi AS, Guarda G, Riteau N, Drexler SK, Tardivel A, Couillin I, et al. Nanoparticles activate the NLR pyrin domain containing 3 (Nlrp3) inflammasome and cause pulmonary inflammation through release of IL-1alpha and IL-1beta. Proc Natl Acad Sci U S A. 2010;107:19449–54.

Zupke O, Distler E, Jurchott A, Paiphansiri U, Dass M, Thomas S, et al. Nanoparticles and antigen-specific T-cell therapeutics: a comprehensive study on uptake and release. Nanomedicine (Lond). 2015;10:1063–76.

Index

J. C. Bonner, J. M. Brown (eds.), *Interaction of Nanomaterials with the Immune System*, Molecular and Integrative Toxicology,
https://doi.org/10.1007/978-3-030-33962-3

The manufacturer's authorised representative in the EU is Springer
Nature Customer Service Centre GmbH, Europaplatz 3, 69115 Heidelberg,
Germany. If you have any concerns regarding our products, please
contact ProductSafety@springernature.com

Printed and bound by CPI Group (UK) Ltd, Croydon, CR0 4YY
30/09/2026
02225449-0007